MATHEMATICS RESEARCH DEVELOPMENTS

ZHANG NEURAL NETWORKS AND NEURAL-DYNAMIC METHOD

MATHEMATICS RESEARCH DEVELOPMENTS

Additional books in this series can be found on Nova's website
under the Series tab.

Additional E-books in this series can be found on Nova's website
under the E-books tab.

MATHEMATICS RESEARCH DEVELOPMENTS

ZHANG NEURAL NETWORKS AND NEURAL-DYNAMIC METHOD

YUNONG ZHANG
AND
CHENFU YI

Nova Science Publishers, Inc.
New York

Copyright © 2011 by Nova Science Publishers, Inc.

All rights reserved. No part of this book may be reproduced, stored in a retrieval system or transmitted in any form or by any means: electronic, electrostatic, magnetic, tape, mechanical photocopying, recording or otherwise without the written permission of the Publisher.

For permission to use material from this book please contact us:
Telephone 631-231-7269; Fax 631-231-8175
Web Site: http://www.novapublishers.com

NOTICE TO THE READER

The Publisher has taken reasonable care in the preparation of this book, but makes no expressed or implied warranty of any kind and assumes no responsibility for any errors or omissions. No liability is assumed for incidental or consequential damages in connection with or arising out of information contained in this book. The Publisher shall not be liable for any special, consequential, or exemplary damages resulting, in whole or in part, from the readers' use of, or reliance upon, this material.

Independent verification should be sought for any data, advice or recommendations contained in this book. In addition, no responsibility is assumed by the publisher for any injury and/or damage to persons or property arising from any methods, products, instructions, ideas or otherwise contained in this publication.

This publication is designed to provide accurate and authoritative information with regard to the subject matter covered herein. It is sold with the clear understanding that the Publisher is not engaged in rendering legal or any other professional services. If legal or any other expert assistance is required, the services of a competent person should be sought. FROM A DECLARATION OF PARTICIPANTS JOINTLY ADOPTED BY A COMMITTEE OF THE AMERICAN BAR ASSOCIATION AND A COMMITTEE OF PUBLISHERS.

Full color presentation of graphics is available in the E-book.

LIBRARY OF CONGRESS CATALOGING-IN-PUBLICATION DATA

Zhang, Yunong.
  Zhang neural networks and neural-dynamic method / Yunong Zhang and Chenfu Yi.
      p. cm.
    Includes index.
    ISBN 978-1-61668-839-4 (hardcover)
  1. Neural networks (Computer science)--Mathematics. I. Yi, Chenfu. II. Title.
    QA76.87.Z4755 2009
    006.3'2--dc22
                            2010016082

Published by Nova Science Publishers, Inc. ✛ *New York*

Contents

List of Figures   xi

List of Tables   xvii

Preface   xix

1 Introduction   **1**
  1.1. Artificial Neural Network 1
  1.2. Recurrent Neural Networks 3
     1.2.1. Overview of Recurrent Neural Networks 3
     1.2.2. Models of Recurrent Neural Networks 5
  1.3. Time-Varying Problems 6
  1.4. Definitions of Convergence and Stability 7
  1.5. Activation Functions Used in Recurrent Neural Networks 8
  1.6. Conclusion 11
  References 12

I Matrix-Vector Problems   **17**

2 ZNN Solving the Time-Varying Linear System $A(t)x(t) = b(t)$ **19**
  2.1. Introduction 19
  2.2. Problem Formulation and Solvers 20
     2.2.1. Zhang Neural Networks 21
     2.2.2. Gradient-Based Neural Networks 22
  2.3. Theoretical Results 23
  2.4. MATLAB-Coding Simulation Techniques 25
     2.4.1. Coding of Activation-Functions 25
     2.4.2. ODE with Mass Matrix 26
     2.4.3. Obtaining Matrix Derivatives 27
     2.4.4. RNN Right-Hand Side 27
     2.4.5. Illustrative Examples 28
  2.5. Simulink-Modeling Techniques 32
     2.5.1. Basic Function Blocks 32
     2.5.2. Generating $\Phi(\cdot)$ 32

v

2.5.3.	Parameter Settings	34
2.5.4.	Illustrative Examples	34
2.6.	An Application: Inverse Kinematic Control	36
2.7.	Conclusion	38
	References	38

3 ZNN for Time-Varying Convex Quadratic Optimization   **41**

3.1.	Introduction	41
3.2.	Time-Varying Quadratic Minimization	42
	3.2.1. Zhang Neural Networks	43
	3.2.2. Gradient Neural Networks	44
	3.2.3. Theoretical Results and Comparisons	45
	3.2.4. Simulation Studies	46
3.3.	Time-Varying Quadratic Program	47
	3.3.1. Neural Network Solvers and Comparison	49
	3.3.2. Convergence Analysis and Results	52
	3.3.3. Illustrative Examples	54
3.4.	Modeling Techniques and Illustrative Examples	57
3.5.	Conclusion	61
	References	61

II Pure Matrix Problems   **67**

4 ZNN for Time-Varying Matrix Inversion and Pseudoinverse-Solving   **69**

4.1.	Introduction	69
4.2.	Problem Formulation and Neural Solvers	70
	4.2.1. ZNN Models	73
	4.2.2. GNN Models	73
4.3.	Theoretical Results	74
4.4.	MATLAB-Coding Simulation Techniques	80
	4.4.1. Kronecker Product and Vectorization	81
	4.4.2. ODE with Mass Matrix	82
	4.4.3. Obtaining Matrix Derivative	83
	4.4.4. Illustrative Examples	84
4.5.	Simulink-Modeling Approach	92
	4.5.1. ZNN-Convergence Simulation	93
	4.5.2. ZNN-Robustness Simulation	96
4.6.	Extension to Time-Varying Matrix Pseudoinverse Solving	97
	4.6.1. Problem Formulation and ZNN Solver	97
	4.6.2. ZNN Simulink Modeling	98
	4.6.3. Generating $A(t)$	99
	4.6.4. Other Parameters Setting	100
	4.6.5. Modeling Results	101
	4.6.6. Comparisons with GNN	101

| Contents | vii |

4.7. Application to Robot Kinematic Control 102
    4.7.1. Preliminaries on Inverse Kinematics 102
    4.7.2. Simulation Based on PUMA560 Robot Arm 104
    4.7.3. Conclusion 106
References 107

5 ZNN for Linear Time-Varying Matrix Equation Solving   113

5.1. Introduction 113
5.2. ZNN for Time-Varying Sylvester Equation Solving 114
    5.2.1. Problem Formulation and Solvers 114
    5.2.2. Theoretical Results 117
    5.2.3. MATLAB-Coding Simulation Techniques 118
    5.2.4. Illustrative Examples by MATLAB Code 119
    5.2.5. Simulink-Modeling Studies 122
5.3. ZNN for Time-Varying Lyapunov Equation Solving 126
    5.3.1. Problem Formulation and Solvers 127
    5.3.2. MATLAB-Coding Simulation Techniques 129
    5.3.3. Illustrative Examples by MATLAB Code 130
    5.3.4. Simulink Modeling and Verification 133
    5.3.5. An Illustrative Example by Simulink-Modeling Approach 135
5.4. ZNN for Time-Varying Linear Matrix Equation $AXB = C$ Solving 137
    5.4.1. Problem Formulation and Solvers 138
    5.4.2. MATLAB-Coding Simulation Techniques 140
    5.4.3. Illustrative Examples by MATLAB Code 142
    5.4.4. Simulink-Modeling Approach 144
5.5. Conclusion 150
References 150

6 ZNN for Time-Varying Matrix Square Roots Finding   155

6.1. Introduction 155
6.2. Problem Formulation and Neural Solvers 156
    6.2.1. Problem Formulation 156
    6.2.2. Zhang Neural Network 156
    6.2.3. Gradient Neural Network 158
    6.2.4. Models and Methods Comparison 158
6.3. MALATB-Coding Simulation Techniques 159
    6.3.1. ODE with Mass Matrix 159
    6.3.2. Obtaining Matrix Derivatives 160
    6.3.3. RNN Right-Hand Side 161
    6.3.4. An Illustrative Verification Example 161
6.4. Simulink-Modeling Techniques 164
    6.4.1. Basic Blocks 164
    6.4.2. Parameters and Options Setting 166
    6.4.3. Modeling and Verification Results 167
6.5. Conclusion 169

viii Contents

References 169

III Return to Scalar or Constant Problems 173

7 Zhang Dynamics for the Scalar-Valued Nonlinear Problems 175
7.1. Introduction 175
7.2. Time-Varying Scalar-Valued Nonlinear Equations 176
    7.2.1. Gradient-Based Dynamics 176
    7.2.2. Zhang Dynamics 177
    7.2.3. Convergence Analysis 177
    7.2.4. Simulation Studies 181
7.3. The Constant Nonlinear Equations 183
    7.3.1. Gradient Dynamics 183
    7.3.2. Zhang Dynamics 183
    7.3.3. Theoretical Analysis 185
    7.3.4. Computer-Simulation Studies 188
7.4. Conclusion 196
References 196

8 ZNN and GNN Models for Linear Constant Problems Solving 201
8.1. Introduction 201
8.2. ZNN for Linear Time-Invariant Equations Solving 202
    8.2.1. ZNN Model for Linear Time-Varying (LTV) Equations Solving . . 202
    8.2.2. Motivation for Simplification 203
    8.2.3. Linear Time-Invariant (LTI) Equations Solving 204
    8.2.4. Illustrative Example 205
8.3. GNN for Linear Equations Solving 206
    8.3.1. Problem Formulation and GNN Solver 207
    8.3.2. Theoretical Analysis 208
    8.3.3. Illustrative Examples 211
8.4. GNN for Constant Matrix Inversion 212
    8.4.1. Main Theoretical Results 213
    8.4.2. Kronecker Product and Vectorization 215
    8.4.3. Illustrative Examples 216
    8.4.4. The Singular Case 220
    8.4.5. An Application to Inverse Kinematics 222
8.5. Conclusion 224
References 224

IV Final Comparisons and Discussions 229

9 Unified Neural-Network Models 231
9.1. Introduction 231
9.2. Unified Neural-Network Models 232

	9.2.1.	Zhang Neural Networks	232
	9.2.2.	Gradient Neural Networks	233
9.3.	Comparisons and Differences		234
9.4.	An Illustrative Application		235
	9.4.1.	QP-Based Scheme-Formulation and Solver	235
	9.4.2.	Computer Simulation and Verification	236
	9.4.3.	CMG Performance-Index Analysis	240
9.5.	Conclusion		242
	References		243

A Appendix **249**

A.1.	Computation of Matrix and Vector	249
	A.1.1. Notation of Matrix and Vector	249
	A.1.2. Matrix/Vector Basic Operation	250
A.2.	Matrix Inversion, Trace and Kronecker Product	251
A.3.	Vector/Matrix Norms	253
A.4.	Matrix Differentiation	254
	References	255

Index **259**

List of Figures

1.1   A new neural network with hidden neurons activated by Bernoulli polynomials. 2

1.2   Architecture and connection of HNN (1.2) solving $Ax = b$. 4

1.3   Activation functions $\phi(\cdot)$ used in recurrent neural networks. 9

2.1   Block diagram which realizes ZNN (2.3). 22

2.2   Circuit schematics which realizes ZNN (2.3). 23

2.3   Online solution of linear time-varying equations by recurrent neural networks with $\gamma = 1$, where theoretical solution is denoted by dotted lines. ... 28

2.4   Computational error $\|x(t) - x^*(t)\|$ of ZNN (2.3) with different γ. 29

2.5   Computational error $\|x(t) - x^*(t)\|$ of GNN (2.6) with different γ. 29

2.6   Solution of linear time-varying equations by imprecisely-constructed ZNN (2.8) with different γ, where dotted lines denote $x^*(t)$. 30

2.7   Computational errors $\|x(t) - x^*(t)\|$ of solving linear time-varying equations by ZNN in the presence of large realization errors. 31

2.8   Computational errors $\|x(t) - x^*(t)\|$ of solving linear time-varying equations by GNN in the presence of realization errors. 31

2.9   Parallel-processing activation-function arrays. 33

2.10 Simulink model of ZNN-model (2.3). 35

2.11 "Scope" output $x(t)$ of ZNN (2.3). 36

2.12 "Scope" output $\|e(t)\|$ of ZNN (2.3). 36

2.13 "Scope" output $\|e(t)\|$ of the perturbed ZNN (2.8). 37

2.14 PUMA560 manipulator tracking a circle. 38

2.15 PUMA560 positioning errors on X, Y and Z axes. 38

3.1   "Moving" minimum of time-varying quadratic function $f(x)$ at different time instants. 43

3.2   Time-varying quadratic minimization (3.1) synthesized online by ZNN (3.6) and GNN (3.9) with design parameter $\gamma = 1$ and using power-sigmoid activation functions, where dotted curves correspond to the theoretical time-varying minimum solution $x^*(t)$. 45

3.3   Time-varying quadratic-function value $f(x)$ minimized online by ZNN (3.6) and GNN (3.9) with design parameter $\gamma = 1$ and using power-sigmoid activation functions, where dotted curves correspond to the theoretical time-varying minimum value $f(x^*(t))$. 46

xii List of Figures

3.4 Residual error $|f(x) - f(x^*)|$ synthesized by ZNN and GNN models with $\gamma = 1$. 47

3.5 "Moving" quadratic function, "Moving" linear constraint and "Moving" optimal solution of time-varying quadratic program (3.10)-(3.11). 48

3.6 Online solution of time-varying quadratic program (3.13)-(3.14) by ZNN and GNN models with $\gamma = 1$, where dashed-dotted blue curves correspond to time-varying theoretical solution $\tilde{x}^*(t)$ and solid red curves correspond to neural-network solution $\tilde{x}(t)$. 55

3.7 Residual errors $\|\tilde{P}(t)\tilde{x}(t) + \tilde{q}(t)\|_2$ of ZNN and GNN models with $\gamma = 1$ for online solution of time-varying quadratic program (3.13)-(3.14). 56

3.8 Residual error $\|\tilde{P}(t)\tilde{x}(t) + \tilde{q}(t)\|_2$ of ZNN (3.17) with $\gamma = 10$ for online solution of time-varying quadratic program (3.13)-(3.14). 57

3.9 A unified overall model of ZNN (3.6) and (3.17) for time-varying QP solving. 58

3.10 A unified overall model of GNN (3.9) and (3.19) for time-varying QP solving. 59

3.11 Output $\tilde{x}(t)$ of Scope 1 exploited in ZNN and GNN modeling. 59

3.12 Output $\|\tilde{P}(t)\tilde{x}(t) + \tilde{q}(t)\|_2$ of Scope 2 exploited in ZNN and GNN modeling. 60

3.13 RNN-synthesized scalar-values of equality-constraint satisfaction $A(t)x(t) - b(t)$. 60

4.1 Block diagram of ZNN model (4.7) for online time-varying matrix inversion. 72

4.2 Online inversion of time-varying sinusoidal matrix $A(t)$ by ZNN model (4.7). 85

4.3 Convergence of computational error $\|X(t) - A^{-1}(t)\|$ by ZNN model (4.7). 86

4.4 Inversion of time-varying sinusoidal matrix $A(t)$ by GNN (4.9). 87

4.5 Convergence of solution error $\|X(t) - A^{-1}(t)\|$ as of GNN model (4.9). .. 87

4.6 Online inversion of sinusoidal matrix $A(t)$ by perturbed ZNN model (4.14). 88

4.7 Convergence of computational error $\|X(t) - A^{-1}(t)\|$ of perturbed ZNN (4.14). 89

4.8 Entry trajectories of time-varying Toeplitz matrix $A(t) = [a_{ij}(t)] \in \mathbb{R}^{n \times n}$. .. 90

4.9 Online inversion of time-varying Toeplitz matrix $A(t)$ by ZNN model (4.7). 90

4.10 Convergence of residual error $\|A(t)X(t) - I\|$ by ZNN (4.7) with $\gamma = 1$. .. 91

4.11 Inversion of time-varying matrix $A(t)$ by model (4.8) using power-sigmoid activation function array and $\gamma = 1$, where dashed-dotted curves denote the theoretical time-varying inverse $A^{-1}(t)$. 91

4.12 Computational error $\|X(t) - A^{-1}(t)\|_F$ by model (4.8) using power-sigmoid activation-function array and different values of parameter γ. 92

4.13 Circuit schematics of the jth neuron column used in ZNN (4.7). 93

4.14 Overall Simulink modeling of ZNN (4.7) for time-varying matrix inversion. 94

4.15 "Scope" outputs of ZNN (4.7) with design parameter $\gamma = 1$. 95

4.16 Overall Simulink modeling of imprecisely-constructed ZNN (4.14) used in online time-varying matrix inversion. 96

4.17 "Scope" outputs of computational error $\|E(t)\|$ synthesized by the imprecisely-constructed ZNN (4.14). 97

4.18 ZNN block diagram for online time-varying matrix pseudoinverse solving. .. 99

4.19 Overall ZNN Simulink model which solves online for the time-varying matrix pseudoinverse. 100

List of Figures

xiii

4.20 Solution-error $\|X(t) - A^+(t)\|$ of ZNN (4.33) solving for time-varying matrix pseudoinverse (4.29a) using different activation functions and with $\gamma = 4$. 101

4.21 State trajectories of ZNN (4.33) using power-sigmoid activation functions, where dotted curves correspond to theoretical time-varying pseudoinverse $A^+(t)$. 102

4.22 Comparison of ZNN and GNN models for online time-varying pseudoinverse solving using linear activation functions with design parameter $\gamma=1$, where dotted curves correspond to theoretical time-varying pseudoinverse $A^+(t)$. 103

4.23 Motion trajectories of PUMA560 manipulator synthesized by ZNN (4.7). . 104

4.24 End-effector errors of PUMA560 manipulator tracking a 10cm-radius circle. 104

4.25 States of ZNN (4.7) for kinematic control of PUMA560, where $x_{ij} = x_{ji}$ shows the symmetry. 105

4.26 States of the traditional GNN model for kinematic control of PUMA560. . . 105

4.27 End-effector motion trajectory of PUMA560 synthesized by the traditional GNN model, where the dotted curve denotes the desired circular path. . . . 106

5.1 Online solution of Sylvester equation (5.1) by ZNN model (5.3). 120

5.2 Convergence of computational error $\|X(t) - X^*(t)\|$ of ZNN model (5.3). . 121

5.3 Online solution of Sylvester equation (5.1) by perturbed ZNN model (5.5) with large differentiation and implementation errors. 122

5.4 Convergence of $\|X(t) - X^*(t)\|$ of perturbed ZNN model (5.5). 122

5.5 Overall Simulink model of ZNN (5.3) for online time-varying Sylvester equation solving. 123

5.6 Scope outputs of ZNN (5.3) when parameter $\gamma = 1$. 124

5.7 Overall Simulink model of imprecisely-constructed ZNN (5.5) for online time-varying Sylvester equation solving. 125

5.8 Scope outputs of imprecisely-constructed ZNN (5.5) when $\gamma = 10$. 126

5.9 Online ZNN and GNN solution of time-varying Lyapunov equation (5.7), where dash-dotted red lines correspond to theoretical solution $P^*(t)$, and solid blue lines correspond to neural-solution $P(t)$ with $\gamma = 1$. 131

5.10 Convergence performance of ZNN and GNN solution errors $\|P(t) - P^*(t)\|_F$. 132

5.11 Overall Simulink model of ZNN (5.9) for online time-varying Lyapunov equation solving. 135

5.12 Overall Simulink model of GNN (5.10) for online time-varying Lyapunov equation solving. 136

5.13 Scope outputs of $P(t)$. 137

5.14 Scope outputs of $\|P(t) - P^*(t)\|_F$. 137

5.15 Online solution of linear time-varying matrix equation (5.13) by GNN (5.19) and ZNN (5.17) with $\gamma = 1$ and using power-sigmoid functions. . . . 143

5.16 Convergence of computational error $\|X(t) - X^*(t)\|$ synthesized by ZNN (5.17) with different values of design parameter γ and using power-sigmoid functions. 144

5.17 Overall model of ZNN model (5.19) for online equation $A(t)X(t)B(t) = C(t)$ solving. 145

List of Figures

5.18 Overall model of GNN (5.17) exploited for online equation $A(t)X(t)B(t) = C(t)$ solving. 145

5.19 Matrix stream $A(t)$ constructed by adding $A_1(t)$ and $A_2(t)$. 146

5.20 State trajectories of ZNN (5.17) using power-sigmoid functions with $\gamma = 1$ and 10, where dotted curves correspond to theoretical solution $X^*(t)$. 148

5.21 Residual error $\|A(t)X(t)B(t) - C(t)\|_F$ of ZNN (5.17) with $\gamma = 1$. 149

5.22 State trajectories of GNN (5.19) using activation functions (1.7) with $\gamma = 1$, where dotted curves correspond to $X^*(t)$. 149

5.23 Convergence of residual error $\|A(t)X(t)B(t) - C(t)\|_F$ of GNN (5.19) using activation functions (1.7) with $\gamma = 1$. 150

6.1 Online solution of time-varying matrix square root $A^{1/2}(t)$ by RNN [i.e., ZNN (6.3) and GNN (6.5)] with $\gamma = 1$, where the theoretical solution $A^{1/2}(t)$ is denoted in dash-dotted curves, while the neural-network solutions are denoted in solid curves. 163

6.2 Convergence of computational error $\|X(t) - X^*(t)\|_F$ of ZNN (6.3) and GNN (6.5) applied to time-varying matrix square roots finding. 163

6.3 Overall ZNN Simulink model of (6.3) applied to the TVMSR problem solving. 165

6.4 Overall GNN Simulink model of (6.5) applied to the TVMSR problem solving. 166

6.5 Necessary "StopFcn" code of "Callbacks" in dialog box "Model Properties". 167

6.6 Online solution of time-varying matrix square root $A^{1/2}(t)$ by ZNN (6.3) and GNN (6.5) with $\gamma = 1$, where the theoretical solution is denoted in dash-dotted curves. 168

6.7 Residual-error profile of RNN models during time-varying square roots finding. 169

7.1 Solution-performance comparison of GD (7.2) and ZD (7.4) with design parameter $\gamma = 1$ for solving nonlinear time-varying equation (7.6), where dash lines in red denote the theoretical solutions. 181

7.2 Residual error $\|x^2 - 2\sin(1.8t)x + \sin^2(1.8t) - 1\|$ of GD (7.2) and ZD (7.4) with design parameter $\gamma = 1$ for solving nonlinear time-varying equation (7.6). 182

7.3 Residual error $\|x^2 - 2\sin(1.8t)x + \sin^2(1.8t) - 1\|$ of ZD (7.4) with design parameter $\gamma = 10$ for solving nonlinear time-varying equation (7.6). 182

7.4 Block diagrams of neural-dynamic solvers for online solution of $f(x) = 0$. . 184

7.5 Simulink-based models representation about GD (7.8) and ZD (7.10). . . . 184

7.6 Convergence behavior of error $e(t)$ with different activation functions. . . . 187

7.7 Online solution of nonlinear equation (7.12) by randomly-initialized GD (7.8) and ZD (7.10) with $\gamma = 1$, where dotted lines in red denote the theoretical solutions. 189

7.8 Online solution to $(x - 4)^{10}(x - 1) = 0$ by ZD (7.10) and GD (7.8) with design parameter $\gamma = 1$. Left: ZD (7.10); Right: GD (7.8). 190

7.9 3-dimensional convergent performance of ZD (7.10) and GD (7.8) with design parameter $\gamma = 1$ for online solution of $(x - 4)^{10}(x - 1) = 0$. 190

List of Figures

xv

7.10 Online solution to $(x-4)^3(x-1) = 0$ by ZD (7.10) and GD (7.8) with design parameter $\gamma = 1$. Left: ZD (7.10); Right: GD (7.8). 191

7.11 Online solution to $(x-4)^2(x-1) = 0$ by ZD (7.10) and GD (7.8) with design parameter $\gamma = 1$. Left: ZD (7.10); Right: GD (7.8). 191

7.12 3-dimensional convergent performance of ZD (7.10) and GD (7.8) with design parameter $\gamma = 1$ for online solution of $(x-4)^3(x-1) = 0$. 192

7.13 3-dimensional convergent performance of ZD (7.10) and GD (7.8) with design parameter $\gamma = 1$ for online solution of $(x-4)^2(x-1) = 0$. 193

7.14 Convergent performance of the neural state of ZD (7.10) for online solution of nonlinear equation (7.14) starting from different initial states. 194

7.15 Online solution of $0.01(x+7)(x-1)(x-8) + \sin x + 2.4 = 0$ by ZD (7.10) and GD (7.8) with parameter $\gamma = 1$. Left: ZD (7.10); Right: GD (7.8). . . . 194

7.16 Online solution of ZD (7.10) (left) and GD (7.8) (right) with $\gamma = 1$ for nonlinear equation $\cos x + 3 = 0$ which evidently has no theoretical roots. . . . 195

8.1 Block diagram of network architecture of ZNN models (8.2) and (8.5). . . . 203

8.2 Circuit schematics of ZNN model (8.5) as derived from (8.6). 205

8.3 Trajectories of $x(t)$ and $\|Ax(t) - b\|_2$ synthesized by ZNN model (8.5). . . . 206

8.4 Comparison between asymptotical convergence and exponential convergence. 207

8.5 Block diagram of GNN model (8.8) solving $Ax = b$. 208

8.6 Solving $Ax = b$ by GNN (8.8) using different activation-functions with $\gamma = 10^6$ and starting from random initial states $x(0)$. 210

8.7 Solution error $\|x(t) - x^*\|_2$ of GNN (8.8) using linear activation-functions: in left graph $\gamma = 10^6$ and in right graph $\gamma = 10^7$. 211

8.8 Solution error $\|x(t) - x^*\|_2$ of GNN (8.8) using power-sigmoid functions: in left graph $\gamma = 10^6$ and in right graph $\gamma = 10^7$. 211

8.9 Energy function convergence of Wang neural network implying global stability for both no-solution case and multiple-solution case. 212

8.10 Online matrix inversion by GNN model (8.17). 217

8.11 Convergence of $\|X(t) - A^{-1}\|_F$ using power-sigmoid activation-function. . 218

8.12 Online matrix inversion by GNN (8.19) with large implementation errors. . 219

8.13 Convergence of computational error $\|X(t) - A^{-1}\|_F$ by perturbed GNN (8.19). 220

8.14 $\|X(t) - PINV(A)\|$ synthesized by (8.17) in the case of A being singular, where solid curves correspond to using the power-sigmoid function, while dashed-dotted curves correspond to the linear function. 221

8.15 The end-effector moving along a circle of radius 30cm. 222

8.16 States $X(t)$ by (8.17), corresponding to $(JJ^T)^{-1}$, for robot. 223

8.17 Positioning error on X, Y and Z axes at the end-effector. 223

9.1 Noncyclic motion trajectories of 4-link planar robot arm performing a square path synthesized without considering CMG criterion (i.e., $\beta = 0$). . . 237

9.2 Cyclic motion trajectories of 4-link planar robot arm performing a square path synthesized by CMG scheme (9.3)-(9.5) with coefficient $\beta = 4$. 238

9.3 End-effector positioning error of 4-link planar robot arm when tracking a square path synthesized with the CMG criterion considered. 239

9.4	Noncyclic motion trajectories of 5-link planar robot arm performing a square path synthesized without considering CMG criterion (i.e., $\beta = 0$).	240
9.5	Cyclic motion trajectories of 5-link planar robot arm performing a square path synthesized by CMG scheme (9.3)-(9.5) with coefficient $\beta = 4$.	242
9.6	Noncyclic motion trajectories of 6-link planar robot arm performing a square path synthesized without considering CMG criterion (i.e., $\beta = 0$).	243
9.7	Cyclic motion trajectories of 6-link planar robot arm performing a square path synthesized by CMG scheme (9.3)-(9.5) with coefficient $\beta = 4$.	243

List of Tables

9.1   Joint-position change of 4-link robot arm when tracking a square path without considering the CMG criterion. 237

9.2   Joint-position change of 4-link robot arm when tracking a square path with the CMG criterion considered. 238

9.3   Joint-position change of 5-link robot arm when tracking a square path without considering the CMG criterion. 241

9.4   Joint-position change of 5-link robot arm when tracking a square path with the CMG criterion considered. 241

Preface

The real-time solution to a mathematical problem arises in numerous fields of science, engineering, and business. It is usually an essential part of many solutions, e.g., matrix/vector computation, optimization, control theory, kinematics, signal processing, and pattern recognition. In recent years, due to the in-depth research on neural networks, numerous recurrent neural networks (RNN) based on the gradient-based method have been developed and investigated. Particularly, some simple neural networks were proposed to solve linear programming problems in real time and implemented on analog circuits. The neural approach is now regarded as a powerful alternative for online computation and optimization because of its parallel distributed nature, self-adaptation ability and convenience of hardware implementation.

However, the conventional gradient-based neural networks (GNN) were designed intrinsically for constant (or termed, time-invariant, static) coefficient matrices rather than time-varying (or termed, time-variant, nonstationary) ones. Then, time-varying problems were traditionally handled by approximating them as static problems via a short-time invariance hypothesis. In other words, the effect of time-variation is often ignored in dealing with time-varying systems arising in practice. Therefore, the obtained results can not fit exactly well with the practical and engineering requirements. This is because GNN does not make full use of the important time-derivative information of the coefficient(s). So, there always exist lagging-behind errors between the GNN-obtained solutions and the theoretical solutions. As a result, GNN belongs to a tracking approach, which could only adapt to the change of coefficients in a posterior passive manner.

As we know, time-varying problems are frequently encountered in scientific and engineering applications, such as aerodynamic coefficients in high-speed aircraft, circuit parameters in electronic circuits, and mechanical parameters in machinery. In order to achieve better performance and higher accuracy for such time-varying systems, it is necessary for us to investigate the effect of time-variation. It follows from a variety of computer simulation results that, GNN approaches could not always solve efficiently time-varying problems, even if very stringent restrictions on design parameters are imposed.

To the authors' knowledge, there is quite little work dealing with such time-varying problems in the neural-network literature at the present stage. Thus, it would be practical and theoretically significant if a new type of RNN is proposed for solving online the time-varying problems exactly and efficiently. In the context of these requirements (e.g., robot-arm motion planning and/or time-varying pole placement), since 12 March 2001, Zhang *et al.* have formally proposed, investigated and developed a special type of recurrent neural networks (or termed, Zhang neural networks, ZNN), which have been substantiated comparatively and theoretically to solve online such time-varying problems exactly and efficiently. In this book, the resultant Zhang neural networks (including Zhang dynamics, ZD, and Zhang neural dynamics, ZND) are designed, developed, modeled, analyzed, compared and simulated for online solution of time-varying problems, such as time-varying Sylvester matrix-equation solving, time-varying Lyapunov matrix-equation solving, time-varying matrix inversion, linear time-varying system solving, nonlinear time-varying matrix-equation (e.g., matrix-square-roots) solving, time-varying nonlinear equation, time-varying optimization and resolution of redundant robot manipulators. In view of making good use of the time-derivative information of coefficients, ZNN belongs to a predictive approach, which is more effective on the system convergence to a time-varying theoretical solution in comparison with GNN models.

In a word, we live in a time-varying world, and each static solution made (even right now) for a problem may have already lagged behind the time. Finding a solution ahead of time (or right for the time) may thus be quite complex. In this book, ZNN, ZD or ZND theory formalizes these problems and solutions in the time-varying context and often provides compact models that could solve those dynamic problems. With the elegant formalization of the problems and solutions made by us, the ZNN, ZD or ZND theory may promise to become a major inspiration for studies in the time-varying problems solving.

Structure of Book

In addition to Appendix A presenting some fundamentals of mathematics related to this research program, this book is divided into ten chapters. Most of the materials of these chapters (excluding Chapters 1, 10, and Appendix A) are derived from the authors' papers published in journals and proceedings of the international conferences, which are also shown evidently in the Authors' Related Publication List part.

Chapter 1 - In view of the remarkable features such as potential high-speed parallel-processing capability, distributed storage and adaptive self-learning, artificial neural networks (ANN) could be widely applied to many science and engineering fields. Being one of the most important neural models, recurrent neural networks (RNN) have been involved widely in many theoretical and practical areas. As a result, the neural-dynamic approach based on RNN has been regarded as a powerful alternative for online computation (e.g., matrix inversion) and optimization. In this chapter, we firstly recall with simplicity the development of ANN (especially RNN). Then, an overview of time-varying problems is presented. For investigation and illustration, some fundamental definitions of convergence/stability are introduced, in addition to four basic types of activation functions used.

Chapter 2 - Differing from GNN models, a special kind of recurrent neural networks has recently been proposed by Zhang *et al.* for time-varying matrix inversion and equations solving. As compared to GNN models, ZNN models are designed based on matrix-valued or vector-valued error functions, instead of scalar-valued energy functions based on matrix norm. In addition, ZNN models are depicted in implicit dynamics instead of explicit dynamics. Furthermore, Zhang neural network globally exponentially converges to the exact solution of linear time-varying equations. In this chapter, we simulate and compare ZNN and GNN models for the online solution of linear time-varying equations using MATLAB-coding approach and Simulink-modeling approach. Computer-simulation results, including the application to robot kinematic control, substantiate the theoretical analysis and demonstrate the efficacy of Zhang neural network on linear time-varying equation solving, especially when using power-sigmoid activation functions [1-3].

Chapter 3 - In this chapter, by following Zhang *et al.*'s method, a recurrent neural network (i.e., ZNN) is developed and analyzed for solving online the time-varying convex quadratic-minimization (QM) and quadratic-programming (QP) problems (of which the latter is subject to a time-varying linear-equality constraint as an example). Differing from conventional GNN, such a ZNN model makes full use of the time-derivative information of time-varying coefficients. The resultant ZNN model is theoretically proved to have global exponential convergence to the time-varying theoretical optimal solution of the investigated time-varying convex quadratic problems. Thus, such a ZNN model can be unified as a superior approach for solving online the time-varying quadratic problems. For the purpose of time-varying quadratic-problems solving, this chapter investigates comparatively both ZNN and GNN solvers, and then their unified modeling techniques. Computer-simulation results substantiate well the efficacy of such ZNN models on solving online the time-varying convex quadratic problems (i.e., QM and QP problems) [4-6].

Chapter 4 - Differing from GNN, a special kind of recurrent neural networks has recently been proposed by Zhang *et al.* for time-varying matrix inversion and pseudoinverse solving. As compared with GNN model, such ZNN models are designed based on a matrix-valued error function, instead of a scalar-valued norm-based energy function. In addition, in this chapter, we simulate and compare ZNN and GNN models for the online solution of time-varying matrix inverse and pseudoinverse using MATLAB-coding and Simulink-modeling techniques as well. Computer-simulation results, including an application to robot kinematic control, substantiate the theoretical analysis and demonstrate the efficacy of ZNN models on linear time-varying inverse and/or pseudoinverse solving, especially when using power-sigmoid activation functions [7-13].

Chapter 5 - For solving online the linear matrix equation (e.g., Sylvester equation, Lyapunov equation, $AXB = C$) with time-varying coefficients, this chapter presents a special kind of recurrent neural networks by using the design method recently proposed by Zhang *et al.*. The resultant ZNN model is deliberately developed in the way that its trajectory could be guaranteed to globally exponentially converge to the time-varying theoretical solution of a given linear matrix equation. For comparison, we develop and simulate the GNN as well, which is exploited to solve online the same time-varying linear matrix equation. In addition, towards the final purpose of hardware realization, this chapter highlights the model building and convergence illustration of ZNN model in comparison with GNN. Computer-simulation results substantiate the theoretical efficacy and superior performance of ZNN for the online solution of time-varying linear matrix equation, especially when using a power-sigmoid activation function array [14-21].

Chapter 6 - ZNN model has been proposed for online time-varying problems solving. Such a ZNN model is usually depicted in implicit dynamics rather than explicit dynamics. In this chapter, we generalize, develop, compare and simulate the ZNN and GNN models for online solution of time-varying matrix square roots. Important simulation and modeling techniques are thus investigated to facilitate the models' verification. Computer-verification results based on power-sigmoid activation functions further substantiate the superior ZNN convergence and efficacy on time-varying problems solving, as compared to the GNN model [22, 23].

Chapter 7 - Different from gradient-based dynamics (GD), a special class of neural dynamics has been found, developed, generalized and investigated by Zhang *et al.*, e.g., for online solution of time-varying nonlinear equations (in form of $f(x(t),t) = 0$) and/or static nonlinear equations (in form of $f(x) = 0$). The resultant Zhang dynamics (ZD) is designed based on the elimination of an indefinite error-function (instead of the elimination of a square-based positive or at least lower-bounded energy-function usually associated with GD and/or Hopfield-type neural networks). For comparative purposes, the GD model is also developed and exploited for online solving such two forms of nonlinear equations. Computer-simulation results substantiate further the theoretical analysis and efficacy of the ZD models for online solution of nonlinear time-varying and/or static equations [24-27].

Chapter 8 - In this chapter, two recurrent neural networks, Zhang neural network (ZNN) and gradient neural network (GNN), are generalized and developed to solve online a set of simultaneous linear constant problems. Firstly, we investigate the convergence, network architecture and electronic implementation of a ZNN model for linear equations solving. Secondly, rather than previously-presented asymptotical convergence, global exponential convergence is proved for GNN models used to solve linear equations. Moreover, superior convergence could be achieved by using power-sigmoid activation-functions, as compared with the situation of using linear activation-functions. Finally, this chapter investigates the simulation of GNN for online solution of the matrix-inverse problem. Several important techniques are employed to simulate such a neural-network system. In addition to investigating the singular case, this chapter also presents an application example on inverse-kinematic control of redundant manipulators via online pseudoinverse solution. Computer-simulation results substantiate further the analysis and efficacy of ZNN and GNN models for online linear constant problems solving [28-32].

Chapter 9 - ZNN and GNN models are unified and investigated in this chapter. Differing from the design of GNN model based on a nonnegative (or lower-bounded at least) scalar-valued norm-based energy function, the ZNN model is designed based on an indefinite matrix/vector-valued error function. For illustration and comparison of such two neural-network models, this chapter presents a cyclic-motion-generation (CMG) scheme (9.3)-(9.5) for redundant robot manipulators. Computer-simulation results based on three types of planar robot arms tracking a square path have substantiated again the efficacy of such a CMG scheme; moreover, theoretical analysis based on both gradient-descent and Zhang *et al.*'s neural-dynamic approaches have further shown the common effectiveness of the design methodologies [33].

Chapter 10 - The preceding chapters have detailedly addressed the design, analysis, applications and comparisons (with GNN/GD/GND) of ZNN (ZD/ZND) for time-varying and time-invariant problems solving. In this concluding chapter, we summarize what has been established in this research program as well as the book; and then we describe some potential future research topics and directions to extend the present contents of this book to more generalized and complicated problems in science and engineering.

This book, as a creative work of the authors' RNN research on the real-time solution of time-varying problems, is intended primarily for researchers, engineers and postgraduates studying in neural networks, intelligent control, signal processing, and so on.

Acknowledgements

This book basically comprises the results of many original research papers of the authors' research group, in which many authors of these original papers have done a great deal of detailed and creative research-work. Therefore, we are much obliged to our contributing authors for their high-quality work.

In particular, for the publication of this book, I would like to thank sincerely the vice president and editor of Nova Science Publishers, Frank Columbus, for his kind reminding, advice and help; and, I would also like to express my gratitude to the IEEE Intellectual Property Rights Office and Springer Verlag Heidelberg Rights and Permissions Department for their generously granting the permission on our reusing, compiling and reprinting some copyrighted materials (which are shown in the Authors' Related Publication List part) in this book.

Yours Sincerely, Respectfully and Gratefully,
Yunong Zhang, PhD, Professor, PhD-Supervisor
School of Information Science & Technology
Sun Yat-Sen University, Guangzhou 510006, China
Emails: zhynong@mail.sysu.edu.cn, ynzhang@ieee.org
Web-page: http://sist.sysu.edu.cn/%7Ezhynong/

Authors' Related Publication List

[1] Zhang, Y; Peng, H. Zhang neural network for linear time-varying equation solving and its robotic application. *Proceedings of the 6th International Conference on Machine Learning and Cybernetics*, 2007, pp. 3543-3548.

[2] Zhang, Y; Chen, K; Ma, W. MATLAB simulation and comparison of Zhang neural network and gradient neural network for online solution of linear time-varying equations. *DCDIS Proceedings of International Conference on Life System Modeling and Simulation*, 2007, pp. 450-454.

[3] Zhang, Y; Guo, X; Ma, W. Modeling and simulation of Zhang neural network for online linear time-varying equations solving based on MATLAB Simulink. *Proceedings of the 7th International Conference on Machine Learning and Cybernetics*, 2008, pp. 805-810.

[4] Zhang, Y; Li, X; Li, Z. Modeling and verification of Zhang neural networks for online solution of time-varying quadratic minimization and programming. *LNCS Proceedings of International Symposium on Intelligence Computation and Applications*, 2009, vol. 5821, pp. 101-110.

[5] Zhang, Y; Li, Z; Yi, C; Chen, K. Zhang neural network versus gradient neural network for online time-varying quadratic function minimization. *LNAI Proceedings of International Conference on Intelligent Computing*, 2008, vol. 5227, pp. 807-814.

[6] Zhang, Y; Li, Z. Zhang neural network for online solution of time-varying convex quadratic program subject to time-varying linear-equality constraints. *Physics Letters A*, 2009, vol. 373, no. 18-19, pp. 1639-1643.

[7] Zhang, Y; Ge, SS. A general recurrent neural network model for time-varying matrix inversion. *Proceedings of the 42nd IEEE Conference on Decision and Control*, 2003, vol. 6, pp. 6169-6174.

[8] Zhang, Y; Ge, SS. Design and analysis of a general recurrent neural network model for time-varying matrix inversion. *IEEE Transactions on Neural Networks*, 2005, vol. 16, no. 6, pp. 1477-1490.

[9] Zhang, Y; Li, Z; Fan, Z; Wang, G. Matrix-inverse primal neural network with application to robotics. *Dynamics of Continuous, Discrete and Impulsive Systems*, Series A, 2007, vol. 14, pp. 400-407.

[10] Zhang, Y; Guo, X; Ma, W; Chen, K; Cai, B. MATLAB Simulink modeling and simulation of Zhang neural network for online time-varying matrix inversion. *Proceedings of the 5th IEEE International Conference on Networking, Sensing and Control*, 2008, pp. 1480-1485.

[11] Zhang, Y; Tan, N; Cai, B; Chen, Z. MATLAB Simulink modeling of Zhang neural network solving for time-varying pseudoinverse in comparison with gradient neural network. *Proceedings of International Symposium on Intelligent Information Technology Application*, 2008, vol. 1, pp. 39-43.

[12] Zhang, Y; Chen, Z; Chen, K; Cai, B. Zhang neural network without using time-derivative information for constant and time-varying matrix inversion. *Proceedings of International Joint Conference on Neural Networks*, 2008, pp. 142-146.

[13] Zhang, Y; Yi, C; Ma, W. Simulation and verification of Zhang neural network for online time-varying matrix inversion. *Simulation Modelling Practice and Theory*, 2009, vol. 17, no. 10, pp. 1603-1617.

[14] Chen, K; Yue, S; Zhang, Y. MATLAB simulation and comparison of Zhang neural network and gradient neural network for online time-varying matrix equation $AXB - C = 0$. *LNAI Proceedings of International Conference on Intelligent Computing*, 2008, pp. 68-75.

[15] Zhang, Y; Chen, K. Comparison on Zhang neural network and gradient neural network for time-varying linear matrix equation $AXB = C$ solving. *Proceedings of IEEE International Conference on Industrial Technology*, 2008, pp. 1-6.

[16] Tan, N; Chen, K; Shi, Y; Zhang, Y. Modeling, verification and comparison of Zhang neural net and gradient neural net for online solution of time-varying linear matrix equation. *Proceedings of the 4th IEEE Conference on Industrial Electronics and Applications*, 2009, pp. 3698-3703.

[17] Zhang, Y; Yue, S; Chen, K; Yi, C. MATLAB simulation and comparison of Zhang neural network and gradient neural network for time-varying Lyapunov equation solving. *LNCS Proceedings of the 5th International Symposium on Neural Networks*, 2008, vol. 5263, pp. 117-127.

[18] Zhang, Y; Chen, K; Li, X; Yi, C; Zhu, H. Simulink modeling and comparison of Zhang neural networks and gradient neural networks for time-varying Lyapunov equation solving. *Proceedings of the Fourth International Conference on Natural Computation*, 2008, pp. 521-525.

[19] Zhang, Y; Fan, Z; Li, Z. Zhang neural network for online solution of time-varying Sylvester equation. *LNCS Proceedings of International Symposium on Intelligence Computation and Applications*, 2007, vol. 4683, pp. 276-285.

[20] Ma, W; Zhang, Y; Wang, J. MATLAB Simulink modeling and simulation of Zhang neural networks for online time-varying Sylvester equation solving. *Proceedings of International Joint Conference on Neural Networks*, 2008, pp. 286-290.

[21] Zhang, Y; Chen, K; Ma, W; Li, P. MATLAB simulation of Zhang neural networks for time-varying Sylvester equation solving. *Proceedings of the International Conference on Information Computing and Automation*, 2007, pp. 392-395.

[22] Zhang, Y; Yang, Y. Simulation and comparison of Zhang neural network and gradient neural network solving for time-varying matrix square roots. *Proceedings of International Symposium on Intelligent Information Technology Application*, 2008, vol. 2, pp. 966-970.

[23] Zhang, Y; Yang, Y; Tan, N. Time-varying matrix square roots solving via Zhang neural network and gradient neural network: modeling, verification and comparison. *LNCS Proceedings of the Sixth International Symposium on Neural Networks*, 2009, vol. 5551, pp. 11-20.

[24] Zhang, Y; Yi, C; Ma, W. Comparison on gradient-based neural dynamics and Zhang neural dynamics for online solution of nonlinear equations. *LNCS Proceedings of International Symposium on Intelligence Computation and Applications*, 2008, vol. 5370, pp. 269-279.

[25] Zhang, Y; Xu, P; Tan, N. Further studies on Zhang neural-dynamics and gradient dynamics for online nonlinear equations solving. *Proceedings of IEEE International Conference on Automation and Logistics*, 2009, pp. 566-571.

[26] Zhang, Y; Yi, C; Zheng, J. Comparison on Zhang neural dynamics and gradient-based neural dynamics for online solution of nonlinear time-varying equation. *Neural Computing and Applications*, 2010, in press.

[27] Zhang, Y; Xu, P; Tan, N. Solution of nonlinear equations by continuous- and discrete-time Zhang dynamics and more importantly their links to Newton iteration. *Proceedings of IEEE International Conference Information, Communications and Signal Processing*, 2009, pp. 1-5.

[28] Zhang, Y. Revisit the analog computer and gradient-based neural system for matrix inversion. *Proceedings of IEEE International Symposium on Intelligent Control*, 2005, pp. 1411-1416.

[29] Zhang, Y; Chen, K; Ma, W; Li, X. MATLAB simulation of gradient-based neural network for online matrix inversion. *LNAI Proceedings of International Conference on Intelligent Computing*, 2007, vol. 4682, pp. 98-109.

[30] Zhang, Y; Chen, K. Global exponential convergence and stability of Wang neural network for solving online linear equations. *Electronics Letters*, 2008, vol. 44, no. 2, pp. 145-146.

[31] Yi, C; Zhang, Y. Analogue recurrent neural network for linear algebraic equation solving. *Electronics Letters*, 2008, vol. 44, no. 18, pp. 1078-1079.

[32] Zhang, Y; Chen, Z; Chen, K. Convergence properties analysis of gradient neural network for solving online linear equations. *ACTA Automatica Sinica*, vol. 35, no. 8, pp. 1136-1139.

[33] Chen, K; Zhang, L; Zhang, Y. Cyclic motion generation of multi-link planar robot performing square end-effector trajectory analyzed via gradient-descent and Zhang et al's neural-dynamic methods. *Proceedings of the 2nd International Symposium on Systems and Control in Aeronautics and Astronautics*, 2008, pp. 1-6.

Chapter 1

Introduction

Abstract

In view of the remarkable features such as potential high-speed parallel-processing capability, distributed storage and adaptive self-learning, artificial neural networks (ANN) could be widely applied to many science and engineering fields. Being one of the most important neural models, recurrent neural networks (RNN) have been involved widely in many theoretical and practical areas. As a result, the neural dynamic approach based on RNN has been regarded as a powerful alternative for online computation (e.g., matrix inversion) and optimization. In this chapter, we firstly recall with simplicity the development of ANN (especially RNN). Then, an overview of time-varying problems is presented. For investigation and illustration, some fundamental definitions of convergence/stability are introduced, in addition to four basic types of activation functions.

1.1. Artificial Neural Network

In recent years, neural-network's researchers have focused on solving a variety of tasks requiring artificial neural networks in different areas, for instance, predicting and evaluating in the financial industry [1–3], solving monotone variational inequalities and related optimization problems [4, 5], and solving linear matrix equations such as matrix inversion [6–10]. An artificial neural network is an interconnected group of artificial neurons that uses a mathematical or computational model for information processing based on a connectionistic approach to computation [6]. In general, ANN could also be termed as simulated neural networks (SNN) or simply neural networks (NN). As a branch of artificial intelligence (AI), artificial neural networks are playing an important role in many theoretical and practical areas [1–10]. In addition, owing to high-speed parallel processing capability, inherent nonlinearity, distributed storage and adaptive self-learning capability, artificial neural networks have been widely applied to many science and engineering fields such as in control system design, signal processing, and robot inverse kinematics [6–11].

Due to in-depth research on neural networks, more and more ANN models have been developed and investigated [1, 3–6, 8]. Generally speaking, the area of ANN is nowadays considered from two main perspectives. That is, cognitive science which is an interdisci-

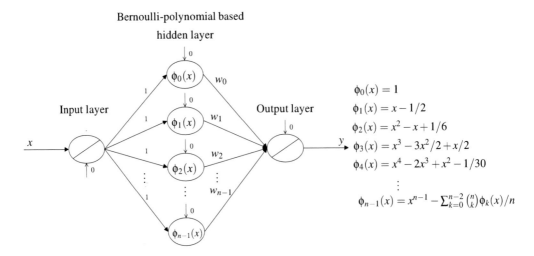

Figure 1.1. A new neural network with hidden neurons activated by Bernoulli polynomials.
Reproduced from Y. Zhang and Q. Ruan, Bernoulli neural network with weights directly determined and with the number of hidden-layer neurons automatically determined, Figure 1, W. Yu et al. (Eds.): ISNN 2009, Part I, LNCS 5551, pp. 36-45, 2009. ©Springer-Verlag Berlin Heidelberg 2009. With kind permission of Springer Science+Business Media.

plinary study of the mind; and connectionism which is a theory of information processing [3]. According to the definition of ANN, the combination of inputs, outputs, neurons, and connection weights constitute the architecture of a neural network [6]. Therefore, artificial neural networks can be classified in different categories in terms of architectures [6]. More precisely, neural networks can be classified, in terms of how the synaptic weights are obtained, into three categories: fixed-weight networks, unsupervised networks, and supervised networks [12]. There is no explicit learning required for the fixed-weight networks in the usual concept that learning is to determine the weights so that for inputs, the network generates desirable outputs [6]. On the other hand, according to the nature of connectivity, neural networks can also be categorized as feedforward neural networks and feedback neural networks (e.g., recurrent neural networks) [6]. Note that, feedforward neural networks have no loops; while feedback neural networks have loops because of feedback connections [3]. In other words, working-signal is not fed back between neurons of feedforward neural networks; while working-signal is fed back between neurons in recurrent neural networks.

As the basic units of ANN, artificial neurons (e.g., M-P model) were first proposed in 1943 by Warren McCulloch who was a neurophysiologist, and Walter Pitts who was an MIT logician [13]. From then on, neural networks have attracted considerable attention as candidates for novel computational systems [12, 14]. Thus lots of artificial neuromimes {e.g., parallel distributed processing (PDP) model [15]} were proposed to improve the performance of neural networks so as to achieve the purpose of human-brain simulation. Besides, the performance of neural networks has been analyzed via the SVD-based methods, in addition to simple circuit/hardware implementation [16]. In the past sixty years, many ANN models have been proposed and studied for science, as well as engineering application purposes. For instance, one classical type of ANN named Hopfield neural network (or termed,

Hopfield net) was proposed by John Hopfield in the early 1980s [17]. It is worth mentioning here that Hopfield neural network has been implemented due to its circuit/hardware realization capability. Another classical type of ANN based on error back-propagation (BP) algorithm (i.e., BP neural network) was proposed by Rumelhart, McClelland and others in the mid-1980s [15]. Nowadays, BP neural network is one of the most widely-used neural networks in the computational-intelligence research and engineering fields [18, 19]. However, there exist some inherent weaknesses in BP-type neural networks, such as relatively slow convergence, local-minima existence, and uncertainties about the optimal number of hidden-layer neurons. In order to overcome the weaknesses, many improved BP algorithms (e.g., algorithmic, activation-function and network-structure improvement) have been proposed [19,20]. For example, a so-called Bernoulli neural network (BNN, as shown in Figure 1.1) was constructed elegantly with hidden-layer neurons activated by a group of Bernoulli polynomials [20].

1.2. Recurrent Neural Networks

Being one of the most important neural models, recurrent neural networks (RNN) have been widely used in many theoretical and practical areas [6–10, 17]. Recently, due to its potential high-speed parallel-processing capability, the neural dynamic approach based on RNN has been regarded as a powerful alternative for online mathematical/algebraic computation (e.g., matrix inversion) [6–10].

1.2.1. Overview of Recurrent Neural Networks

A recurrent neural network (RNN), also called a feed-back neural network, is a class of artificial neural networks where connections between units form a directed loop (or termed, a directed cycle, a feed-back loop) [3,6]. Note that, the inherent loop in RNN creates an internal state of the network which allows it to exhibit dynamical behavior [6]. In other words, there exist feedback connections among neurons in the architecture of a recurrent neural network. Thus a recurrent neural network exhibits dynamical behavior [6]. Given an initial state, the state of a recurrent neural network evolves as time elapses. If the recurrent neural network is stable without oscillation, an equilibrium state can eventually be reached. Roughly speaking, a recurrent neural network may be viewed as the implementation of a dynamical system in physical devices (e.g., analog VLSI circuits) [6, 21–23].

As probably we know, in a feedforward neural network (e.g., BP-type neural network), signals propagate in only one direction, from an input stage through intermediate neurons to an output stage. Therefore, data from neurons of a lower layer are sent to neurons of an upper layer by feedforward connection networks. However, in a recurrent neural network (e.g., Hopfield-type neural network), signals propagate from the output of any neuron to the input of any neuron, and thus they bring data from neurons of an upper layer back to neurons of a lower layer [6]. Different from feedforward neural networks which are static (i.e., a given input can produce only one set of outputs, and hence carry no memory), recurrent neural networks not only operate on an input space but also on an internal state space, and hence the networks' architectures enable the information to be temporally memorised in the networks [3].

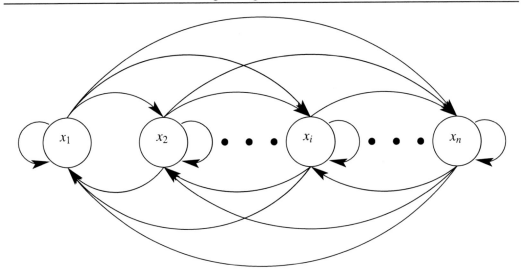

Figure 1.2. Architecture and connection of HNN (1.2) solving $Ax = b$.

Because of the existence of state feedback (i.e., feedback connections exist among neurons), a recurrent neural network exhibits dynamical behavior [6]. Generally speaking, recurrent neural networks could be described as dynamic equations for online problems solving. For such RNN, dynamical systems theory is usually used to model and analyze them. On the other hand, the dynamic equations could be formulated as the ordinary differential equation (ODE) [24]. In mathematics, the dynamic equations of RNN for online problem solving could generally be written in the form of [25]

$$M\dot{X}(t) = f(X(t),t), \quad (1.1)$$

where coefficient M is a mass matrix which could be defined accordingly for a specific problem solving, $\dot{X}(t)$ is the time-derivative of $X(t)$, and $f(\cdot)$ is a continuous linear and/or nonlinear mapping.

Take Hopfield-neural-network (HNN) as an example, it could be used to solve online a system of linear equations [6,26]. The linear-equations problem is generally formulated as: $Ax = b$, where coefficient matrix $A \in \mathbb{R}^{n \times n}$, coefficient vector $b \in \mathbb{R}^{n \times 1}$, and $x \in \mathbb{R}^{n \times 1}$ is the unknown vector to be obtained. The HNN's design procedure based on gradient algorithm is summarized as follows. Firstly, to obtain online solutions of the linear system, a norm-based scalar-valued energy function $\mathcal{E}(t) = \|Ax(t) - b\|_2^2/2$ could be defined and minimized. Secondly, researchers usually exploit the negative of the gradient $\partial \mathcal{E}/\partial x = A^T(Ax(t) - b)$ as the descent direction to minimize $\mathcal{E}(t)$. Thus, the following classic linear Hopfield-neural-network (or termed, gradient neural network, GNN) model could be derived to solve $Ax = b$ in real-time t:

$$\dot{x}(t) = dx/dt = -\gamma A^T (Ax(t) - b), \quad (1.2)$$

where $\gamma > 0$ is a design parameter used to scale the HNN convergence rate, and its value should be set as large as hardware permits or selected appropriately for experiments [6, 26, 27]. As seen from the HNN (1.2), it is exactly an ordinary differential equation (or simply termed, a dynamic equation), which can be viewed as a basic form of recurrent neural networks. The neural-network architecture of HNN model (1.2) is shown in Figure 1.2.

Before ending this subsection, it is worth mentioning that recurrent neural network can simulate as associative memories or pattern recognition, in addition to their applications in algebra and optimization (e.g., Hopfield-type recurrent neural networks have been developed to solve systems of linear equations and linearly-constrained quadratic real programs) [6, 28–31]. Moreover, the theoretical analysis for recurrent neural networks have also been intensively studied during the last decade [6, 32], since guaranteeing stability and convergence is very important.

1.2.2. Models of Recurrent Neural Networks

Unlike unsupervised and supervised networks, recurrent neural networks (RNN) do not usually need learning due to their fixed weight structures [6]. In 1982, Tank and Hopfield [6, 17, 21] introduced a linear programming neural network implemented by using an analog circuit which are well suited for applications that require on-line optimization. They showed that the state of the network evolved in such a way that the energy function of the network was monotonically nonincreasing with time. The Tank-Hopfield network has a shortcoming in that its equilibrium point may not be a solution to the original problem because it fails to satisfy the Kuhn-Tucker optimality conditions for a minimizer. However, their seminal work has inspired many researchers to investigate alternative neural networks for solving a wide variety of algebra and optimization problems. For instance, Hopfield-type recurrent neural networks have been developed and investigated to solve online problems of linear equations and linearly-constrained quadratic real programs [6, 29–31].

Since then, recurrent neural networks have attracted considerable attention as candidates for novel computational systems. In 1986, the Jordan neural network was proposed by Jordan for control of robots [3]. This network consists of a multilayer perceptron with one hidden layer and a feedback loop from the output layer to an additional input called the context layer, in which there are self-recurrent loops. In Jordan's network, past values of network outputs are fed back into hidden units. Another recurrent neural network was proposed in 1989 by Williams and Zipser for nonlinear adaptive filtering and pattern recognition [3]. This network consists of three layers: the input layer, the processing layer and the output layer. In the Williams-Zipser architecture, the neural network is fully connected, having one hidden layer. In 1990, the Elman network was proposed by Elman for problems in linguistics [3]. In Elman's network, past values of the outputs of hidden units are fed back into themselves. In addition, hybrid neural network architectures are proposed, developed and investigated [3, 33, 34]. These networks consist of a cascade of a neural network and a linear adaptive filter.

Note that, since 12 March 2001, Zhang *et al* have formally proposed a special class of recurrent neural network for online time-varying problems solving (e.g., matrix inversion) [6–10]. Differing from the scalar-valued norm-based energy function used in conventional gradient-based neural networks (GNN), the design of Zhang neural network (ZNN) is based on a matrix-valued error function. Such ZNN design method could make every entry of the matrix-valued error function converge to zero. In addition, ZNN is depicted in implicit dynamics, instead of explicit dynamics usually associated with GNN and/or Hopfield-type neural networks. More detailed description and important theoretical results about ZNN for time-varying problems solving would be shown in the ensuing chapters.

1.3. Time-Varying Problems

In recent years, many studies have been reported on real-time solutions of algebraic equations including matrix inversion and Sylvester equation [7, 29, 35–38]. Generally speaking, the methods reported in these references are related to the gradient descent method in optimization, which can be summarized as follows: first, construct a cost function such that its minimal point is the solution to the equation; then, a recurrent neural network is developed to evolve along a descent direction of this cost function until a minimum of the cost function is reached. A typical descent direction is defined by the negative gradient.

However, if the coefficients of the equation are time-varying, then a gradient-based method may not work well. Because of the effects of the time-varying coefficients, the negative gradient direction can no longer guarantee the decrease of the cost function. Usually a neural network of much faster convergence in comparison to the time-varying coefficients is required for a real-time solution if the gradient-based method adopted. The shortcomings of applying such a method to time-varying cases are two-fold: the much faster convergence is usually at cost of the precision or with stringent restrictions on design parameters, and such method is not applicable to the case where the coefficients vary quickly or the case of large-scale complex control systems.

To illustrate clearly and lay a basis for further discussing about the time-varying problems, here we also take the system of linear equations shown in Section 1.2.1. as an example. Evidently, the coefficients matrix $A \in \mathbb{R}^{n \times n}$ and vector $b \in \mathbb{R}^{n \times 1}$ are constant/static since they have no relevance with time t. However, if the coefficients of the linear system are functions of time t [i.e., $A(t)$ and $b(t)$], the linear matrix equation would be reformulated as: $A(t)x(t) = b(t)$, which is one of time-varying problems. The so-called time-varying problems could thus be described as a class of problems of linear and/or nonlinear matrix equations of which the coefficients change as a function of time [e.g., the coefficient matrix A and vector b in (1.2) are time-varying/nonstationary] [6]. Such time-varying systems are widely encountered in practical and engineering fields, such as aerodynamic coefficients in high-speed aircraft, circuit parameters in electronic circuits, mechanical parameters in machinery, and diffusion coefficients in chemical processes [39].

To the best of our acknowledge, there is a little work dealing with such time-varying problems solving in the neural-network literature at present stage. In general, the effects of time-variation arising in such time-varying systems are often ignored in practical application and scientific research. However, to achieve better performance and higher accuracy for time-varying systems, time-variation needs to be included in the model used to analyze a given system. Following the idea of using first-order time derivatives, a special kind of recurrent neural network for online time-varying problems solving was proposed by Zhang *at al* via utilizing the time derivatives of the coefficients [6–10]. The neural dynamics are elegantly introduced by defining a matrix/vector/scalar-valued error-monitoring function (rather than the usual norm-based scalar-valued energy function) such that the computation error can be made decreasing to zero globally and exponentially. In addition, Zhang neural network (ZNN) is depicted in an implicit dynamics, instead of an explicit dynamics usually associated with gradient-based neural networks and/or Hopfield-type neural networks [7, 29, 35–38]. It is worth mentioning that such a ZNN model could be reduced to a discrete-time model, which is depicted by a system of difference equations for the purposes of hardware (e.g., digital-circuit) implementation [27].

1.4. Definitions of Convergence and Stability

For a better understanding of the solution characteristics of recurrent neural networks, the following definitions are presented which might have been scattered in the literature (see [6, 40–43] and the references therein).

Definition 1.4.1. A recurrent neural network is said to be globally convergent, if starting from any initial point $x(t_0)$ taken in the whole associated Euclidean space, every state trajectory of the recurrent neural network converges to an equilibrium point x^* that depends on the initial state of the trajectory. Note that the initial time instant $t_0 \geq 0$, and there might exist many non-isolated equilibrium points x^*.

Definition 1.4.2. A recurrent neural network is said to be globally exponentially convergent, if every trajectory starting from any initial point $x(t_0)$ satisfies $\forall t \geq t_0 \geq 0$,

$$\|x(t) - x^*\| \leq \eta \|x(t_0) - x^*\| \exp(-\lambda(t - t_0)),$$

where η and λ are positive constants, x^* is an equilibrium point depending on initial states $x(t_0)$, and the symbol $\|\cdot\|$ hereafter denotes the Euclidean norm of a matrix or vector unless specified otherwise. Note that there might exist many non-isolated equilibrium points x^*, and the exponential convergence implies the arbitrarily fast convergence of the recurrent neural network.

Definition 1.4.3. An equilibrium point x^* of a recurrent neural network is locally stable if every trajectory, starting from the initial condition $x(t_0)$ near the equilibrium point x^*, stays near that equilibrium point x^*. On the other hand, an equilibrium point x^* of a recurrent neural network is globally stable if every trajectory starting from any initial condition $x(t_0)$ stays near that equilibrium point x^*.

Definition 1.4.4. An equilibrium point x^* of a recurrent neural network is locally asymptotically stable if it is locally stable and, in addition, the state $x(t)$ of the recurrent neural network, starting from any initial condition $x(t_0)$ near the equilibrium point x^*, converges to the equilibrium point x^* as time t increases. Note that the equilibrium point x^* is assumed here to be isolated.

Definition 1.4.5. An equilibrium point x^* of a recurrent neural network is globally asymptotically stable if it is globally stable and, in addition, all the state $x(t)$ of the recurrent neural network converges to the equilibrium point x^* as time t increases. Note that the equilibrium point x^* is assumed here to be unique.

Definition 1.4.6. An equilibrium point x^* of a recurrent neural network is locally exponentially stable if it is locally asymptotically stable and, in addition, the state $x(t)$ of the recurrent neural network, starting from any initial condition $x(t_0)$ near the equilibrium point x^*, converges as follows to the equilibrium point x^* as time t increases: $\forall t \geq t_0 \geq 0$,

$$\|x(t) - x^*\| \leq \eta \|x(t_0) - x^*\| \exp(-\lambda(t - t_0)),$$

where η and λ are positive constants, x^* is an isolated equilibrium point.

Definition 1.4.7. An equilibrium point x^* of a recurrent neural network is globally exponentially stable if it is globally asymptotically stable and, in addition, all the state $x(t)$ of the recurrent neural network converges as follows to the equilibrium point x^* as time t increases: $\forall t \geq t_0 \geq 0$,

$$\|x(t) - x^*\| \leq \eta \|x(t_0) - x^*\| \exp(-\lambda(t - t_0)),$$

where η and λ are positive constants, x^* is a unique equilibrium point.

Remark 1.4.1. Note that in the context of recurrent neural networks, the global (exponential) convergence of a neural-network system does not imply the global (exponential) stability of such a neural-network system in a conventional Lyapunov sense. This is because usually there might exist an infinite number of non-isolated solutions (or to say, equilibrium points) in a neural-network system. If all the state trajectory of a recurrent neural network converges to one of such non-isolated solutions/equilibriums, we call the recurrent neural network globally convergent. On the other hand, the global stability of a system normally requires the equilibrium point to be unique (and also isolated). □

Remark 1.4.2. Strictly speaking, we seldom say that a recurrent neural network (being a dynamic system) is stable or unstable. Instead, we talk about an equilibrium point (of such a neural network system) being stable or not. Thus, we first find the equilibrium points of such a neural network system, and then determine whether those equilibrium points are stable or not. In the case that a neural network system has a single equilibrium point that is globally attractive (meaning that all solutions converge to that equilibrium point from all initial conditions), we sometimes say that the recurrent neural network system is (globally) stable. But we really should say that the equilibrium point is globally stable. On the other hand, this comment is quite different from the usage of "global (exponential) convergence of a recurrent neural network", which is discussed in Definitions 1.4.1 and 1.4.2, and Remark 1.4.1. Furthermore, the concept of global stability could be extended if necessary. □

1.5. Activation Functions Used in Recurrent Neural Networks

In this section, we introduce some common activation functions used in recurrent neural networks. As we know, nonlinearity and errors always exist. Even if a linear activation function is used, the nonlinear phenomenon may appear in its hardware implementation. In addition, the nonlinearity inherent in the network is due to the overall action of all the activation functions of the neurons within the structure. Therefore, the choice of activation functions has a key influence on the complexity and performance of recurrent neural networks [3].

As is mentioned above, a neural network model is composed of a set of inputs, a set of outputs, and a set of simple processing elements. Each element is called a neuron. Each

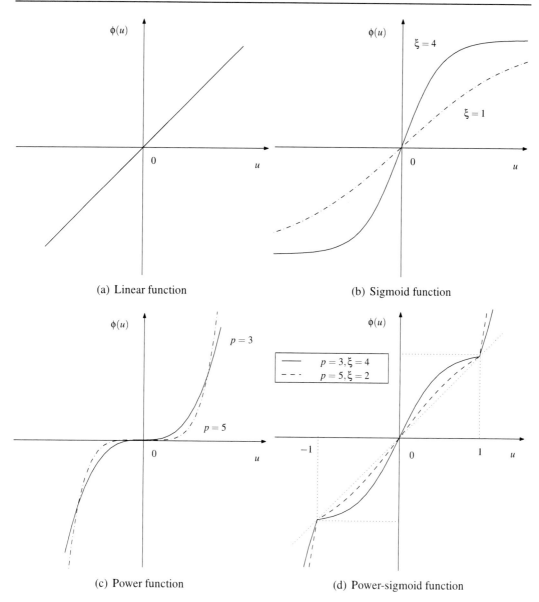

Figure 1.3. Activation functions $\phi(\cdot)$ used in recurrent neural networks.

of the neurons can send a signal to the other neurons in the neural network. These signals are rescaled by the so-called connection weights. These neurons usually sum a number of the weighted input variables and pass the result through a nonlinearity, which is the input-output relationship of the transformation. This relationship is characterized by a real-valued continuous function and usually called activation function [6]. Note that, the activation function is the mapping from the neural networks' input provided by a combination function to the output of a neuron [3]. Each neuron performs a simple transformation concurrently in a parallel distributed manner. The strength of interaction among neurons is reflected by the connection weights which determine the functional behavior of the neural network [6].

Take the previously presented continuous-time HNN (1.2) as an example, by adding an activation function array, its [i.e., the HNN (1.2)] general nonlinear form could be formulated as

$$\dot{x}(t) = dx/dt = -\gamma A^T \Phi(Ax(t) - b), \qquad (1.3)$$

where $\Phi(\cdot) : \mathbb{R}^{n \times 1} \to \mathbb{R}^{n \times 1}$ denotes a matrix-valued activation-function array of neural networks. The processing array $\Phi(\cdot)$ is made of n monotonically-increasing odd activation-functions $\phi(\cdot)$. Generally speaking, any monotonically-increasing odd activation-function $\phi(\cdot)$, being the ith element of matrix array $\Phi(\cdot)$, could be used for recurrent neural networks' construction. For instance, the following four basic types of activation functions are adopted and investigated [6, 27].

- linear activation function:

$$\phi(u) = u, \qquad (1.4)$$

- bipolar-sigmoid activation function:

$$\phi(u) = \frac{1 - \exp(-\xi u)}{1 + \exp(-\xi u)} \qquad (1.5)$$

with $\xi \geq 1$,

- power activation function:

$$\phi(u) = u^p \qquad (1.6)$$

with integer $p \geq 3$, and

- power-sigmoid activation function:

$$\phi(u) = \begin{cases} u^p, & \text{if } |u| \geq 1 \\ \frac{1 + \exp(-\xi)}{1 - \exp(-\xi)} \cdot \frac{1 - \exp(-\xi u)}{1 + \exp(-\xi u)}, & \text{otherwise} \end{cases} \qquad (1.7)$$

with suitable design parameters $\xi \geq 1$ and $p \geq 3$,

which could be shown in Figure 1.3. Other types of activation functions can thus be generated based on these basic types. It is worth mentioning that different choices of activation function arrays $\Phi(\cdot)$ [as seen in (1.3)] may lead to different performance [6–10, 27]. For example, for (1.3), if a linear or power-sigmoid activation-function array $\Phi(\cdot)$ is used, then the state vector $x(t)$ of HNN model (1.3) starting from any initial state $x_0 \in \mathbb{R}^n$ will exponentially converge to the theoretical solution $x^* = A^{-1}b$ of linear equation $Ax(t) = b$. In addition, the exponential-convergence rate is at least $\alpha\gamma$ with α denoting the minimum eigenvalue of $A^T A$. Moreover, the HNN model (1.3) using power-sigmoid activation functions has better convergence than that using linear activation functions [equivalently, HNN (1.2)].

In view of the fact that nonlinear phenomena may appear even in the hardware implementation of a linear activation function; e.g., in the form of saturation and/or inconsistency of the linear slope, and in the form of truncation and roundoff errors in digital realization [27], it may be necessary to investigate the impact of different activation functions in the neural network. The investigation of different activation functions (such as the sigmoid, power, and power-sigmoid functions) may show us an answer to the problems and side-effects of imprecise implementation and nonlinearities occurring to linear activation functions [27].

1.6. Conclusion

This chapter simply introduces artificial neural networks (especially RNN). As for time-varying problem solving, some definitions of convergence/stability have been given, including four basic types of activation function arrays.

References

[1] Kwon, YK; Moon, BR. A hybrid neurogenetic approach for stock forecasting. *IEEE Transactions on Neural Networks,* 2007, vol. 18, no. 3, pp. 851-864.

[2] Lee, RST. iJADE stock advisor: an intelligent agent based stock prediction system using hybrid RBF recurrent network. *IEEE Transactions on Systems, Man and Cybernetics,* Part A, 2004, vol. 34, no. 3, pp. 421-428.

[3] Mandic, DP; Chambers, JA. *Recurrent Neural Network for Prediction.* Singapore: Wiley, 2001.

[4] He, B; Yang, H. A neural-network model for monotone linear asymmetric variational inequalities. *IEEE Transactions on Neural Networks,* 2000, vol. 11, no. 1, pp. 3-16.

[5] Xia, Y; Wang, J. A general projection neural network for solving monotone variational inequalities and related optimization problems. *IEEE Transactions on Neural Networks,* 2004, vol. 15, no. 2, pp. 318-328.

[6] Zhang, Y. *Analysis and Design of Recurrent Neural Networks and their Applications to Control and Robotic Systems.* Ph.D. Dissertation, Chinese University of Hong Kong; 2002.

[7] Zhang, Y; Wang, J. Recurrent neural networks for nonlinear output regulation. *Automatica,* 2001, vol. 37, no. 8, pp. 1161-1173.

[8] Zhang, Y; Jiang, D; Wang, J. A recurrent neural network for solving Sylvester equation with time-varying coefficients. *IEEE Transactions on Neural Networks,* 2002, vol. 13, no. 5, pp. 1053-1063.

[9] Zhang, Y; Ge, SS. Design and analysis of a general recurrent neural network model for time-varying matrix inversion. *IEEE Transactions on Neural Networks,* 2005, vol. 16, no. 6, pp. 1477-1490.

[10] Zhang, Y; Ge, SS; Lee, TH. A unified quadratic-programmingbased dynamical system approach to joint torque optimization of physically constrained redundant manipulators. *IEEE Transactions on Systems, Man, and Cybernetics,* Part B, 2004, vol. 34, no. 5, pp. 2126-2132.

[11] Steriti, RJ; Fiddy, MA. Regularized image reconstruction using SVD and a neural network method for matrix inversion. *IEEE Transactions on Signal Processing,* 1993, vol. 41, no. 10, pp. 3074-3077.

[12] Kung, SY. *Digital Neural Network.* New Jersey: Prentice Hall, Englewood Cliffs, 1993.

[13] McCulloch, MS; Pitts, W. A logical calculus of the ideas immanent in nervous activity. *Bulletin of Mathematical Biology,* 1943, vol. 5, no. 4, pp. 115-133.

[14] Wilde, PD. *Neural Network Models,* New York: Springer-Verlag, 1996.

[15] Rumelhart, DE; McClelland, JL; PDP Research Group. *Parallel Distributed Processing,* Cambridge, USA: MIT Press, 1986.

[16] Manherz, RK; Jordan, BW; Hakimi, SL. Analog methods for computation of the generalized inverse. *IEEE Transactions on Automatic Control,* 1968, vol. 13, no. 5, pp. 582-585.

[17] Hopfield, JJ. Neural networks and physical systems with emergent collective computational abilities. *Proceedings of National Academy of Sciences, USA, Biophysics,* 1982, vol. 79, pp. 2554-2558.

[18] Guo, W; Qiao, Y; Hou, H. BP neural network optimized with PSO algorithm and its application in forecasting. *Proceedings of IEEE International Conference on Information Acquisition,* 2006, pp. 617-621.

[19] Jenkins, WM. Neural network weight training by mutation. *Computers and Structures,* 2006, vol. 84, no. 31/32, pp. 2107-2112.

[20] Zhang, Y; Ruan, G. Bernoulli neural network with weights directly determined and with the number of hidden-layer neurons automatically determined. *Proceedings of the Sixth International Symposium on Neural Networks,* 2009, pp. 36-45.

[21] Tank, DW; Hopfield, JJ. Simple neural optimization networks: an A/D converter, signal decision circuit, and a linear programming circuit. *IEEE Transactions on Circuits and Systems,* 1986, vol. 33, no. 5, pp. 533-541.

[22] Botros, NM; Abdul-Aziz, M. Hardware implementation of an artificial neural network using field programmable gate arrays (FPGAs). *IEEE Transactions on Industrial Electronics,* 1994, vol. 41, no. 6, pp. 665-667.

[23] Diorio, C; Rao, RPN. Neural circuits in silicon. *Nature,* 2000, vol. 405, no. 6789, pp. 891-892.

[24] Morris, T; Harry, P. *Ordinary Differential Equations.* New York: Dover Publications, 2003.

[25] The MathWorks. *Inc., MATLAB 7.0.* MA: Natick, 2004.

[26] Zhang, Y; Li, Z; Chen, K; Cai, B. Common nature of learning exemplified by BP and Hopfield neural networks for solving online a system of linear equations. *Proceedings of IEEE International Conference on Networking, Sensing and Control,* 2008, pp. 832-836.

References

[27] Zhang, Y; Ma, W; Cai, B. From Zhang neural network to Newton iteration for matrix inversion. *IEEE Transactions on Circuits and Systems I: Regular Papers,* 2009, vol. 56, no. 7, pp. 1405-1415.

[28] Sudharsanan, SI; Sunareshan, MK. Equilibrium characterization of dynamical neural networks and a systematic synthesis procedure for associative memories. *IEEE Transactions on Neural Networks,* 1991, vol 2, no. 5, pp. 509-521.

[29] Cichocki, A; Unbehauen, R. Neural networks for solving systems of linear equation and related problems. *IEEE Transactions on Circuits and Systems I: Fundamental Theory and Applications,* 1992, vol. 39, no. 2, pp. 124-138.

[30] Maa, CY; Shanblatt, MA. Linear and quadratic programming neural network analysis. *IEEE Transactions on Neural Networks,* 1992, vol. 3, no. 4, pp. 580-594.

[31] Liang, X; Wang, J. A recurrent neural network for nonlinear optimization with a continuously differentiable objective function and bound constraints. *IEEE Transactions on Neural Networks,* 2000, vol. 11, no. 6, pp. 1251-1262.

[32] Zhang, Y; Wang, J. Global exponential stability of recurrent neural networks for synthesizing linear feedback control systems via pole assignment. *IEEE Transactions on Neural Networks,* 2002, vol. 13, no. 3, pp. 633-644.

[33] Khalaf, AAM; Nakayama, K. A hybrid nonlinear predictor: analysis of learning process and predictability for noisy time series. *IEICE Transections on Fundamentals,* 1999, vol. 82, no. 8, pp. 1420-1427.

[34] Khotanzad, A; Lu, J. Non-parametric prediction of AR processes using neural networks. *Proceedings of International Conference on Acoustics, Speech, and Signal Processing,* 1990, vol. 5, pp. 2551-2554.

[35] Wang, J. A recurrent neural network for real-time matrix inversion. *Applied Mathematics and Computation,* 1993, vol. 55, no. 1, pp. 89-100.

[36] Wang, J; Wu, G. A multilayer recurrent neural network for solving continuous-time algebraic Riccati equations. *Neural Networks,* 1998, vol. 11, no. 5, pp. 939-950.

[37] Song, Y; Yam, Y. Complex recurrent neural network for computing the inverse and pseudo-inverse of the complex matrix. *Applied Mathematics and Computation,* 1998, vol. 93, no. 2-3, pp. 195-205.

[38] Xia, Y; Wang, J; Hung, DL. Recurrent neural networks for solving linear inequalities and equations. *IEEE Transactions on Circuits and Systems I: Fundamental Theory and Applications,* 1999, vol. 46, no. 4, pp. 452-462.

[39] Lee, HC; Choi, JW. Linear time-varying eigenstructure assignment with flight control application. *IEEE Transactions on Aerospace and Electronic Systems,* 2004, vol. 40, no. 1, pp. 145-157.

[40] Zhang, Y; Wang, J; Xu, Y. A dual neural network for bi-criteria kinematic control of redundant manipulators. *IEEE Transactions on Robotics and Automation,* 2002, vol. 18, no. 6, pp. 923-931.

[41] Zhang, Y; Wang, J; Xia, Y. A dual neural network for redundancy resolution of kinematically redundant manipulators subject to joint limits and joint velocity limits. *IEEE Transactions on Neural Networks,* 2003, vol. 14, no. 3, pp. 658-667.

[42] Zhang, Y; Wang, J. A dual neural network for constrained joint torque optimization of kinematically redundant manipulators. *IEEE Transactions on Systems, Man, and Cybernetics,* part B, 2002, vol. 32, no. 5, pp. 654-662.

[43] Zhang, Y. *Dual Neural Networks: Design, Analysis, and Application to Redundant Robotics, in: Progress in Neurocomputing Research.* New York: Nova Science Publishers, 2007.

Part I

Matrix-Vector Problems

Chapter 2

ZNN Solving the Time-Varying Linear System $A(t)x(t) = b(t)$

Abstract

Different from gradient-based neural networks (in short, gradient neural networks), a special kind of recurrent neural networks has recently been proposed by Zhang *et al.* for time-varying matrix inversion and equation solving. As compared to gradient neural networks (GNN), Zhang neural networks (ZNN) are designed based on matrix-valued or vector-valued error functions, instead of scalar-valued error functions based on matrix norm. In addition, Zhang neural networks are depicted in implicit dynamics instead of explicit dynamics. Furthermore, the Zhang neural network globally exponentially converges to the exact solution of linear time-varying equations. In this chapter, we simulate and compare the Zhang neural network and gradient neural network for the online solution of linear time-varying equations using the MATLAB-coding approach and Simulink-modeling approach. Computer-simulation results, including the application to robot kinematic control, substantiate the theoretical analysis and demonstrate the efficacy of the Zhang neural network on linear time-varying equation solving, especially when using power-sigmoid activation functions.

2.1. Introduction

The problem of linear equation solving (including matrix-inverse problems as a closely-related topic) is considered to be one of the basic problems widely encountered in science and engineering. It is usually an essential part of many solutions; e.g., as preliminary steps for optimization [1], signal-processing [2], electromagnetic systems [3], and robot inverse kinematics [4]. In mathematics, the problem is formulated as $Ax = b$, where coefficient matrix $A \in \mathbb{R}^{n \times n}$, coefficient vector $b \in \mathbb{R}^n$, and $x \in \mathbb{R}^n$ is the unknown vector to be obtained.

There are two general types of solutions to the problem of linear equations. One is the numerical algorithms performed on digital computers (i.e., on today's computers). Usually, the minimal arithmetic operations for a numerical algorithm are proportional to the cube of the coefficient-matrix dimension n, i.e., $O(n^3)$ operations. Evidently, such serial-processing numerical algorithms may not be efficient enough for large-scale online or real-time applications. Thus, some $O(n^2)$-operation algorithms were proposed in order to remedy this computational problem, e.g., in [5, 6]. However, they may still not be fast enough. For example, in [5], it takes on average around one hour to solve a set of linear equations with dimension $n = 60000$. Being the second general type of solution, many parallel-processing computational methods have been developed, analyzed, and implemented on specific architectures [2, 4, 7–16]. The dynamic-system approach is one of the important parallel-processing methods for solving a set of linear equations [2, 7–13]. Recently, because of the in-depth research on neural networks, numerous dynamic and analog solvers based on recurrent neural networks (RNN) have been developed and investigated [2, 8–13]. The neural-dynamic approach is now regarded as a powerful alternative to online computation because of its parallel distributed nature and convenience of hardware implementation [4, 7, 10, 16, 17].

Different from gradient neural networks (GNN) for time-invariant problem solving [7–10, 16, 18], a special kind of recurrent neural networks has recently been proposed by Zhang *et al.* [10–13] for real-time solution of time-varying problems. In other words, in our context of $Ax = b$, coefficients A and b could respectively be $A(t)$ and $b(t)$, time-varying ones. The design method of Zhang neural networks (ZNN) is completely different from that of gradient neural networks. In this chapter, we generalize such a design method to solving online a set of linear time-varying equations, namely, $A(t)x(t) = b(t)$ over time t. We also simulate the resultant Zhang neural network and compare it with gradient neural network on the solution of linear time-varying equations.

Nowadays, more research and development tools have been provided, which could translate different levels of design languages and finally implement a system/algorithm onto hardware [19]. MATLAB Simulink® is one of such useful tools for modeling the dynamic systems at a high level by exploiting many built-in toolboxes, which could further generate an FPGA (field programmable gate array) realization of the systems [20, 21]. For the final purpose of neural-network hardware-implementation, the modeling and simulation of ZNN based on MATLAB Simulink is investigated in this chapter as our first "hardwarizing" step.

2.2. Problem Formulation and Solvers

Let us consider the smoothly time-varying nonsingular matrix $A(t) \in \mathbb{R}^{n \times n}$ and vector $b(t) \in \mathbb{R}^n$. We are to find $x(t) \in \mathbb{R}^n$ such that the following set of linear time-varying equations holds true:

$$A(t)x(t) - b(t) = 0, \quad t \in [0, +\infty). \tag{2.1}$$

Without loss of generality, $A(t)$ and $b(t)$ are assumed to be known, while their time derivatives $\dot{A}(t)$ and $\dot{b}(t)$ are assumed to be known or could be estimated. The existence of the inverse $A^{-1}(t)$ at any time instant t is theoretically assumed in this chapter but might not needed in practice. In the ensuing subsections, ZNN and GNN are developed, which are exploited comparatively to solve (2.1) in real time t.

2.2.1. Zhang Neural Networks

A theoretical solution to (2.1) is $x(t) = A^{-1}(t)b(t)$, which actually results in the matrix-inverse problem and its online solvers [10, 13]. As we know [3, 5], solving directly for a vector-valued term, such as the direct solution $x(t)$ of (2.1), is generally more efficient and accurate than the way of solving firstly for the inverse A^{-1} and then $x = A^{-1}b$. So, in this chapter, the direct solution of linear time-varying equation (2.1) is preferred. By following the design method of Zhang *et al* [11–13], we can generalize a recurrent neural network to solving $A(t)x(t) = b(t)$ in real time t as follows.

Step 1. To monitor the solution process, instead of scalar-valued error functions, the following vector-valued error function is defined: $e(t) = A(t)x(t) - b(t) \in \mathbb{R}^n$.

Step 2. The error-function derivative $\dot{e}(t) \in \mathbb{R}^n$ should be made such that every element $e_i(t) \in \mathbb{R}$ of $e(t) \in \mathbb{R}^n$ converges to zero, $i = 1, \cdots, n$.

Step 3. A general form of $\dot{e}(t)$ could thus be

$$\frac{de(t)}{dt} = -\Gamma\Phi\big(e(t)\big), \tag{2.2}$$

where design parameter Γ and activation-function array $\Phi(\cdot)$ are described as follows.

- $\Gamma \in \mathbb{R}^{n \times n}$ is a positive-definite matrix used to scale the convergence of the solution. Γ could simply be γI, with scalar $\gamma > 0$ and I denoting the identity matrix.

- $\Phi(\cdot) : \mathbb{R}^n \to \mathbb{R}^n$ denotes a vector-form activation-function array (or termed, mapping) of neural networks. A simple example of $\Phi(\cdot)$ is the linear one, i.e., $\Phi(A(t)x(t) - b(t)) = A(t)x(t) - b(t)$ (Note that, other activation function arrays could be seen in Section 1.5.).

Step 4. Expanding the design formula (2.2) leads to the following implicit dynamic equation of Zhang neural network for solving a set of linear time-varying equations, (2.1).

$$A(t)\dot{x}(t) = -\dot{A}(t)x(t) - \gamma\Phi\big(A(t)x(t) - b(t)\big) + \dot{b}(t), \tag{2.3}$$

where $x(t)$, starting from an initial condition $x(0) = x_0 \in \mathbb{R}^n$, is the activation state-vector corresponding to the theoretical solution $x^*(t) := A^{-1}(t)b(t)$ of (2.1). It is worth mentioning that when using the linear activation-function array $\Phi(e) = e$, ZNN (2.3) reduces to the following classic linear one [11, 12, 22]:

$$A(t)\dot{x}(t) = -\big(\dot{A}(t) + \gamma A(t)\big)x(t) + \big(\dot{b}(t) + \gamma b(t)\big).$$

For modeling and realization purposes, we could firstly transform (2.3) into the following dynamics:

$$\dot{x}(t) = -\dot{A}(t)x(t) - A(t)\dot{x}(t) - \gamma\Phi\big(A(t)x(t) - b(t)\big) + \dot{b}(t) + \dot{x}(t). \tag{2.4}$$

Therefore, the corresponding block diagram of ZNN (2.3) is shown in Figure 2.1 by actually building (2.4). In addition, the ith-neuron dynamic equation of ZNN (2.3) is

$$\dot{x}_i = -\sum_{k=1}^{n} \dot{a}_{ik}x_k - \sum_{k=1}^{n} a_{ik}\dot{x}_k - \gamma\phi\left(\sum_{k=1}^{n} a_{ik}x_k - b_i\right) + \dot{b}_i + \dot{x}_i, \tag{2.5}$$

where

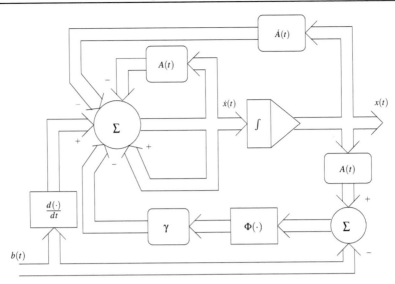

Figure 2.1. Block diagram which realizes ZNN (2.3). *Reproduced from Y. Zhang, X. Guo et al., Modeling and simulation of Zhang neural network for online linear time-varying equations solving based on MATLAB Simulink, Figure 1, Proceedings of the Seventh International Conference on Machine Learning and Cybernetics, pp. 455-460. ©[2008] IEEE. Reprinted, with permission.*

- x_i denotes the ith neuron of ZNN (2.3) corresponding to the ith element of state vector $x(t)$, $i = 1, 2, \ldots, n$;

- time-varying weights a_{ik} and $\dot{a}_{ik}$ are defined respectively as the ikth entries of matrix $A(t)$ and its time-derivative $\dot{A}(t)$;

- b_i and $\dot{b}_i$ denotes the ith elements of vector $b(t)$ and its time-derivative $\dot{b}(t)$.

According to this neuron equation, the circuit schematics of ZNN model (2.3) could thus be depicted in Figure 2.2.

2.2.2. Gradient-Based Neural Networks

For comparison, it is worth mentioning that almost all numerical algorithms and neural-dynamic computational schemes were designed intrinsically for constant coefficients $A \in \mathbb{R}^{n \times n}$ and $b \in \mathbb{R}^n$, rather than time-varying ones. The aforementioned neural-dynamic computational schemes [7–9, 12] are in general related to the gradient-descent method in optimization [23]. By using such a design method, we could have the classic linear gradient-based neural network (in short, gradient neural network), $\dot{x}(t) = -\gamma A^T (Ax(t) - b)$. This could be generalized to the following nonlinear form:

$$\dot{x}(t) = -\gamma A^T \Phi (Ax(t) - b). \tag{2.6}$$

Note that the GNN models generated by the negative-gradient method could be accurate and effective only if coefficients A and b are constant. When A and b are time-varying, the GNN solution could only approximately approach the theoretical solution $x^*(t)$.

ZNN Solving the Time-Varying Linear System $A(t)x(t) = b(t)$

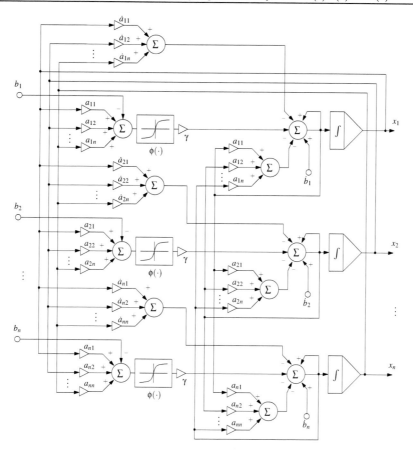

Figure 2.2. Circuit schematics which realizes ZNN (2.3). *Reproduced from Y. Zhang, X. Guo et al., Modeling and simulation of Zhang neural network for online linear time-varying equations solving based on MATLAB Simulink, Figure 2, Proceedings of the Seventh International Conference on Machine Learning and Cybernetics, pp. 455-460. ©[2008] IEEE. Reprinted, with permission.*

For the model-implementation error due to the imprecise implementation of system dynamics, we could consider the following perturbed model of GNN (2.6).

$$\dot{x} = -\gamma A^T \Phi(Ax - b) + \Delta_R(t), \tag{2.7}$$

where additive term $\Delta_R(t)$, being the model-implementation error, satisfies $\|\Delta_R(t)\| < +\infty$ for any $t \in [0, +\infty)$. Note that, in this situation, the steady-state residual error $\lim_{t \to +\infty} \|x(t) - x^*(t)\|$ of GNN (2.6) and (2.7) could be decreased by increasing γ.

2.3. Theoretical Results

For ZNN (2.3) solving linear time-varying equation (2.1), we have the following theorems on its global convergence and exponential convergence.

Theorem 2.3.1. *Given smoothly time-varying nonsingular matrix $A(t) \in \mathbb{R}^{n \times n}$ and vector $b(t) \in \mathbb{R}^n$, if a monotonically-increasing odd activation-function array $\Phi(\cdot)$ is used, then* *the state vector $x(t)$ of ZNN (2.3), starting from any initial state $x_0 \in \mathbb{R}^n$, always converges to the time-varying theoretical solution $x^*(t) = A^{-1}(t)b(t)$ of equation (2.1). In addition, ZNN (2.3) possesses the following properties.*

- *If the linear activation function (1.4) is used, then exponential convergence with the rate γ is achieved for (2.3).*

- *If the bipolar sigmoid activation-function (1.5) is used, then the superior convergence can be achieved for (2.3) for error range $[-\varepsilon, \varepsilon]$, $\exists \varepsilon > 0$, as compared to the case of using the linear activation function. This is because the error signal $e_i = [Ax - b]_i$ in (2.3) is amplified by the bipolar sigmoid function for such an error range.*

- *If the power activation function (1.6) is used, then the superior convergence can be achieved for (2.3) for error ranges $(-\infty, -1]$ and $[1, +\infty)$, as compared to the linear case. This is because error $e_i = [Ax - b]_i$ in (2.3) is amplified by the power function for such error ranges.*

- *If the power-sigmoid activation function (1.7) is used, then the superior convergence can be achieved for the whole error range $(-\infty, +\infty)$, as compared to the linear case. This is in view of the above mentioned properties.* □

In the analog implementation or simulation of recurrent neural networks, we usually assume it to be under ideal conditions. However, there always exist some realization errors. The implementation errors of $A(t)$ and $b(t)$, the differentiation errors of $A(t)$ and vector $b(t)$, and the model-implementation error appear most frequently in hardware realization. For these realization errors possibly appearing in ZNN model (2.3), we have the following analysis results.

Theorem 2.3.2. *Consider the following perturbed ZNN model with imprecise matrix implementation $\hat{A} = A + \Delta_A(t)$ and imprecise vector implementation $\hat{b} = b + \Delta_b(t)$:*

$$\hat{A}\dot{x} = -\dot{\hat{A}}x - \gamma\Phi(\hat{A}x - \hat{b}) + \dot{\hat{b}}.$$

If $\|\Delta_A(t)\| \leq \varepsilon_0$ and $\|\Delta_b(t)\| \leq \varepsilon_1$ for any time instant $t \in [0, \infty)$, $\exists 0 \leq \varepsilon_0, \varepsilon_1 < +\infty$, then the computational error $\|x(t) - x^*(t)\|$ over time t and the steady-state residual error $\lim_{t \to \infty} \|x(t) - x^*(t)\|$ will always be uniformly upper bounded by some positive scalars, provided that the invertibility condition of matrix $\hat{A}$ still holds true. □

For the differentiation errors and the model implementation error, the following perturbed ZNN dynamics is considered, as compared to the original dynamic equation (2.3).

$$A\dot{x} = -\left(\dot{A} + \Delta_{\dot{A}}(t)\right)x - \gamma\Phi(Ax - b) + \dot{b} + \Delta_R(t), \tag{2.8}$$

where $\Delta_{\dot{A}}(t) \in \mathbb{R}^{n \times n}$ denotes the differentiation error of matrix $A(t)$, and $\Delta_R(t) \in \mathbb{R}^n$ denotes the model-implementation error (including the differentiation error of vector $b(t)$ as a part). These errors may result from truncating/roundoff errors in digital realization and/or high-order residual errors of circuit components in analog realization.

Theorem 2.3.3. *Consider the perturbed ZNN model with implementation errors $\Delta_{\dot{A}}(t)$ and $\Delta_R(t)$, which is finally depicted in (2.8). If $\|\Delta_{\dot{A}}(t)\| \leq \varepsilon_2$ and $\|\Delta_R(t)\| \leq \varepsilon_3$ for any $t \in [0, \infty)$, $\exists 0 \leq \varepsilon_2$, $\varepsilon_3 < +\infty$, then the computational error $\|x(t) - x^*(t)\|$ over time t and the steady-state residual error $\lim_{t \to \infty} \|x(t) - x^*(t)\|$ will always be uniformly upper bounded by some positive scalars, provided that the design parameter $\gamma > 0$ is large enough. More importantly, as the design parameter γ tends to positive infinity, the steady-state residual error vanishes to zero.* □

Theorem 2.3.4. *In addition to the general robustness results in Theorems 2.3.2 and 2.3.3, the imprecisely-constructed ZNN (2.8) possesses the following properties.*

- *If the sigmoid activation function is used, then the steady-state residual error can be made smaller by increasing γ or ξ, as compared to the linear case. In addition, if the sigmoid function is used, then superior convergence and robustness properties exist for error range $|e_i(t)| \leq \varepsilon$, $\exists \varepsilon > 0$, compared to the linear case.*

- *If the power activation function is used, then we could remove the requirement of γ being large enough. In addition, if the power function is used, then superior convergence and robustness properties exist for error range $|e_i(t)| > 1$, as compared to the liner case.*

- *If the power-sigmoid activation function is used, then we could remove the requirement of γ being large enough. In addition, if the power-sigmoid function is used, then superior convergence and robustness properties exist for the whole error range $e_i(t) \in (-\infty, +\infty)$, as compared to the linear case.* □

2.4. MATLAB-Coding Simulation Techniques

In this section, two important simulation techniques are employed and briefed as follows. i) MATLAB routine "ode45" with a mass-matrix property is introduced firstly to solve the initial-value ordinary-differential-equation (ODE) problem, where $A(t)$ on the left-hand side of ZNN (2.3) can be viewed as a mass matrix. ii) Secondly, matrix derivatives [e.g., $\dot{A}(t)$] could be obtained by using MATLAB routine "diff" and symbolic math toolbox.

2.4.1. Coding of Activation-Functions

To simulate the implicit-dynamics systems such as (2.3) and (2.8), the activation-functions are to be defined firstly in MATLAB. Inside the body of a user-defined function, the MATLAB routine "nargin" returns the number of input arguments which are used to call the function. By using "nargin", different kinds of activation-functions can be generated at least with their default input argument(s). The linear activation-function defined in equation (1.4) can be generated simply by using the following MATLAB code.

```
function output=AFMlinear(X)
output=X;
```

The sigmoid activation-function defined in equation (1.5) with $\xi = 4$ as its default input value can be generated by using the following MATLAB code.

```
function output=AFMsigmoid(X,xi)
if nargin==1, xi=4; end
output=(1-exp(-xi*X))./(1+exp(-xi*X));
```

The power activation-function defined in equation (1.6) with $p = 3$ as its default input value can be generated by using the following MATLAB code.

```
function output=AFMpower(X,p)
if nargin==1, p=3; end
output=X.^p;
```

The power-sigmoid activation-function defined in equation (1.7) with $\xi = 4$ and $p = 3$ being its default values can be generated below.

```
function output=AFMpowersigmoid(X,xi,p)
if nargin==1, xi=4; p=3;
elseif nargin==2, p=3;
end
output=(1+exp(-xi))/(1-exp(-xi))*(1-exp(-xi*X))./ ...
    (1+exp(-xi*X));
i=find(abs(X)>=1);
output(i)=X(i).^p;
```

2.4.2. ODE with Mass Matrix

In the simulation of ZNN (2.3), the MATLAB routine "ode45" is preferred because "ode45" can solve the initial-value ODE problem with a mass matrix, $A(t,x)\dot{x} = g(t,x)$, $x(0) = x_0$, where the nonsingular matrix $A(t,x)$ on the left-hand side of such an equation is termed the mass matrix, and $g(t,x) := -\dot{A}x - \gamma\Phi(Ax - b) + \dot{b}$ for our case. The following code is used to generate such a mass matrix, where a time-varying matrix $A(t)$ is defined to be $[\sin t, \cos t; -\cos t, \sin t]$ as an example.

```
function M=mA(t,x)
M=[sin(t) cos(t); -cos(t) sin(t)];
```

To solve the ODE with a mass matrix, the MATLAB routine "odeset" should also be used. Its "Mass" property could be assigned to be the function "mA", which returns the value of the mass matrix $A(t,x)$. If $A(t,x)$ does not depend on the state variable x and the function "mA" is to be invoked with only one input argument t, then the value of the "MStateDep" property of "odeset" should be set to "none". For example, the following MATLAB code can be used to solve an ODE problem resulting from ZNN (2.3) with random initial state x_0.

```
tspan=[0 10]; x0=4*(rand(4,1)-0.5*ones(4,1));
options=odeset('Mass',@mA,'MStateDep','none');
[t,x]=ode45(@ZNNrside,tspan,x0,options);
```

The function "ZNNrside", to be shown in Subsection 2.4.4., returns the evaluation of the right-hand side of ZNN (2.3).

It is worth mentioning here that GNN need not use this mass-matrix technique, because it appears to have no mass matrix on the left hand side of its dynamic equation (or say, it actually has an identity matrix on the left hand side as the mass matrix). Thus, the code to solve an initial-value ODE problem resulting from GNN (2.6) could be written as follows.

```
tspan=[0 10]; options=odeset();
x0=4*(rand(4,1)-0.5*ones(4,1));
[t,x]=ode45(@GNNrside,tspan,x0,options);
```

2.4.3. Obtaining Matrix Derivatives

While matrix $A(t)$ can be generated by the user-defined function "mA", its time derivative $\dot{A}(t)$ required by equation (2.3) should also be obtained. Without loss of generality, we can use MATLAB routine "diff" to generate $\dot{A}(t)$ and $\dot{b}(t)$ required in ZNN (2.3).

In order to get the time derivative of $A(t)$, a symbolic object "u" is constructed firstly, then code "D=diff(mA(u))" is used to generate the analytical form of $\dot{A}(t)$. Finally, evaluating such an analytical form of $\dot{A}(t)$ with a numerical t will generate the required $\dot{A}(t)$. The code for generating $\dot{A}(t)$ is as follows, and we can generate $\dot{b}(t)$ similarly.

```
syms u; D=diff(mA(u)); u=t; diffA=eval(D);
```

Note that, without using the above symbolic "u", the command "diff(mA(t))" will return the row difference of $A(t)$, which is not the desired time derivative of $A(t)$.

2.4.4. RNN Right-Hand Side

The following code is used to define a function which evaluates the right-hand side of ZNN (2.3). In other words, it returns the evaluation of $g(t,x)$ in equation $A(t)\dot{x}(t) = g(t,x)$, where $g(t,x) := -\dot{A}(t)x(t) - \gamma\Phi(A(t)x(t) - b(t)) + \dot{b}(t)$ with power-sigmoid function.

```
function y=ZNNrside(t,x,gamma)
if nargin==2, gamma=1; end
syms u; A=diff(mA(u)); B=diff(vB(u));
u=t; dA=eval(A); dB=eval(B);
% using power-sigmoid function (recommended)
y=-dA*x-gamma*AFMpowersigmoid(mA(t,x)*x-...
    vB(t,x))+dB;
```

The MATLAB code simulating GNN (2.6) is presented below for comparison. It returns the evaluation of $g(t,x)$ in equation $\dot{x}(t) = g(t,x)$, where, in this case, $g(t,x) := -\gamma A^T \Phi(Ax(t) - b)$ with power-sigmoid activation functions.

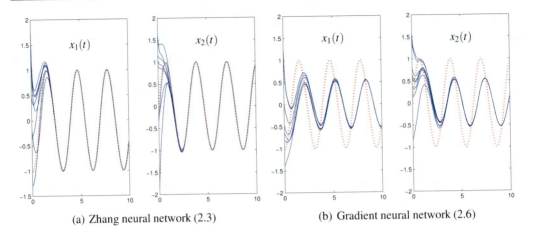

(a) Zhang neural network (2.3) (b) Gradient neural network (2.6)

Figure 2.3. Online solution of linear time-varying equations by recurrent neural networks with $\gamma = 1$, where theoretical solution is denoted by dotted lines.

```
function y=GNNrside(t,x,gamma)
if nargin==2, gamma=1; end
A=mA(t,x);
% using power-sigmoid function (recommended)
y=-gamma*A'*AFMpowersigmoid(A*x-vB(t,x));
```

2.4.5. Illustrative Examples

For illustration and comparison, let us consider the real-time solution of linear time-varying equation (2.1) with the following coefficients:

$$A(t) = \begin{bmatrix} \sin t & \cos t \\ -\cos t & \sin t \end{bmatrix}, \quad b(t) = \begin{bmatrix} \sin t \\ \cos t \end{bmatrix}.$$

In addition, the derivatives of the coefficients are assumed to be measurable or given analytically:

$$\dot{A}(t) = \begin{bmatrix} \cos t & -\sin t \\ \sin t & \cos t \end{bmatrix}, \quad \dot{b}(t) = \begin{bmatrix} \cos t \\ -\sin t \end{bmatrix}.$$

We choose $A(t)$ as such, because simple algebraic manipulations can verify that $A^{-1}(t) = A^T(t)$ and

$$x^*(t) = A^{-1}(t)b(t) = \begin{bmatrix} \sin^2 t - \cos^2 t \\ 2\cos t \sin t \end{bmatrix}.$$

The neural network (2.3) is thus in the specific form,

$$A \begin{bmatrix} \dot{x}_1(t) \\ \dot{x}_2(t) \end{bmatrix} = -\dot{A} \begin{bmatrix} x_1(t) \\ x_2(t) \end{bmatrix} - \gamma \Phi \left(A \begin{bmatrix} x_1(t) \\ x_2(t) \end{bmatrix} - b \right) + \dot{b}$$

where the activation-function array $\Phi(\cdot)$ is constituted by power-sigmoid activation functions (1.7) with $\xi = 4$ and $p = 3$, and the design parameter $\gamma = 1$ for illustrative purposes.

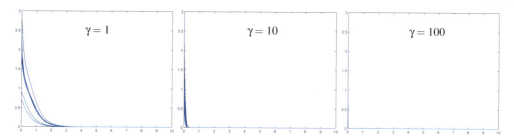

Figure 2.4. Computational error $\|x(t)-x^*(t)\|$ of ZNN (2.3) with different γ.

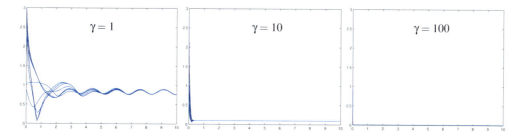

Figure 2.5. Computational error $\|x(t)-x^*(t)\|$ of GNN (2.6) with different γ.

2.4.5.1. Simulation of Convergence

By using the above MATLAB simulation techniques, Figure 2.3(a) can be generated which shows the global exponential convergence of ZNN model (2.3). Similarly, GNN model (2.6) can be simulated, and thus we generate Figure 2.3(b) which shows that the GNN solution is appreciably lag behind the theoretical solution $x^*(t)$.

To monitor the network convergence, we can also show the norm of the computational error, $\|x(t)-x^*(t)\|$, over time t. For example, Figure 2.4 shows that starting from random initial conditions, the computational errors $\|x(t)-x^*(t)\|$ all converge to zero. Note that such a convergence can be expedited by increasing γ. For example, if γ is increased to 10^3, the convergence time is within 4 milliseconds; and, if γ is increased to 10^6, the convergence time is within 4 microseconds. For comparison, the error-norm of GNN (2.6) is shown in Figure 2.5 under the same design parameters and random initial conditions. Evidently, such an error is considerably large. This is because the time-derivative information of $A(t)$ and $b(t)$ has not been utilized in traditional gradient-based computational schemes.

2.4.5.2. Simulation of Robustness

To show the robustness characteristics of the presented ZNN, the following differentiation error $\Delta_{\dot{A}}(t)$ and model-implementation error $\Delta_R(t)$ (including the $b(t)$-differentiation error as a part) are added:

$$\Delta_{\dot{A}}(t) = \varepsilon_1 \begin{bmatrix} \cos 3t & -\sin 3t \\ \sin 3t & \cos 3t \end{bmatrix}, \quad \Delta_R(t) = \varepsilon_2 \begin{bmatrix} \sin 3t \\ \cos 3t \end{bmatrix}$$

with $\varepsilon_1 = \varepsilon_2 = 0.5$. The following code defines the function "ZNNimprecise" for ODE solvers, which evaluates the right-hand side of the perturbed ZNN (2.8). Then, using the

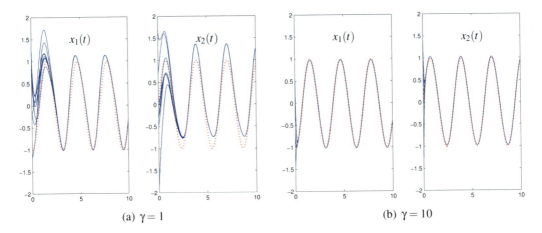

Figure 2.6. Solution of linear time-varying equations by imprecisely-constructed ZNN (2.8) with different γ, where dotted lines denote $x^*(t)$.

MATLAB commands "ZNNrobust(1)" and "ZNNrobust(10)" leads to Figure 2.6(a) and (b), respectively.

```
function y=ZNNimprecise(t,x,gamma)
if nargin==2, gamma=1; end
syms u; A=diff(mA(u)); B=diff(vB(u));
u=t; dA=eval(A); dB=eval(B);
e1=.5; e2=.5;
DA=e1*[cos(3*t) -sin(3*t);sin(3*t) cos(3*t)];
deltaR=e2*[sin(3*t);cos(3*t)];
dA=dA+DA; dB=dB+deltaR;
y=-dA*x-gamma*AFMpowersigmoid(mA(t,x)*x-...
    vB(t,x))+dB;
```

```
function ZNNrobust(gamma)
tspan=[0 10];
options=odeset('Mass',@mA,'MStateDep','none');
for iter=1:8
    x0=4*(rand(2,1)-0.5*ones(2,1));
    [t,x]=ode45(@ZNNimprecise,tspan,x0,...
        options,gamma);
    subplot(1,2,1);plot(t,x(:,1)); hold on
    subplot(1,2,2);plot(t,x(:,2)); hold on
end
subplot(1,2,1); xStar1=sin(t).^2-cos(t).^2;
plot(t,xStar1,'k:'); hold on
subplot(1,2,2); xStar2=2*sin(t).*cos(t);
plot(t,xStar2,'k:'); hold on
```

Similarly, we can show the computational error $\|x(t)-x^*(t)\|$ of the perturbed ZNN (2.8) in the presence of large differentiation and model-implementation errors. As seen from Figure 2.7, the computational error $\|x(t)-x^*(t)\|$ of the perturbed ZNN (2.8) is still bounded and very small.

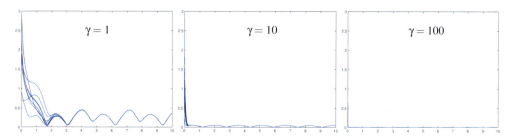

Figure 2.7. Computational errors $\|x(t)-x^*(t)\|$ of solving linear time-varying equations by ZNN in the presence of large realization errors.

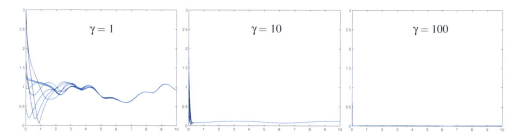

Figure 2.8. Computational errors $\|x(t)-x^*(t)\|$ of solving linear time-varying equations by GNN in the presence of realization errors.

To investigate the robustness of GNN, the following model-implementation error $\Delta_R(t)$ is added in a higher-frequency sinusoidal form (with $\varepsilon_2 = 0.5$): $\Delta_R(t) = \varepsilon_2[\sin 3t; \cos 3t]$. The following MATLAB code is used to define the function "GNNimprecise" for ODE solvers, which returns the evaluation of the right-hand side of perturbed GNN (2.7). Note that, to use the linear activation function, we only need to change "AFMpowersigmoid" to "AFMlinear" in the above MATLAB codes. We can then show the computational error $\|x(t) - x^*(t)\|$ of GNN (2.7) in the presence of large realization error $\Delta_R(t)$. See Figure 2.8. With imprecise implementation, the perturbed GNN works well, and its computational error is relatively small.

```
function y=GNNimprecise(t,x,gamma)
if nargin==2, gamma=1; end
e2=.5; deltaR=e2*[sin(3*t);cos(3*t)];
A=mA(t,x);
% using power-sigmoid function (recommended)
y=-gamma*A'*AFMpowersigmoid(A*x-vB(t,x))...
    +deltaR;
```

Finally, it is worth pointing out that, by using ZNN (2.3) to solve linear time-varying equation (2.1), the maximum steady-state computational errors $\lim_{t\to\infty}\|x(t)-x^*(t)\|$ are about 8×10^{-14}, 2.5×10^{-14}, and 2×10^{-15} for $\gamma=1$, $\gamma=10$, and $\gamma=100$, respectively. In contrast, by using GNN (2.6), the maximum steady-state computational errors are about 0.9, 0.1, and 0.01 for $\gamma=1$, $\gamma=10$, and $\gamma=100$, respectively.

2.5. Simulink-Modeling Techniques

While Section 2.3. presents theoretical results of ZNN (2.3), in this section, we investigate the MATLAB Simulink modeling techniques.

2.5.1. Basic Function Blocks

As we may know, MATLAB Simulink includes a comprehensive block library of sinks, sources, linear and nonlinear components, connectors, and so on. These basic function blocks we used to construct the dynamic system (2.3) are briefed as follows.

- The *Constant* block could generate a constant scalar or matrix by directly specifying the "Constant value" parameter. For example, if ZNN is used to solve the linear time-invariant equations, we could use the *Constant* block to generate matrix A and vector b in (2.3).

- The *Sine Wave* block can generate the sine wave "O(t)=Amplitude*Sin(Frequency*t +Phase)+Bias" by specifying the parameters "Amplitude", "Frequency", "Phase" and "Bias".

- The *Product* block provides two types of multiplication, either element-wise or matrix-wise product.

- The *Subsystem* block is used to encapsulate the blocks for constructing the sigmoid and power-sigmoid function, which makes the whole system more readable.

- The *MATLAB Function* block here computes the matrix norm. We could also use it to generate matrix $A(t)$ and $b(t)$ with the *Clock* block's output as its input.

- The *Math Function* block can represent mathematical functions including logarithmic, exponential, power and modulus functions.

- The *Switch* block is used to construct the power-sigmoid subsystem in this chapter. "u2>=Threshold" is chosen for the option "Criteria for passing first input", and the value of "Threshold" is set to 1. Simply put, if the value of $|e_i(t)|$ (i.e. the 2nd input of the *Switch* block) is not smaller than 1, then the 1st input of the *Switch* block (i.e. the power activation-function value) will be passed to the output. Otherwise, the 3rd input of the *Switch* block (i.e. the "sigmoid" function value) will be passed to the output.

- The *Integrator* block makes continuous-time integration on the input signals. In this section, we set the "Initial condition" as "4*(rand(2,1)-0.5*ones(2,1))" to restrict x_0 randomly between -2 and 2.

2.5.2. Generating $\Phi(\cdot)$

Four types of activation functions and their processing-arrays are investigated here.

ZNN Solving the Time-Varying Linear System $A(t)x(t) = b(t)$ 33

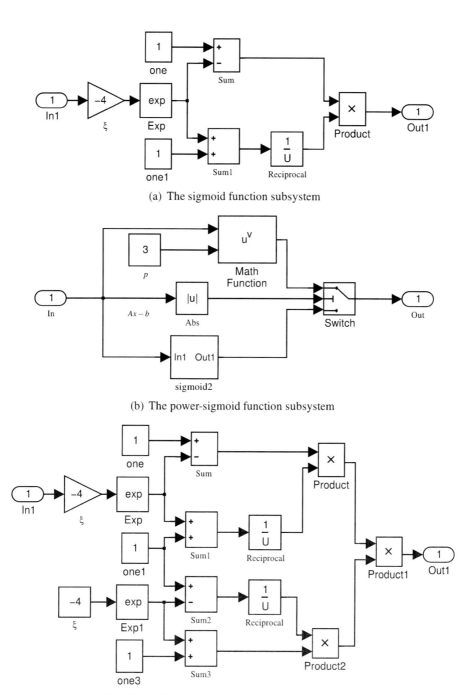

(a) The sigmoid function subsystem

(b) The power-sigmoid function subsystem

(c) The sigmoid2 subsystem in the power-sigmoid subsystem

Figure 2.9. Parallel-processing activation-function arrays. *Reproduced from Y. Zhang, X. Guo et al., Modeling and simulation of Zhang neural network for online linear time-varying equations solving based on MATLAB Simulink, Figure 3, Proceedings of the Seventh International Conference on Machine Learning and Cybernetics, pp. 455-460. ©[2008] IEEE. Reprinted, with permission.*

- For using the linear-activation-function array, there is no need to use any block since its output equals input.

- For using the power-activation-function array, we can use the *Math Function* block by choosing "pow" in its function list with input parameter p set as 3.

- For the final FPGA/ASIC realization of ZNN model (2.3), we are here constructing the subsystems of bipolar-sigmoid and power-sigmoid activation-function-arrays, which are shown in Figure 2.9(a) and (b). Note that, in the power-sigmoid case, we employ subsystem "sigmoid2" to construct the sigmoid part of equation (1.7).

2.5.3. Parameter Settings

Other important parameters and options related to the blocks we used in this section could be specified as follows.

- For the blocks which have the default option "Collapse 2-D results to 1-D" or "Interpret vector parameters as 1-D", we need deselect them.

- The default option "Element-wise" of *Product* blocks has to be changed to "Matrix" so as to perform the standard matrix multiplication, except the ones inside the activation-function processing-arrays.

2.5.4. Illustrative Examples

Based on the MATLAB Simulink modeling techniques investigated in the preceding subsections, we are to simulate ZNN (2.3) for solving online linear time-varying equations in this subsection. For illustrative and comparative purposes, let us consider the real-time solution of equation $A(t)x(t) - b(t) = 0$ with coefficients defined as:

$$A(t) = \begin{bmatrix} \cos t & \sin t \\ -\sin t & \cos t \end{bmatrix}, \quad b(t) = \begin{bmatrix} \sin 2t \\ \cos 2t \end{bmatrix},$$

where the theoretical solution $x^*(t) = [\sin t; \cos t]$.

2.5.4.1. Convergence Simulation

The overall Simulink model of ZNN (2.3) is depicted in Figure 2.10, where $p = 3$ and $\xi = 4$ are the default parameter-values used in the activation-function-arrays. Note that, in Figure 2.10, the power-sigmoid array is chosen, since superior performance could be achieved in general as compared with other cases. We could replace the "power-sigmoid" subsystem with other subsystems as well for alternative testing.

Since the entries of the coefficients $A(t)$ and $b(t)$ in our simulation example are all sine/cosine functions, we use the *Sine Wave* block to generate them. For $A(t)$, the "Phase" parameter is set as "[pi/2 0; pi pi/2]", "Amplitude" as "1", "Frequency" as "1", and "Bias" as "0". Similarly, the parameters for $b(t)$ are set as "[0; pi/2]", "1", "2", and "0", respectively, in the same order as those for $A(t)$.

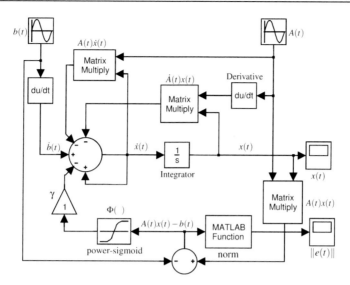

Figure 2.10. Simulink model of ZNN-model (2.3). *Reproduced from Y. Zhang, X. Guo et al., Modeling and simulation of Zhang neural network for online linear time-varying equations solving based on MATLAB Simulink, Figure 4, Proceedings of the Seventh International Conference on Machine Learning and Cybernetics, pp. 455-460. ©[2008] IEEE. Reprinted, with permission.*

After the Simulink model of ZNN (2.3) shown in Figure 2.10 has been established, we click the "Start simulation" button to run it. When the simulation completes, the output of $x(t)$ could be seen [i.e., Figure 2.11(a)] by double-clicking the corresponding *Scope* block. Figure 2.11 shows that, started from a randomly-generated initial state $x_0 \in [-2,2]^2$, $x(t)$ always converges to the theoretical solution $x^*(t) = [\sin t; -\cos t]$. For comparison, we also show the "scope" output for the situation of $\gamma = 10$ in Figure 2.11(b). Evidently, when $\gamma=10$, the convergence is much faster than that of $\gamma = 1$. To make a better monitoring of neural-network convergence, we could also show the error-norm "scope" (i.e., $\|e(t)\|$) over time t) in Figure 2.12.

By zooming in the "scope" outputs of ZNN (2.3) when $\gamma=1$, we can find that the computational error converges to zero (without appreciable error) in 3.3 seconds. Some other observations show that, if γ is increased to 10^2, the convergence time is within 0.035 seconds; if γ is 10^3, the convergence time is within 3.5 milliseconds; and if $\gamma = 10^4$, the convergence time is within 0.35 milliseconds. Furthermore, superior convergence has been observed by using the power-sigmoid activation-function-array. These substantiate the theoretical results presented in Section 2.3..

2.5.4.2. Robustness Simulation

To show ZNN-robustness characteristics, the following differentiation error $\Delta_{\dot{A}}(t)$ and model-implementation error $\Delta_R(t)$ are added (with $\varepsilon_2 = \varepsilon_3 = 0.3$):

$$\Delta_{\dot{A}}(t) = \varepsilon_2 \begin{bmatrix} \cos 6t & -\sin 6t \\ \sin 6t & \cos 6t \end{bmatrix}, \Delta_R(t) = \varepsilon_3 \begin{bmatrix} \sin 6t \\ \cos 6t \end{bmatrix}.$$

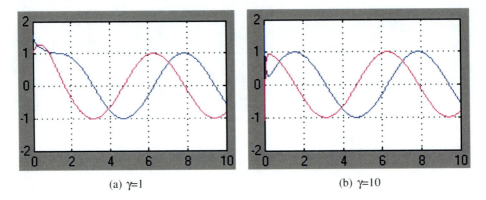

(a) γ=1 (b) γ=10

Figure 2.11. "Scope" output $x(t)$ of ZNN (2.3). *Reproduced from Y. Zhang, X. Guo et al., Modeling and simulation of Zhang neural network for online linear time-varying equations solving based on MATLAB Simulink, Figure 5, Proceedings of the Seventh International Conference on Machine Learning and Cybernetics, pp. 455-460. ©[2008] IEEE. Reprinted, with permission.*

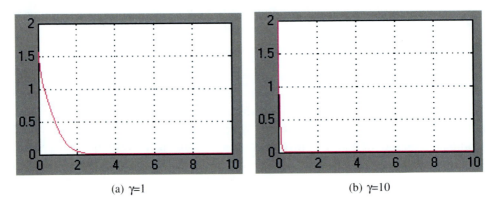

(a) γ=1 (b) γ=10

Figure 2.12. "Scope" output $\|e(t)\|$ of ZNN (2.3). *Reproduced from Y. Zhang, X. Guo et al., Modeling and simulation of Zhang neural network for online linear time-varying equations solving based on MATLAB Simulink, Figure 6, Proceedings of the Seventh International Conference on Machine Learning and Cybernetics, pp. 455-460. ©[2008] IEEE. Reprinted, with permission.*

By considering the above "realization error" imposed onto ZNN (2.3), the scope output of $\|e(t)\|$ is shown in Figure 2.13. As we analyzed previously and illustrate here, the computational error of the imprecisely-constructed ZNN model (2.8) is always upper bounded by some small positive scalar, provided that the design parameter γ is large enough. Moreover, when design parameter γ increases, such a computational error decreases. The modeling and simulation results substantiate the theoretical analysis.

2.6. An Application: Inverse Kinematic Control

The inverse-kinematic problem is one of the fundamental tasks in operating robot manipulators [4, 10, 12, 13, 18]. Specifically, as shown in Figure 2.14, we need find the joint trajectories $\theta(t) \in \mathbb{R}^n$ in real time t, given the end-effector trajectories $r(t) \in \mathbb{R}^n$. The

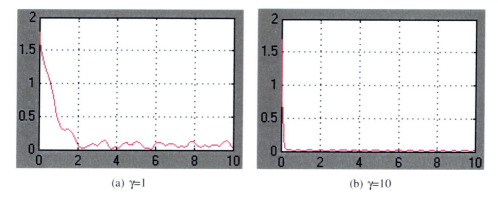

Figure 2.13. "Scope" output $\|e(t)\|$ of the perturbed ZNN (2.8). *Reproduced from Y. Zhang, X. Guo et al., Modeling and simulation of Zhang neural network for online linear time-varying equations solving based on MATLAB Simulink, Figure 7, Proceedings of the Seventh International Conference on Machine Learning and Cybernetics, pp. 455-460. ©[2008] IEEE. Reprinted, with permission.*

inverse-kinematic problem is usually solved at velocity level, where the end-effector velocity $\dot{r}(t) \in \mathbb{R}^n$ and joint velocity $\dot{\theta}(t) \in \mathbb{R}^n$ satisfy

$$J(\theta)\dot{\theta}(t) = \dot{r}(t), \qquad (2.9)$$

where $J(\theta) \in \mathbb{R}^{n \times n}$ is the Jacobian matrix given analytically. Evidently, the velocity-level inverse-kinematic problem (2.9) is depicted (or can be viewed) as a set of linear time-varying equation, (2.1), with coefficients $A(t) := J(\theta,t)$ and $b(t) := \dot{r}(t)$, and $x(t) := \dot{\theta}(t)$ which is to be solved online.

Now, we exploit ZNN (2.3) to solve (2.9) for $\dot{\theta}(t)$ in real time. The time derivatives, $\dot{J}$ and $\ddot{r}$, can be approximated by using finite-difference or given analytically. For example, to perform an orientation-consistent circular path-following task as shown in Figure 2.14, $\dot{r}(t)$ and $\ddot{r}(t)$ could be

$$\begin{bmatrix} -v\omega \sin(\omega t) \\ v\omega \cos(\alpha)\cos(\omega t) \\ -v\omega \sin(\alpha)\cos(\omega t) \\ 0 \in \mathbb{R}^3 \end{bmatrix}, \begin{bmatrix} -v\omega^2 \cos(\omega t) \\ -v\omega^2 \cos(\alpha)\sin(\omega t) \\ v\omega^2 \sin(\alpha)\sin(\omega t) \\ 0 \in \mathbb{R}^3 \end{bmatrix},$$

where the task duration $T = 10.0$s, $\omega = 2\pi/T$, radius $v = 0.1$, and the revolute angle α about the X axis is $\pi/6$rad. The initial state $\theta(0) = [\pi, -\pi, \pi, -\pi, \pi, -\pi]^T/4$ in radians. Design parameter $\gamma = 10^3$. Figure 2.14 illustrates the simulated motion of the robot arm in the 3-dimensional work space, which is sufficiently close to the desired trajectory $r(t)$. Specifically, as shown in Figure 2.15, the Cartesian positioning errors at the PUMA560 end-effector are less than 8×10^{-5}m. In addition, the maximal steady-state velocity error is also very small (around 2×10^{-5}m/s).

This application example has substantiated the efficacy of Zhang neural network (2.3) on the issues of time-varying equation solving and robot inverse-kinematic control.

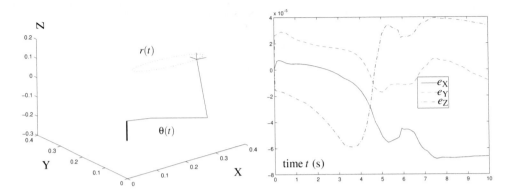

Figure 2.14. PUMA560 manipulator tracking a circle. *Reproduced from Y. Zhang and H. Peng, Zhang neural network for linear time-varying equation solving and its robotic application, Figure 5, Proceedings of the Sixth International Conference on Machine Learning and Cybernetics, pp. 3543-3548. ©[2007] IEEE. Reprinted, with permission.*

Figure 2.15. PUMA560 positioning errors on X, Y and Z axes. *Reproduced from Y. Zhang and H. Peng, Zhang neural network for linear time-varying equation solving and its robotic application, Figure 6, Proceedings of the Sixth International Conference on Machine Learning and Cybernetics, pp. 3543-3548. ©[2007] IEEE. Reprinted, with permission.*

2.7. Conclusion

By following the neural design approach of Zhang *et al.* [11–13], a special recurrent neural network has been proposed to solve online a set of linear time-varying equations. Such a ZNN could globally exponentially converge to the exact solution of linear time-varying equations, especially when using the power-sigmoid activation functions. Compared to GNN, ZNN has much superior convergence and robustness. In addition, several important MATLAB-coding techniques have been introduced for the simulation of both recurrent neural networks; e.g., "ode45" with a mass-matrix property, and the obtaining of matrix derivatives. Simulation examples have substantiated the theoretical analysis of the neural network. On the other hand, the MATLAB Simulink modeling techniques and results have been investigated and showed towards the final FPGA/ASIC circuits implementation of such a neural network. It is also substantiated that the ZNN model is effective and efficient for solving online time-varying problems. In Section 2.6., we have applied ZNN to the robot inverse-kinematic control. The results have demonstrated the possibility and feasibility of ZNN for time-varying problems solving and engineering applications. Future research directions may lie in its hardware implementation and/or numerical-algorithm development.

References

[1] Zhang, Y. Towards piecewise-linear primal neural networks for optimization and redundant robotics. *Proceedings of IEEE International Conference on Networking, Sensing and Control*, 2006, pp. 374-379.

[2] Steriti, RJ; Fiddy, MA. Regularized image reconstruction using SVD and a neural network method for matrix inversion. *IEEE Transactions on Signal Processing*, 1993, vol. 41, no. 10, pp. 3074-3077.

[3] Sarkar, T; Siarkiewicz, K; Stratton, R. Survey of numerical methods for solution of large systems of linear equations for electromagnetic field problems. *IEEE Transactions on Antennas and Propagation*, 1981, vol. 29, no. 6, pp. 847-856.

[4] Sturges, Jr RH. Analog matrix inversion (robot kinematics). *IEEE Journal of Robotics and Automation*, 1988, vol. 4, no. 2, pp. 157-162.

[5] Zhang, Y; Leithead, WE; Leith, DJ. Time-series Gaussian process regression based on Toeplitz computation of $O(N^2)$ operations and $O(N)$-level storage. *Proceedings of the 44th IEEE Conference on Decision and Control*, 2005, pp. 3711-3716.

[6] Leithead, WE; Zhang, Y. $O(N^2)$-operation approximation of covariance matrix inverse in Gaussian process regression based on quasi-Newton BFGS methods. *Communications in Statistics-Simulation and Computation*, 2007, vol. 36, no. 2, pp. 367-380.

[7] Manherz, RK; Jordan, BW; Hakimi, SL. Analog methods for computation of the generalized inverse. *IEEE Transactions on Automatic Control*, 1968, vol. 13, no. 5, pp. 582-585.

[8] Jang, J; Lee, S; Shin, S. An optimization network for matrix inversion. *Neural Information Processing Systems, American Institute of Physics*, 1988, pp. 397-401.

[9] Wang, J. A recurrent neural network for real-time matrix inversion. *Applied Mathematics and Computation*, 1993, vol. 55, no. 1, pp. 89-100.

[10] Zhang, Y. Revisit the analog computer and gradient-based neural system for matrix inversion. *Proceedings of IEEE International Symposium on Intelligent Control*, 2005, pp. 1411-1416.

[11] Zhang, Y; Jiang, D; Wang, J. A recurrent neural network for solving Sylvester equation with time-varying coefficients. *IEEE Transactions on Neural Networks*, 2002, vol. 13, no. 5, pp. 1053-1063.

[12] Zhang, Y; Ge, SS. A general recurrent neural network model for time-varying matrix inversion. *Proceedings of the 42nd IEEE Conference on Decision and Control*, 2003, pp. 6169-6174.

[13] Zhang, Y; Ge, SS. Design and analysis of a general recurrent neural network model for time-varying matrix inversion. *IEEE Transactions on Neural Networks*, 2005, vol. 16, no. 6, pp. 1477-1490.

[14] Yeung, KS; Kumbi, F. Symbolic matrix inversion with application to electronic circuits. *IEEE Transactions on Circuits and Systems*, 1988, vol. 35, no. 2, pp. 235-238.

[15] El-Amawy, A. A systolic architecture for fast dense matrix inversion. *IEEE Transactions on Computers*, 1989, vol. 38, no. 3, pp. 449-455.

[16] Carneiro, NCF; Caloba, LP. A new algorithm for analog matrix inversion. *Proceedings of the 38th Midwest Symposium on Circuits and Systems*, 1995, vol. 1, pp. 401-404.

[17] Mead, C. *Analog VLSI and Neural Systems*. Reading, MA: Addison-Wesley, 1989.

[18] Zhang, Y. A set of nonlinear equations and inequalities arising in robotics and its online solution via a primal neural network. *Neurocomputing*, 2006, vol. 70, pp. 513-524.

[19] Shanblatt, MA; Foulds, B. A Simulink-to-FPGA implementation tool for enhanced design flow. *Proceedings of IEEE International Conference on Microelectronic Systems Education*, 2005, pp. 89-90.

[20] Grout, IA. Modeling, simulation and synthesis: from Simulink to VHDL generated hardware. *Proceedings of the 5th World Multi-Conference on Systemics, Cybernetics and Informatic*, 2001, vol. 15, pp. 443-448.

[21] Grout, IA; Keane, K. A MATLAB to VHDL conversion toolbox for digital control. *Proceedings of IFAC Symposium on Computer Aided Control Systems Design*, 2000, pp. 164-169.

[22] Jiang, D. Analog computing for real-time solution of time-varying linear equations. *Proceedings of International Conference on Communications, Circuits and Systems*, 2004, vol. 2, pp. 1367-1371.

[23] Bazaraa, MS; Sherali HD; Shetty, CM. *Nonlinear Programming-Theory and Algorithms*. New York: Wiley, 1993.

Chapter 3

ZNN for Time-Varying Convex Quadratic Optimization

Abstract

In this chapter, by following Zhang *et al.*'s method (which has been proposed formally since March 2001), a recurrent neural network (termed as Zhang neural network, ZNN) is developed and analyzed for solving online the time-varying convex quadratic-minimization (QM) and quadratic-programming (QP) problems (of which the latter is subject to a time-varying linear-equality constraint as an example). Different from conventional gradient-based neural networks (GNN), such a ZNN model makes full use of the time-derivative information of time-varying coefficients. The resultant ZNN model is theoretically proved to have global exponential convergence to the time-varying theoretical optimal solution of the investigated time-varying convex quadratic problems. Thus, such a ZNN model can be unified as a superior approach for solving online the time-varying quadratic problems. For the purpose of time-varying quadratic-problem solving, this chapter investigates comparatively both ZNN and GNN solvers, and then their unified modeling techniques. Computer-simulation results substantiate well the efficacy of such ZNN models on solving online the time-varying convex quadratic problems (i.e., QM and QP problems).

3.1. Introduction

The online solution of (equality-constrained) quadratic programs (including the quadratic-minimization problems solving as its special case) are widely encountered in various areas; e.g., optimal controller design [1], power-scheduling [2], robot-arm motion planning [3], and digital signal processing [4]. A well-accepted approach to solving linear-equality constrained quadratic-programs is the numerical algorithms/methods performed on digital computers (e.g., nowadays computers). However, the minimal arithmetic operations of a numerical QP algorithm are proportional to the cube of the related Hessian matrix's dimension, and consequently such a numerical algorithm may not be efficient enough [4].

In the past decades, many neural-dynamic models have been proposed, developed and implemented on specific architectures, e.g., analog RNN solvers [5–10]. The neural-dynamic approach is now regarded as a powerful alternative to real-time computation and optimization in view of its parallel-processing distributed nature, self-adaptation ability, and hardware-implementation convenience. A large number of the aforementioned neural-dynamic models , however, belong to the well-known gradient (or termed, gradient-descent, negative-gradient) method, being feasible and efficient intrinsically for static (or termed, constant, time-invariant) problem solving (e.g., static QM and QP).

Different from conventional design methods of numerical algorithms and gradient neural networks [3–10], in this chapter, Zhang neural networks (ZNN), together with their modeling techniques and results, are presented for online solution of time-varying convex quadratic-minimization and quadratic-programs subject to time-varying linear-equality constraints. Theoretical and modeling results both demonstrate the ZNN superiority on "moving"-problem solving.

3.2. Time-Varying Quadratic Minimization

Consider the following time-varying quadratic minimization (QM) problem:

$$\text{minimize } f(x) := x^T(t)P(t)x(t)/2 + q^T(t)x(t) \in \mathbb{R}, \tag{3.1}$$

where given Hessian matrix $P(t) \in \mathbb{R}^{n \times n}$ is smoothly time-varying and positive-definite for any time instant $t \in [0, +\infty) \subset \mathbb{R}$, and given coefficient vector $q(t) \in \mathbb{R}^n$ is smoothly time-varying as well. In the expression (3.1), unknown vector $x(t) \in \mathbb{R}^n$ is to be solved all over the time so as to make the value of $f(x)$ always smallest.

As we may recognize or know, solving the time-varying quadratic minimization problem (3.1) could be done by zeroing the partial-derivative $\nabla f(x)$ of $f(x)$ [4] at every time instant t; in mathematics,

$$\nabla f(x) := \frac{\partial f(x)}{\partial x} = \mathbf{0} \in \mathbb{R}^n, \ \forall t \in [0, +\infty). \tag{3.2}$$

More specifically, it follows from the above that the theoretical time-varying solution $x^*(t) \in \mathbb{R}^n$ to (3.1), being the minimum point of $f(x)$ at any time instant t, satisfies $P(t)x^*(t) + q(t) = 0$ or here equivalently $x^*(t) = -P^{-1}(t)q(t)$. The theoretical minimum value $f^* := f(x^*)$ of time-varying quadratic function $f(x)$ is thus achieved as $f^* = x^{*T}(t)P(t)x^*(t)/2 + q^T(t)x^*(t)$.

In order to demonstrate the significance of the interesting problem and the visual effects, we could take the following time-varying coefficients $P(t) \in \mathbb{R}^{2 \times 2}$ and $q(t) \in \mathbb{R}^2$ as an example:

$$P(t) = \begin{bmatrix} 0.5\sin(t)+2 & \cos(t) \\ \cos(t) & 0.5\cos(t)+2 \end{bmatrix}, \ q(t) = \begin{bmatrix} \sin(t) \\ \cos(t) \end{bmatrix}. \tag{3.3}$$

Figure 3.1 shows the three-dimensional plots of $f(x)$ with respect to $x = (x_1, x_2) \in \mathbb{R}^{2 \times 2}$, but at different time instants (i.e., $t = 1.60, 4.65, 5.75$ and 7.25 seconds). We can see quite evidently that the shape and minimum value f^* of $f(x)$ together with its minimum solution $x^*(t)$ are all "moving" with time t, so that this time-varying quadratic minimization problem (3.1) could be considered as a "moving minimum" problem.

ZNN for Time-Varying Convex Quadratic Optimization

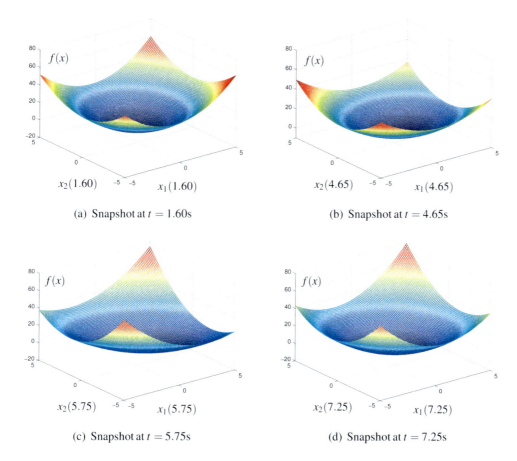

Figure 3.1. "Moving" minimum of time-varying quadratic function $f(x)$ at different time instants. *Reproduced from Y. Zhang, Z. Li et al., Zhang neural network versus gradient neural network for online time-varying quadratic function minimization, Figure 1, D.-S. Huang et al. (Eds.): ICIC 2008, LNAI 5227, pp. 807-814, 2008. ©Springer-Verlag Berlin Heidelberg 2008. With kind permission of Springer Science+Business Media.*

3.2.1. Zhang Neural Networks

In this subsection, Zhang *et al*'s neural-dynamic design method [11–14] is generalized and applied to solving the online time-varying quadratic minimization problem (3.1). The design procedure could be formalized as follows.

Step 1. To track the "moving" minimum-point of $f(x)$, instead of transforming the problem to a scalar-valued squared (or norm-based) function to minimize, we could define the vector-valued error function $e(t) \in \mathbb{R}^n$ as ∇f; in mathematics,

$$e(t) := \nabla f = P(t)x(t) + q(t). \tag{3.4}$$

Step 2. To make every element $e_i(t)$ of error function $e(t) \in \mathbb{R}^n$ converge to zero (i.e., for any $i \in 1, 2, \cdots, n$), the time derivative $\dot{e}(t)$ of error-function $e(t)$ could be constructed better as (or termed, the general ZNN-design formula):

$$\dot{e}(t) := \frac{de(t)}{dt} = \frac{d(\nabla f)}{dt} = -\Gamma \Phi(e(t)) = -\Gamma \Phi(\nabla f), \tag{3.5}$$

where design parameter $\Gamma \in \mathbb{R}^{n \times n}$ is generally a positive-definite matrix but could be replaced here with any $\gamma > 0 \in \mathbb{R}$ directly to scale the ZNN-convergence rate. Note that, Γ (or γ), being a set of reciprocals of capacitance-parameters, should be set as large as hardware permits [15] or set appropriately for simulative/experimental purposes. $\Phi(\cdot) : \mathbb{R}^n \to \mathbb{R}^n$ denotes an activation-function processing-array from $\mathbb{R}^n$ to $\mathbb{R}^n$ but preferably with each element decoupled. In addition, each scalar-valued processing-unit $\phi(\cdot)$ of array $\Phi(\cdot)$ should be a monotonically-increasing odd activation function. In this section, two types of activation function $\phi(\cdot)$ are investigated: 1) linear activation function (1.4), and 2) power-sigmoid activation function (1.7) (for more details, see Section 1.5. in Chapter 1).

Step 3. By expanding the ZNN-design formula (3.5), the following implicit dynamic equation of Zhang neural network could readily be constructed for minimizing online the time-varying quadratic function (3.1):

$$P(t)\dot{x}(t) = -\dot{P}(t)x(t) - \gamma \Phi\left(P(t)x(t) + q(t)\right) - \dot{q}(t), \tag{3.6}$$

where $x(t) \in \mathbb{R}^n$, starting with any initial condition $x(0) \in \mathbb{R}^n$, denotes the neural-state vector which corresponds to the theoretical time-varying minimum solution $x^*(t)$ of non-stationary quadratic function (3.1). In addition, from (3.6), it is worth writing out the following linear ZNN model for the same time-varying quadratic function minimization purposes:

$$P(t)\dot{x}(t) = -\left(\dot{P}(t) + \gamma P(t)\right)x(t) - \left(\gamma q(t)) + \dot{q}(t)\right). \tag{3.7}$$

In summary, the designed ZNN models could solve online the time-varying quadratic function minimization problem depicted in (3.1)! In other words, the "moving" minimum point could be found by the ZNN models in real time and in an error-free manner. For this efficacy, please refer to the ensuing subsections.

3.2.2. Gradient Neural Networks

For comparison, we develop here the conventional gradient-based neural networks to solve online the quadratic minimization problem depicted in (3.1). However, please note that almost all numerical algorithms and neural-dynamic computational schemes (specially, gradient-based neural networks) [4, 10, 16–18] were designed intrinsically for the problems with constant coefficients rather than time-varying ones. According to the gradient-descent design method [10], we could obtain the gradient-based neural networks which minimize the stationary quadratic function $f(x) = x^T P x / 2 + q^T x$. The design procedure is as follows.

Step 1. Let us define a scalar-valued norm-based (or squared) energy function $\mathcal{E}(x) = \|\nabla f\|_2^2 / 2 = \|Px + q\|_2^2 / 2$, where $\| \cdot \|_2$ denotes the two-norm of a vector. Evidently, x is the minimum solution if and only if $\mathcal{E}(x) = 0$ is reached.

Step 2. The following design-formula and dynamics of GNN model could then be adopted to minimize online the stationary quadratic function $x^T P x / 2 + q^T x$:

$$\dot{x}(t) = \frac{dx(t)}{dt} = -\gamma \frac{\partial \|\nabla f\|_2^2 / 2}{\partial x} = -\gamma P^T \left(Px(t) + q\right), \tag{3.8}$$

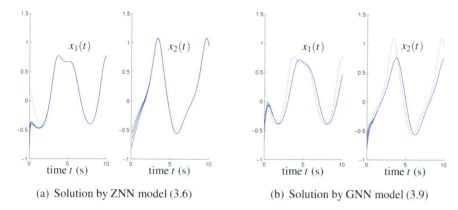

(a) Solution by ZNN model (3.6) (b) Solution by GNN model (3.9)

Figure 3.2. Time-varying quadratic minimization (3.1) synthesized online by ZNN (3.6) and GNN (3.9) with design parameter $\gamma = 1$ and using power-sigmoid activation functions, where dotted curves correspond to the theoretical time-varying minimum solution $x^*(t)$. *Reproduced from Y. Zhang, Z. Li et al., Zhang neural network versus gradient neural network for online time-varying quadratic function minimization, Figure 2, D.-S. Huang et al. (Eds.): ICIC 2008, LNAI 5227, pp. 807-814, 2008. ©Springer-Verlag Berlin Heidelberg 2008. With kind permission of Springer Science+Business Media.*

where design parameter $\gamma > 0$ is defined the same as that in ZNN models for the purpose of scaling the neural-network convergence rate. In addition, the following nonlinear GNN model could be extended from the above by exploiting the aforementioned activation-function processing-array $\Phi(\cdot)$:

$$\dot{x}(t) = -\gamma P^T \Phi\bigl(Px(t) + q\bigr). \tag{3.9}$$

In summary, the designed GNN models could theoretically only solve the stationary quadratic-function minimization problem (and might be extended in practice to handle approximately the time-varying quadratic function minimization problem). For this point, please refer to the ensuing subsections.

3.2.3. Theoretical Results and Comparisons

For ZNN model (3.6) which minimizes time-varying quadratic function (3.1), we could have the following propositions on its global exponential convergence when using the two aforementioned activation-function arrays [10–12, 19].

Proposition 3.2.1. *Given smoothly time-varying positive-definite matrix $P(t) \in \mathbb{R}^{n \times n}$ and vector $q(t) \in \mathbb{R}^n$, if an odd activation-function array $\Phi(\cdot)$ is employed, then the neural-state vector $x(t)$ of ZNN (3.6), starting from any initial state $x(0) \in \mathbb{R}^n$, always converges to the theoretical time-varying minimum solution $x^*(t)$ of non-stationary quadratic function (3.1). In addition, Zhang neural network (3.6) possesses the following properties.*

1) *If the linear-activation-function array is used, then global exponential convergence could be achieved for ZNN (3.6) with error-convergence rate γ.*

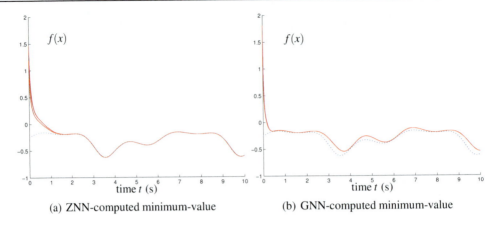

(a) ZNN-computed minimum-value (b) GNN-computed minimum-value

Figure 3.3. Time-varying quadratic-function value $f(x)$ minimized online by ZNN (3.6) and GNN (3.9) with design parameter $\gamma = 1$ and using power-sigmoid activation functions, where dotted curves correspond to the theoretical time-varying minimum value $f(x^*(t))$. Reproduced from Y. Zhang, Z. Li et al., Zhang neural network versus gradient neural network for online time-varying quadratic function minimization, Figure 3, D.-S. Huang et al. (Eds.): ICIC 2008, LNAI 5227, pp. 807-814, 2008. ©Springer-Verlag Berlin Heidelberg 2008. With kind permission of Springer Science+Business Media.

2) *If the power-sigmoid-activation-function array is used, then superior convergence can be achieved for the whole error range $(-\infty, +\infty)$, as compared to the linear-activation-function-array situation.* □

For comparison, the following proposition about gradient neural network (3.9) could be provided but with exactness only for stationary quadratic minimization (i.e., the quadratic minimization with coefficients being constant) [10–12, 19].

Proposition 3.2.2. *Consider the situation with constant positive-definite matrix $P \in \mathbb{R}^{n \times n}$ and constant vector $q \in \mathbb{R}^n$ associated with quadratic minimization (3.1). If the linear activation function array is used, then global exponential convergence to the constant minimum point x^* of stationary $f(x)$ can be achieved by GNN (3.9) with convergence rate proportional to the product of γ and minimum eigenvalue of $P^T P$. If the power-sigmoid activation function array is used, then superior convergence can be achieved for GNN (3.9) over the whole error range $(-\infty, +\infty)$, as compared to the case of using linear activation functions.* □

3.2.4. Simulation Studies

For comparison, both ZNN model (3.6) and GNN model (3.9) are employed to carry out the minimization process and to obtain the time-varying minimum solution $x^*(t)$ of non-stationary quadratic function (3.1). The power-sigmoid activation function array (with parameters $\xi = 4$ and $p = 3$) is used in the model simulation of both ZNN (3.6) and GNN (3.9) with $\gamma = 1$.

- As seen from Figure 3.2(a), starting form randomly-generated initial state $x(0)$, neural-state vector $x(t)$ of ZNN (3.6) could always "elegantly" converge to the time-

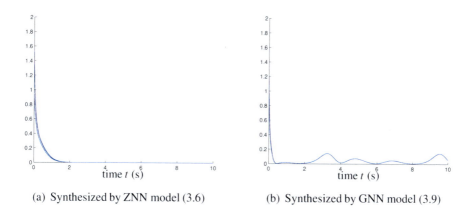

(a) Synthesized by ZNN model (3.6)   (b) Synthesized by GNN model (3.9)

Figure 3.4. Residual error $|f(x)-f(x^*)|$ synthesized by ZNN and GNN models with $\gamma = 1$. *Reproduced from Y. Zhang, Z. Li et al., Zhang neural network versus gradient neural network for online time-varying quadratic function minimization, Figure 4, D.-S. Huang et al. (Eds.): ICIC 2008, LNAI 5227, pp. 807-814, 2008. ©Springer-Verlag Berlin Heidelberg 2008. With kind permission of Springer Science+Business Media.*

varying minimum solution $x^*(t)$. On the other hand, as can be seen from Figure 3.2(b), relatively large steady-state computational errors of gradient-based neural network (3.9) exist!

- The ZNN-solution exactness and GNN-solution deviation could be further shown in Figure 3.3(a) and (b). We could observe from the left graph of the figure that when proceeding to its steady-state (e.g., after $t \geq 2$), ZNN model (3.6) could always compute accurately the minimum value of non-stationary quadratic function $f(x)$ at every time instant t. On the other hand, as seen from the right graph of Figure 3.3, though proceeding earlier to its steady-state (e.g., after $t \geq 1.5$), GNN model (3.9) generates a result, sometimes equal to or sometimes larger than (i.e., in a hit-and-miss manner) the theoretical minimum value of non-stationary quadratic function $f(x)$ over time t.

- To monitor the convergence properties, we could also show the residual error $|f(x) - f(x^*(t))|$, where $|\cdot|$ denotes the absolute value of a scalar. As seen from Figure 3.4(a), $f(x)$ minimized by ZNN (3.6) could converge perfectly to the theoretical time-varying minimum-value $f(x^*(t))$. In comparison, as seen from Figure 3.4(b), $f(x)$ minimized by GNN (3.9) can not achieve $f(x^*(t))$ exactly.

3.3. Time-Varying Quadratic Program

Facing the efficacy of ZNN (3.6) on time-varying quadratic minimization, as a further investigation, we consider the following time-varying convex quadratic programming problem which is subject to the time-varying linear-equality constraints

$$\text{minimize} \quad x^T(t)P(t)x(t)/2 + q^T(t)x(t), \tag{3.10}$$
$$\text{subject to} \quad A(t)x(t) = b(t), \tag{3.11}$$

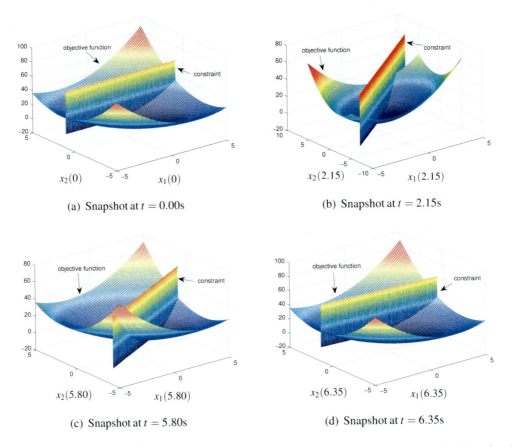

Figure 3.5. "Moving" quadratic function, "Moving" linear constraint and "Moving" optimal solution of time-varying quadratic program (3.10)-(3.11).

where time-varying decision-vector $x(t) \in \mathbb{R}^n$ is unknown and to be solved at any time instant $t \in [0, +\infty)$. In addition to the coefficients' description in Section 3.2. and smoothly time-varying coefficient vector $b(t) \in \mathbb{R}^m$, the smoothly time-varying coefficient matrix $A(t) \in \mathbb{R}^{m \times n}$ in equality constraint (3.11) is assumed to be of full row rank. Likewise, without loss of generality, the coefficient matrices and vectors, together with their time derivatives, are assumed to be known or at least can be estimated accurately. To guarantee the solution uniqueness, such time-varying quadratic program (3.10)-(3.11) should be strictly convex [20, 21] with $P(t) \in \mathbb{R}^{n \times n}$ positive-definite at any time instant $t \in [0, +\infty)$ throughout this section.

Facing the time-varying quadratic program (3.10)-(3.11) and based on the preliminary results on equality-constrained optimization problems [3, 20, 21], we have its related Lagrangian:

$$L(x(t), \lambda(t), t) = x^T(t)P(t)x(t)/2 + q^T(t)x(t) + \lambda^T(t)(A(t)x(t) - b(t)),$$

where $\lambda(t) \in \mathbb{R}^m$ denotes the Lagrange-multiplier vector.

As we may recognize or know, solving the time-varying quadratic program (3.10)-

(3.11) could be done by zeroing the equations below at any time instant $t \in [0, +\infty)$:

$$\begin{cases} \frac{\partial L(x(t),\lambda(t),t)}{\partial x(t)} = P(t)x(t) + q(t) + A^T(t)\lambda(t) = 0, \\ \frac{\partial L(x(t),\lambda(t),t)}{\partial \lambda(t)} = A(t)x(t) - b(t) = 0. \end{cases}$$

The above equations could be further written as

$$\tilde{P}(t)\tilde{x}(t) = -\tilde{q}(t), \tag{3.12}$$

where

$$\tilde{P}(t) := \begin{bmatrix} P(t) & A^T(t) \\ A(t) & \mathbf{0}_{m \times m} \end{bmatrix} \in \mathbb{R}^{(n+m) \times (n+m)},$$

$$\tilde{x}(t) := \begin{bmatrix} x(t) \\ \lambda(t) \end{bmatrix} \in \mathbb{R}^{n+m}, \quad \tilde{q}(t) := \begin{bmatrix} q(t) \\ -b(t) \end{bmatrix} \in \mathbb{R}^{n+m}.$$

Moreover, for the purposes of better understanding and comparison, we know that the time-varying theoretical solution could be written as

$$\tilde{x}^*(t) = [x^{*T}(t), \lambda^{*T}(t)]^T := -\tilde{P}^{-1}(t)\tilde{q}(t) \in \mathbb{R}^{n+m}.$$

To demonstrate the significance of the interesting problem (3.10)-(3.11) as well as for better visual effects and readability, we can take the following specific time-varying quadratic program as an example:

$$\text{minimize} \quad ((\sin t)/4 + 1)x_1^2(t) + ((\cos t)/4 + 1)x_2^2(t)$$
$$+ \cos tx_1(t)x_2(t) + \sin 3tx_1(t) + \cos 3tx_2(t), \tag{3.13}$$
$$\text{subject to} \quad \sin 4tx_1(t) + \cos 4tx_2(t) = \cos 2t. \tag{3.14}$$

Figure 3.5 shows the three-dimension snapshots at different time instants (i.e., $t = 0.00$, 2.15, 5.80 and 6.35s), where horizontal axes correspond to $x_1(t) \in \mathbb{R}$ and $x_2(t) \in \mathbb{R}$. From the figure, we can see evidently that the linear-constraint plane and the objective-function surface are both "moving" with time t. In other words, the problem that we are going to solve [namely, (3.10)-(3.11)] is a quite challenging problem in the sense that the optimal solution is also "moving" with time t due to the "moving" effects of the objective-function surface and linear-constraint plane.

3.3.1. Neural Network Solvers and Comparison

In the literature, conventional gradient-based neural networks and computational algorithms have been developed to compute algebraic and optimization problems with constant coefficients [11, 20–22]. However, when applied to the time-varying situation, a much faster convergence rate is required for these gradient-based approaches as compared to the variational rate of coefficient matrices and vectors. This may thus impose very stringent restrictions on hardware realization and/or sacrifice the solution precision very much (see Figure 1 of [13]). Different from the above gradient-based approaches, the following Zhang neural network is proposed in this subsection for finding the optimal solution of the time-varying quadratic program (3.10)-(3.11).

3.3.1.1. Zhang Neural Networks

To monitor the solving process of time-varying quadratic program (3.10)-(3.11) via time-varying linear system (3.12), we could firstly define the following vector-valued error function (rather than the scalar-valued nonnegative energy functions used in gradient-based neural approaches):

$$\tilde{e}(t) = \tilde{P}(t)\tilde{x}(t) + \tilde{q}(t) \in \mathbb{R}^{n+m}, \tag{3.15}$$

of which each element could be positive, negative, and lower-unbounded. Then, to make each element $\tilde{e}_i(t)$ of error vector $\tilde{e}(t) \in \mathbb{R}^{n+m}$ converge to zero, the following ZNN design formula can be adopted [11–13, 19]:

$$\frac{d\tilde{e}(t)}{dt} = -\gamma \Phi\big(\tilde{e}(t)\big), \tag{3.16}$$

where design-parameter $\gamma > 0$ and $\Phi(\cdot) : \mathbb{R}^{n+m} \to \mathbb{R}^{n+m}$ denote the same as those in Subsection 3.2.1..

Expanding ZNN design formula (3.16) leads to the following ZNN model depicted in an implicit-dynamic equation, which solves online the augmented time-varying linear matrix-vector equation (3.12) as well as the time-varying equality-constrained quadratic-program (3.10)-(3.11):

$$\tilde{P}(t)\dot{\tilde{x}}(t) = -\dot{\tilde{P}}(t)\tilde{x}(t) - \gamma \Phi\big(\tilde{P}(t)\tilde{x}(t) + \tilde{q}(t)\big) - \dot{\tilde{q}}(t), \tag{3.17}$$

or written in a more complete form as below:

$$\begin{bmatrix} P(t) & A^T(t) \\ A(t) & \mathbf{0} \end{bmatrix} \begin{bmatrix} \dot{x}(t) \\ \dot{\lambda}(t) \end{bmatrix} = - \begin{bmatrix} \dot{P}(t) & \dot{A}^T(t) \\ \dot{A}(t) & \mathbf{0} \end{bmatrix} \begin{bmatrix} x(t) \\ \lambda(t) \end{bmatrix}$$
$$- \gamma \Phi \left(\begin{bmatrix} P(t)x(t) + A^T(t)\lambda(t) + q(t) \\ A(t)x(t) - b(t) \end{bmatrix} \right) - \begin{bmatrix} \dot{q}(t) \\ -\dot{b}(t) \end{bmatrix},$$

where state vector $\tilde{x}(t) \in \mathbb{R}^{n+m}$, starting from an initial condition $\tilde{x}(0) \in \mathbb{R}^{n+m}$, corresponds to the theoretical solution of (3.12), of which the first n elements constitute the neural-network solution corresponding to the optimal solution of time-varying quadratic program (3.10)-(3.11). In addition, it is worth writing out the following linear ZNN model [simplified from ZNN (3.17)] for the online solution of time-varying quadratic program (3.10)-(3.11):

$$\tilde{P}(t)\dot{\tilde{x}}(t) = -\left(\dot{\tilde{P}}(t) + \gamma\tilde{P}(t)\right)\tilde{x}(t) - \left(\gamma\tilde{q}(t) + \dot{\tilde{q}}(t)\right), \tag{3.18}$$

or writing out in a more complete form as below:

$$\begin{bmatrix} P(t) & A^T(t) \\ A(t) & \mathbf{0} \end{bmatrix} \begin{bmatrix} \dot{x}(t) \\ \dot{\lambda}(t) \end{bmatrix} = -\left(\begin{bmatrix} \dot{P}(t) + \gamma P(t) & \dot{A}^T(t) + \gamma A^T(t) \\ \dot{A}(t) + \gamma A(t) & \mathbf{0} \end{bmatrix} \right) \begin{bmatrix} x(t) \\ \lambda(t) \end{bmatrix}$$
$$- \begin{bmatrix} \gamma q(t) + \dot{q}(t) \\ -\gamma b(t) - \dot{b}(t) \end{bmatrix}.$$

3.3.1.2. Gradient Neural Networks

For comparison, it is worth pointing out here that we can develop a gradient-based neural network to solve online the quadratic program (3.10)-(3.11). However, similar to almost all numerical algorithms and neural-dynamic schemes mentioned before, the gradient neural networks are designed intrinsically for problems with constant coefficient matrices and/or vectors [e.g., constant coefficients P, A, q and b in (3.10)-(3.11)] [11,19,22]. Now we show the GNN design procedure as the following.

1) Firstly, a scalar-valued norm-based energy function, such as $\|\tilde{P}\tilde{x} + \tilde{q}\|_2^2/2$ with $\|\cdot\|_2$ denoting the two norm of a vector, is constructed such that its minimum point is the solution of linear system $\tilde{P}\tilde{x} = -\tilde{q}$.

2) Secondly, an algorithm is designed to evolve along a descent direction of this energy function until the minimum point is reached. The typical descent direction is the negative of the gradient of energy function $\|\tilde{P}\tilde{x} + \tilde{q}\|_2^2/2$, i.e.,

$$-\frac{\partial \|\tilde{P}\tilde{x} + \tilde{q}\|_2^2/2}{\partial \tilde{x}} = -\tilde{P}^T \left(\tilde{P}\tilde{x} + \tilde{q}\right)$$

3) Thirdly, by using the above negative gradient to construct and apply the neural network to the time-varying situation, we could have a linear GNN model solving (3.10)-(3.11), $\dot{\tilde{x}}(t) = -\gamma \tilde{P}^T(t)\left(\tilde{P}(t)\tilde{x}(t) + \tilde{q}(t)\right)$, and also a generalized nonlinear GNN model,

$$\dot{\tilde{x}}(t) = -\gamma \tilde{P}^T(t)\Phi\left(\tilde{P}(t)\tilde{x}(t) + \tilde{q}(t)\right). \tag{3.19}$$

3.3.1.3. Methods and Models Comparison

In this subsubsection, we would like to compare the above-presented design methods and their resultant models; namely, ZNN (3.17) and GNN (3.19) exploited for online solution of time-varying quadratic program (3.10)-(3.11). The main differences and novelties may lie in the following facts.

1) ZNN model (3.17) is designed based on the elimination of every element of the vector-valued error function $\tilde{e}(t) = \tilde{P}(t)\tilde{x}(t) + \tilde{q}(t)$. In contrast, GNN model (3.19) is designed based on the elimination of the scalar-valued norm-based energy function $\|\tilde{P}\tilde{x} + \tilde{q}\|_2^2$ (note that, generally speaking, the GNN design method could only handle static problems with constant coefficients).

2) ZNN (3.17) is depicted in an implicit dynamics, i.e., $\tilde{P}(t)\dot{\tilde{x}}(t) = \cdots$, which coincides well with systems in nature and in practice (e.g., in analogue electronic circuits and mechanical systems owing to Kirchhoff's and Newton's laws, respectively [8,11,12,23]). In contrast, GNN (3.19) is depicted in an explicit dynamics, i.e., $\dot{\tilde{x}}(t) = \cdots$, which is usually associated with classic Hopfield-type and/or gradient-based neural-network models. Comparing the implicit and explicit dynamics/systems, we can see that the former seem to have higher abilities in representing dynamic systems, as it can preserve physical parameters even in the coefficient matrix on the left-hand side of the system as well, e.g., the so-called mass matrix $\tilde{P}(t)$ in (3.17).

3) ZNN model (3.17) could systematically and methodologically exploit the time-derivative information of coefficient matrices and vectors [namely, $\dot{P}(t)$, $\dot{A}(t)$, $\dot{q}(t)$ and $\dot{b}(t)$] during its real-time solving process. On the other hand, GNN model (3.19) has not exploited

such important information [i.e., $\dot{P}(t), \dot{A}(t), \dot{q}(t)$ and $\dot{b}(t)$], it may be thus less effective on solving the time-varying problem.

4) As shown in the ensuring subsections, ZNN model (3.17) could globally exponentially converge to the theoretical time-varying optimal solution of time-varying quadratic program (3.10)-(3.11). In contrast, GNN model (3.19) could only generate approximate solutions to (3.10)-(3.11) with much larger steady-state errors, thus less effective on the time-varying quadratic program solving.

3.3.2. Convergence Analysis and Results

In this subsection, three theorems about the convergence properties of ZNN model (3.17) are established for online solution of time-varying quadratic program (3.10)-(3.11). The analysis includes the situations of using linear or power-sigmoid activation functions (with their shapes shown in Figure 1.3(a) and (d) in Section 1.5.).

Theorem 3.3.1. *Consider smoothly time-varying strictly-convex quadratic program* (3.10)–(3.11). *If a monotonically-increasing odd activation-function array* $\Phi(\cdot)$ *is used, then the state vector* $\tilde{x}(t)$ *of ZNN model* (3.17), *starting from any initial state* $\tilde{x}(0) \in \mathbb{R}^{n+m}$, *could globally converge to the unique time-varying theoretical solution* $\tilde{x}^*(t) = [x^{*T}(t), \lambda^{*T}(t)]^T$ *of time-varying linear system* (3.12). *In addition, the first n elements of solution* $\tilde{x}^*(t)$ *constitute the time-varying optimal solution* $x^*(t)$ *to the time-varying quadratic program* (3.10)–(3.11).

Proof. For ZNN design formula (3.16), we could define a Lyapunov function candidate $v(t) = \|\tilde{e}(t)\|_2^2/2 = \tilde{e}^T(t)\tilde{e}(t)/2 \geq 0$, and its time-derivative is thus

$$\dot{v}(t) = \frac{dv(t)}{dt} = \tilde{e}^T(t)\frac{d\tilde{e}(t)}{dt}$$
$$= -\gamma\tilde{e}^T(t)\Phi\big(\tilde{e}(t)\big) = -\gamma\sum_{i=1}^{n+m}\tilde{e}_i(t)\phi\big(\tilde{e}_i(t)\big). \tag{3.20}$$

Because $\phi(\tilde{e}_i(t))$ is a monotonically-increasing odd activation function, we have

$$\tilde{e}_i(t)\phi\big(\tilde{e}_i(t)\big) \begin{cases} > 0, \text{if } \tilde{e}_i(t) > 0, \\ = 0, \text{if } \tilde{e}_i(t) = 0, \\ > 0, \text{if } \tilde{e}_i(t) < 0, \end{cases}$$

which guarantees the negative-definiteness of $\dot{v}(t)$. By Lyapunov theory [24], error function $\tilde{e}(t)$ could globally converge to zero. In view of $\tilde{e}(t) = \tilde{P}(t)\tilde{x}(t) + \tilde{q}(t) = \tilde{P}(t)(\tilde{x}(t) - \tilde{x}^*(t))$ and $\tilde{P}(t)$ being nonsingular [20] for any time instant t, we could obtain that $\tilde{x}(t) \to \tilde{x}^*(t)$, as $t \to +\infty$. That is, state vector $\tilde{x}(t)$ of ZNN (3.17) globally converges to the exact time-varying theoretical solution $\tilde{x}^*(t)$ of time-varying linear system (3.12). As seen from Section 3.3., the theoretical solution $\tilde{x}^*(t)$ of time-varying linear system (3.12) contains two parts: the first part is $x^*(t)$ which is the optimal solution to time-varying quadratic program (3.10)-(3.11) and is constituted by the first n elements of $\tilde{x}^*(t)$; and the second part is $\lambda^*(t)$ which is the corresponding Lagrange-multiplier vector for (3.10)-(3.11) and is constituted by the last m elements of $\tilde{x}^*(t)$. The proof is thus complete. $\square$

Theorem 3.3.2. *In addition to Theorem 3.3.1, if the linear activation function $\phi(\tilde{e}_i(t)) = \tilde{e}_i(t)$ is used, then the state vector $\tilde{x}(t)$ of ZNN model (3.17) could globally exponentially converge to the unique theoretical solution $\tilde{x}^*(t) = [x^{*T}(t), \lambda^{*T}(t)]^T$ of time-varying linear system (3.12), with $x^*(t)$ being the time-varying optimal solution to time-varying problem (3.10)–(3.11).*

Proof. Let us review ZNN model (3.17): it could be equivalently rewritten as

$$\tilde{P}(t)\dot{\hat{x}}(t) = -\dot{\tilde{P}}(t)\hat{x}(t) - \gamma\Phi\big(\tilde{P}(t)\hat{x}(t)\big), \tag{3.21}$$

where $\hat{x}(t) := \tilde{x}(t) - \tilde{x}^*(t)$ denotes the difference between the neural-network state-vector $\tilde{x}(t)$ and the theoretical time-varying solution $\tilde{x}^*(t)$. In addition, the relation between $\hat{x}(t)$ and $\tilde{e}(t)$ is

$$\begin{aligned}
\|\hat{x}(t)\|_2 &= \|\tilde{x}(t) - \tilde{x}^*(t)\|_2 \\
&\leq \frac{1}{\sqrt{\lambda_{min}}}\|\tilde{P}(t)(\tilde{x}(t) - \tilde{x}^*(t))\|_2 \\
&\leq \frac{1}{\sqrt{\lambda_{min}}}\|\tilde{e}(t)\|_2.
\end{aligned} \tag{3.22}$$

where $\lambda_{min} > 0$ denotes the minimal eigenvalue of matrix $\tilde{P}^T(t)\tilde{P}(t)$ all over the time $t \in [0, +\infty)$. If the linear activation function array $\Phi(\tilde{e}(t)) = \tilde{e}(t)$ is used, then it follows from ZNN design formula (3.16) that $\tilde{e}(t) = \tilde{e}(0)\exp(-\gamma t)$, or in element form, $\tilde{e}_i(t) = \tilde{e}_i(0)\exp(-\gamma t)$, $i = 1, 2, \cdots, (n+m)$. Thus, we have

$$\|\hat{x}(t)\|_2 \leq \frac{1}{\sqrt{\lambda_{min}}}\|\tilde{e}(t)\|_2 = \sqrt{\sum_{i=1}^{n+m} \frac{\tilde{e}_i^2(0)}{\lambda_{min}}}\exp(-\gamma t),$$

which implies $\hat{x}(t)$ globally exponentially converges to zero with exponential convergence rate γ. That is, ZNN state $\tilde{x}(t)$ globally exponentially converges to theoretical solution $\tilde{x}^*(t)$ with exponential convergence rate γ, of which the first n elements constitute the time-varying optimal solution $x^*(t)$ to time-varying quadratic program (3.10)-(3.11). The proof is thus complete. $\square$

Theorem 3.3.3. *In addition to Theorems 3.3.1 and 3.3.2, if we use the power-sigmoid activation function*

$$\phi\big(\tilde{e}_i(t)\big) = \begin{cases} \tilde{e}_i^{2r-1}(t), & |\tilde{e}_i(t)| \geq 1 \\ \dfrac{1+\exp(-\zeta)}{1-\exp(-\zeta)}\dfrac{1-\exp\big(-\zeta\tilde{e}_i(t)\big)}{1+\exp\big(-\zeta\tilde{e}_i(t)\big)}, & |\tilde{e}_i(t)| \leq 1 \end{cases} \tag{3.23}$$

with suitable design parameters $\zeta \geq 2$ and $r \geq 2$ (being an integer), then state vector $\tilde{x}(t)$ of ZNN model (3.17) is superiorly globally convergent to the theoretical time-varying solution $\tilde{x}^*(t) = [x^{*T}(t), \lambda^{*T}(t)]^T$ with $x^*(t)$ being the optimal solution to time-varying quadratic program (3.10)–(3.11), which is compared with the situation of using linear activation functions as presented in Theorem 3.3.2.

Proof. 1) For error range $|\tilde{e}_i(t)| \geq 1$

The power activation function $\phi(\tilde{e}_i(t)) = \tilde{e}_i^{2r-1}(t)$ with $r \geq 2$ is used specifically for this error range. Review the Lyapunov function candidate $v(t) = \tilde{e}^T(t)\tilde{e}(t)/2$ and equation (3.20). Over such an error range, we have

$$\dot{v}(t)_{\text{ps}} = -\gamma \sum_{i=1}^{n+m} \tilde{e}_i(t)\phi(\tilde{e}_i(t)) = -\gamma \sum_{i=1}^{n+m} \tilde{e}_i^{2r}(t)$$

$$\begin{cases} = -\gamma\sum_{i=1}^{n+m} \tilde{e}_i^2(t) = \dot{v}(t)_{\text{lin}}, & \forall |\tilde{e}_i(t)| = 1, \\ < -\gamma\sum_{i=1}^{n+m} \tilde{e}_i^2(t) = \dot{v}(t)_{\text{lin}}, & \forall |\tilde{e}_i(t)| > 1, \\ \ll -\gamma\sum_{i=1}^{n+m} \tilde{e}_i^2(t) = \dot{v}(t)_{\text{lin}}, & \forall |\tilde{e}_i(t)| \gg 1, \end{cases}$$

where $\dot{v}(t)_{\text{ps}}$ and $\dot{v}(t)_{\text{lin}}$ denote the time-derivative of $v(t)$ activated by power-sigmoid activation functions and linear activation functions, respectively. Evidently, if follows from $v(t) = \tilde{e}^T(t)\tilde{e}(t)/2$ and $\dot{v}(t)_{\text{ps}} \leq \dot{v}(t)_{\text{lin}}$ that $v(t)_{\text{ps}} \leq v(t)_{\text{lin}}$ [25], implying that if we use the power-sigmoid activation function over error range $|\tilde{e}_i(t)| > 1$, superior convergence can be achieved for ZNN (3.17), as compared to the situation of using the linear activation function with exponential convergence rate γ.

2) For error range $|\tilde{e}_i(t)| \leq 1$

The bipolar-sigmoid activation function $\phi(\tilde{e}_i) = \tau(1 - \exp(-\zeta\tilde{e}_i)(1 + \exp(-\zeta\tilde{e}_i))$ is used specifically for this error range, with $\tau := (1 + \exp(-\zeta))/(1 - \exp(-\zeta)) > 0$. Review the Lyapunov function candidate $v(t) = \tilde{e}^T(t)\tilde{e}(t)/2$ and equation (3.20) again. Over the error range $|\tilde{e}_i(t)| \leq 1$ (as seen from Figure 1.3(a) and (b) as well), we can prove with $\zeta \geq 2$ that

$$\dot{v}(t)_{\text{ps}} = -\gamma \sum_{i=1}^{n+m} \tilde{e}_i(t)\phi(\tilde{e}_i(t))$$

$$\begin{cases} < -\gamma\sum_{i=1}^{n+m} \tilde{e}_i^2(t) = \dot{v}(t)_{\text{lin}}, & \forall 0 < |\tilde{e}_i(t)| < 1, \\ = -\gamma\sum_{i=1}^{n+m} \tilde{e}_i^2(t) = \dot{v}(t)_{\text{lin}}, & \forall |\tilde{e}_i(t)| = 0 \text{ or } 1, \end{cases}$$

which, together with $v(t) = \tilde{e}^T(t)\tilde{e}(t)/2$, guarantees that $v(t)_{\text{ps}} \leq v(t)_{\text{lin}}$ [25], implying that if we use the power-sigmoid activation function over error range $0 < |\tilde{e}_i(t)| < 1$, superior convergence can also be achieved for ZNN (3.17), as compared to the situation of using the linear activation function with exponential convergence rate γ.

Summarizing the above analysis of the two sub-cases, if we use the power-sigmoid activation function to construct ZNN (3.17), superior convergence could be achieved, as compared to the situation of using the linear activation function. The proof is now thus complete. $\square$

3.3.3. Illustrative Examples

While Subsections 3.3.1. and 3.3.2. present the models and related analysis results of ZNN (3.17) and GNN (3.19), in this subsection we show an illustrative computer-simulation example so as to demonstrate the characteristics of the neural-network convergence. For illustration and comparison purposes, both ZNN model (3.17) and GNN model (3.19) are employed to solve online such a time-varying quadratic program (3.10)-(3.11), which are

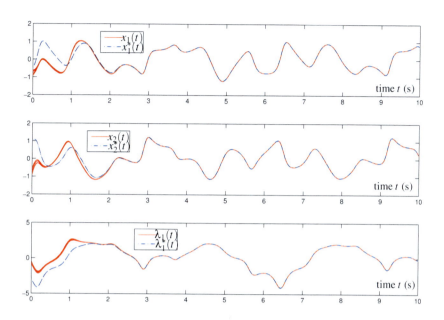

(a) State vector $\tilde{x}(t) \in \mathbb{R}^3$ of ZNN (3.17) solving the time-varying quadratic program (3.13)-(3.14)

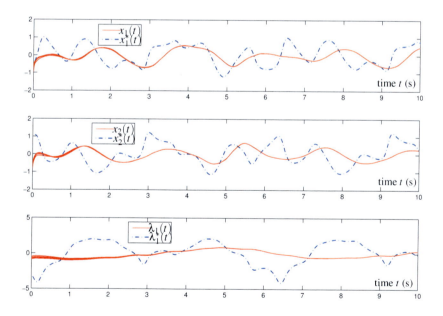

(b) State vector $\tilde{x}(t) \in \mathbb{R}^3$ of GNN (3.19) solving the time-varying quadratic program (3.13)-(3.14)

Figure 3.6. Online solution of time-varying quadratic program (3.13)-(3.14) by ZNN and GNN models with $\gamma = 1$, where dashed-dotted blue curves correspond to time-varying theoretical solution $\tilde{x}^*(t)$ and solid red curves correspond to neural-network solution $\tilde{x}(t)$.

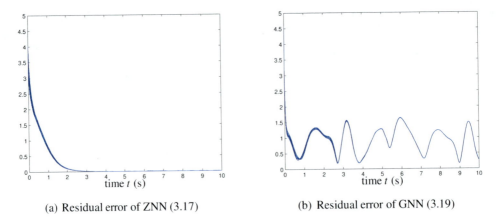

(a) Residual error of ZNN (3.17) (b) Residual error of GNN (3.19)

Figure 3.7. Residual errors $\|\tilde{P}(t)\tilde{x}(t) + \tilde{q}(t)\|_2$ of ZNN and GNN models with $\gamma = 1$ for online solution of time-varying quadratic program (3.13)-(3.14).

based on the use of the power-sigmoid activation function (3.23) with design parameters $r = 2$ and $\zeta = 4$.

Consider the specific time-varying quadratic programming problem depicted in (3.13)-(3.14). Evidently, we can rewrite the time-varying quadratic program (3.13)-(3.14) in the matrix-vector form (3.10)-(3.11) with the following coefficient matrices and vectors

$$P(t) = \begin{bmatrix} 0.5\sin t + 2 & \cos t \\ \cos t & 0.5\cos t + 2 \end{bmatrix}, \quad q(t) = \begin{bmatrix} \sin 3t \\ \cos 3t \end{bmatrix},$$

$$A(t) = [\sin 4t, \cos 4t], \quad b(t) = \cos 2t, \quad x(t) = [x_1(t), x_2(t)]^T.$$

It follows from equation (3.12) and ZNN (3.17) that we have

$$\tilde{P}(t) = \begin{bmatrix} 0.5\sin t + 2 & \cos t & \sin 4t \\ \cos t & 0.5\cos t + 2 & \cos 4t \\ \sin 4t & \cos 4t & 0 \end{bmatrix},$$

$$\tilde{q}(t) = \begin{bmatrix} \sin 3t, & \cos 3t, & -\cos 2t \end{bmatrix}^T,$$

and $\tilde{x}(t) = [x_1(t), x_2(t), \lambda_1(t)]^T$. Then we have the following simulation results.

As seen from Figure 3.6(a), starting form six randomly-generated initial states $\tilde{x}(0) = [x_1(0), x_2(0), \lambda_1(0)]^T \in [-1, 1]^3$, state-vectors $\tilde{x}(t)$ of ZNN model (3.17) could always globally converge to the theoretical time-varying solution $\tilde{x}^*(t) = [x_1^*(t), x_2^*(t), \lambda_1^*(t)]^T$ exactly, of which the $x_1^*(t)$ and $x_2^*(t)$ are the time-varying optimal solution to time-varying quadratic program (3.13)-(3.14). In contrast, as shown in Figure 3.6(b), the state-vectors $\tilde{x}(t)$ of GNN model (3.19) do not fit well with the theoretical solution $\tilde{x}^*(t)$ with quite large computational errors.

Moreover, we could also monitor the residual error $\|\tilde{P}(t)\tilde{x}(t) + \tilde{q}(t)\|_2$ during the problem-solving process by both neural networks. As seen from Figures 3.7(a) and 3.8 as well as other simulation data, by using ZNN model (3.17) to solve time-varying quadratic program (3.13)-(3.14), its residual error $\|\tilde{P}(t)\tilde{x}(t) + \tilde{q}(t)\|_2$ is globally convergent to zero

ZNN for Time-Varying Convex Quadratic Optimization

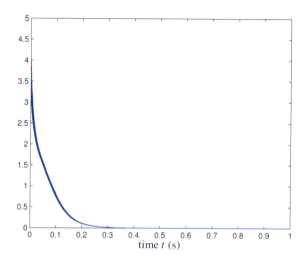

Figure 3.8. Residual error $\|\tilde{P}(t)\tilde{x}(t)+\tilde{q}(t)\|_2$ of ZNN (3.17) with $\gamma = 10$ for online solution of time-varying quadratic program (3.13)-(3.14).

within 5 seconds. More specifically, the steady-state residual-error $\lim_{t\to+\infty}\|\tilde{P}(t)\tilde{x}(t)+\tilde{q}(t)\|_2$ is around 2.842700×10^{-13} and 2.661423×10^{-13} (as measured at $t = 100$s), which correspond to the use of design parameter $\gamma = 1$ and 10, respectively. In contrast, it is seen from Figure 3.7(b) and other simulation data that, by using GNN (3.19) to solve the time-varying quadratic program (3.13)-(3.14) under the same simulation conditions, its residual error $\|\tilde{P}(t)\tilde{x}(t)+\tilde{q}(t)\|_2$ is rather large, and its steady-state residual-error $\lim_{t\to+\infty}\|\tilde{P}(t)\tilde{x}(t)+\tilde{q}(t)\|_2$ is about 1.461293 and 0.814059, which correspond to the use of design parameter $\gamma = 1$ and 10, respectively. In addition, it is worth pointing out that, as shown in Figures 3.7(a) and 3.8, the convergence time for ZNN model (3.17) can be expedited from around 4 seconds to 0.4 second, when design parameter γ is increased from 1 to 10. Furthermore, if $\gamma = 10^4$, the convergence time is around 0.4 millisecond. This may show that ZNN (3.17) has an exponential-convergence property, which could be expedited effectively by increasing the value of design parameter γ.

In summary, this illustrative example substantiates the theoretical results, efficacy and advantages of the ZNN model (3.17) which could solve the time-varying quadratic programs, as compared to the less favorable GNN model (3.19).

3.4. Modeling Techniques and Illustrative Examples

To simulate and/or model the above-presented ZNN and GNN solvers, the following MATLAB Simulink modeling techniques (see [14, 26] and references therein) are investigated in this section to show the RNN-solution characteristics. A Simulink model is a representation of the design and/or implementation of a dynamic system satisfying a set of requirements. The modeling techniques are related to those basic function-blocks and their parameters' setting. By interconnecting the basic function-blocks with appropriate parameters, the diagrams of ZNN (3.6) and (3.17), and GNN (3.9) and (3.19) can be built up, as shown in

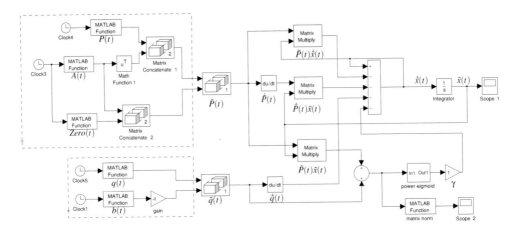

Figure 3.9. A unified overall model of ZNN (3.6) and (3.17) for time-varying QP solving.
Reproduced from Y. Zhang, X. Li et al., Modeling and verification of Zhang neural networks for online solution of time-varying quadratic minimization and programming, Figure 1, Z. Cai et al. (Eds.): ISICA 2009, LNCS 5821, pp. 101-110, 2009. ©Springer-Verlag Berlin Heidelberg 2009. With kind permission of Springer Science+Business Media.

Figures 3.9 and 3.10, which are exploited to solve online the time-varying quadratic program (3.10)-(3.11).

With the overall ZNN and GNN models built up and depicted in Figures 3.9 and 3.10, prior to running them, we may have to change some of the default simulation options by opening the "Configuration Parameters" dialog box and setting

- Solver: "ode45 (Dormand-Prince)";

- Absolute tolerance: "1e-6 (i.e., 10^{-6})";

- Relative tolerance: "1e-6 (i.e., 10^{-6})";

- Algebraic loop: "none".

Then, for illustration and comparison purposes, ZNN (3.17) and GNN (3.19), activated by the power-sigmoid activation function (1.7) with design parameters $p = 3$ and $\xi = 4$, could both be modeled and employed to solve online such a time-varying quadratic-programming problem depicted in (3.10)-(3.11).

More specifically, let us consider the time-varying quadratic-program (3.10)-(3.11) with the following coefficient matrices and vectors (as an example):

$$P(t) = \begin{bmatrix} 0.5\sin t + 2 & \cos t \\ \cos t & 0.5\cos t + 2 \end{bmatrix}, \quad q(t) = \begin{bmatrix} \sin 6t \\ \cos 6t \end{bmatrix},$$

$$A(t) = [\sin 4t, \cos 4t], \quad b(t) = \cos 2t.$$

It follows from equation (3.12), ZNN (3.17) and GNN (3.19) that

$$\tilde{P}(t) = \begin{bmatrix} 0.5\sin t + 2 & \cos t & \sin 4t \\ \cos t & 0.5\cos t + 2 & \cos 4t \\ \sin 4t & \cos 4t & 0 \end{bmatrix},$$

$$\tilde{q}(t) = \begin{bmatrix} \sin 6t, & \cos 6t, & -\cos 2t \end{bmatrix}^T.$$

ZNN for Time-Varying Convex Quadratic Optimization

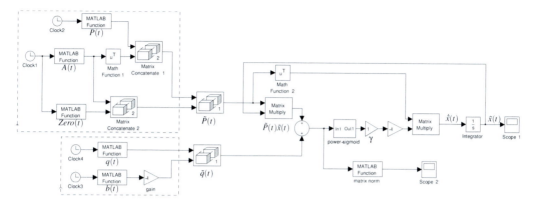

Figure 3.10. A unified overall model of GNN (3.9) and (3.19) for time-varying QP solving. *Reproduced from Y. Zhang, X. Li et al., Modeling and verification of Zhang neural networks for online solution of time-varying quadratic minimization and programming, Figure 2, Z. Cai et al. (Eds.): ISICA 2009, LNCS 5821, pp. 101-110, 2009. ©Springer-Verlag Berlin Heidelberg 2009. With kind permission of Springer Science+Business Media.*

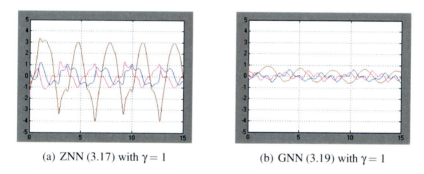

(a) ZNN (3.17) with $\gamma = 1$ (b) GNN (3.19) with $\gamma = 1$

Figure 3.11. Output $\tilde{x}(t)$ of Scope 1 exploited in ZNN and GNN modeling. *Reproduced from Y. Zhang, X. Li et al., Modeling and verification of Zhang neural networks for online solution of time-varying quadratic minimization and programming, Figure 3, Z. Cai et al. (Eds.): ISICA 2009, LNCS 5821, pp. 101-110, 2009. ©Springer-Verlag Berlin Heidelberg 2009. With kind permission of Springer Science+Business Media.*

ZNN (3.17) is thus formulated in the following specific form [so will be GNN (3.19)]:

$$\begin{bmatrix} 0.5\sin t + 2 & \cos t & \sin 4t \\ \cos t & 0.5\cos t + 2 & \cos 4t \\ \sin 4t & \cos 4t & 0 \end{bmatrix} \begin{bmatrix} \dot{x}_1(t) \\ \dot{x}_2(t) \\ \dot{\lambda}(t) \end{bmatrix}$$

$$= - \begin{bmatrix} 0.5\cos t & -\sin t & 4\cos 4t \\ -\sin t & -0.5\sin t & -4\sin 4t \\ 4\cos 4t & -4\sin 4t & 0 \end{bmatrix} \begin{bmatrix} x_1(t) \\ x_2(t) \\ \lambda(t) \end{bmatrix} - \begin{bmatrix} 6\cos 6t \\ -6\sin 6t \\ 2\sin 2t \end{bmatrix}$$

$$- \Phi\left(\begin{bmatrix} 0.5\sin t + 2 & \cos t & \sin 4t \\ \cos t & 0.5\cos t + 2 & \cos 4t \\ \sin 4t & \cos 4t & 0 \end{bmatrix} \begin{bmatrix} x_1(t) \\ x_2(t) \\ \lambda(t) \end{bmatrix} + \begin{bmatrix} \sin 6t \\ \cos 6t \\ -\cos 2t \end{bmatrix} \right).$$

Then we could have the specific ZNN and GNN models built up and depicted in Figures 3.9 and 3.10, respectively. After running them, we can obtain their online solutions

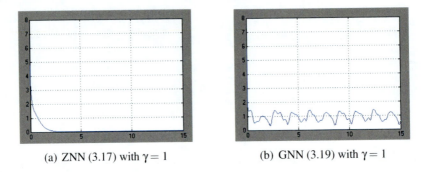

(a) ZNN (3.17) with $\gamma = 1$ (b) GNN (3.19) with $\gamma = 1$

Figure 3.12. Output $\|\tilde{P}(t)\tilde{x}(t) + \tilde{q}(t)\|_2$ of Scope 2 exploited in ZNN and GNN modeling. *Reproduced from Y. Zhang, X. Li et al., Modeling and verification of Zhang neural networks for online solution of time-varying quadratic minimization and programming, Figure 4, Z. Cai et al. (Eds.): ISICA 2009, LNCS 5821, pp. 101-110, 2009. ©Springer-Verlag Berlin Heidelberg 2009. With kind permission of Springer Science+Business Media.*

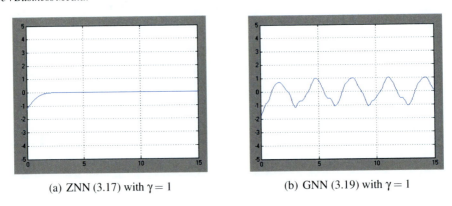

(a) ZNN (3.17) with $\gamma = 1$ (b) GNN (3.19) with $\gamma = 1$

Figure 3.13. RNN-synthesized scalar-values of equality-constraint satisfaction $A(t)x(t) - b(t)$. *Reproduced from Y. Zhang, X. Li et al., Modeling and verification of Zhang neural networks for online solution of time-varying quadratic minimization and programming, Figure 5, Z. Cai et al. (Eds.): ISICA 2009, LNCS 5821, pp. 101-110, 2009. ©Springer-Verlag Berlin Heidelberg 2009. With kind permission of Springer Science+Business Media.*

as shown in Figure 3.11(a) and (b): evidently, the ZNN and GNN solution-curves are quite different. In addition, as seen from Figure 3.12(a), by using ZNN model (3.17) to solve the time-varying quadratic program (3.10)-(3.11), its residual error $\|\tilde{P}(t)\tilde{x}(t) + \tilde{q}(t)\|_2$ decreases to zero fast. For comparison, as seen from Figure 3.12(b), by using GNN model (3.19) to solve the time-varying quadratic program (3.10)-(3.11), under the same modeling conditions, its residual error $\|\tilde{P}(t)\tilde{x}(t) + \tilde{q}(t)\|_2$ is however much larger (than the ZNN one) and is also with very clear fluctuations. Furthermore, corresponding to the above RNN-solution process, we could also show the value of time-varying equality-constraint satisfaction [i.e., the value of $A(t)x(t) - b(t)$] for QP (3.10)-(3.11). The results are in Figure 3.13, where the rapid convergence of $A(t)x(t) - b(t)$ to zero [or to say, the satisfaction of time-varying equality-constraint $A(t)x(t) = b(t)$] has been achieved well by ZNN model (3.17). In contrast, as seen especially from Figure 3.13(b), by using GNN model (3.19) to solve the time-varying quadratic program (3.10)-(3.11), the time-varying equality-constraint $A(t)x(t) = b(t)$ can not be satisfied accurately in view of the observation that the steady-state scalar-value of $A(t)x(t) - b(t)$ is not zero but fluctuating around.

In summary, from the modeling and verification results, we can say that, about online solution of time-varying QP problems, the resultant ZNN models could perform much better than gradient-based models. This may further imply that ZNN method and models might be a powerful alternative to the relevant time-varying optimization problem solving.

3.5.   Conclusion

In this chapter, by following Zhang *et al.*'s design method, a kind of recurrent neural networks (i.e., Zhang neural networks) has been proposed, unified and modeled for solving online the time-varying convex quadratic-minimization and quadratic-program problems (of which the latter is subject to a time-varying linear matrix-vector equality-constraint). Different from conventional gradient-based neural-network methods and models, the Zhang neural network could methodologically utilize the time-derivative information of time-varying problems and thus achieve global exponential convergence to the time-varying optimal solution. In addition, instead of writing simulation-codes, we could build up the neural-network models readily and rapidly by employing, creating and connecting the modeling blocks. Computer-simulation results have demonstrated the feasibility and efficacy of Zhang *et al.*'s neural-dynamic design-method and solvers for handling time-varying problems. Compared to conventional gradient-based solvers, superior performance can be achieved by the ZNN models for time-varying convex quadratic problem solving, especially using the power-sigmoid activation-function array.

References

[1] Johansen, TA; Fsosen, TI; Berge, SP. Constrained nonlinear control allocation with singularity avoidance using sequential quadratic programming. *IEEE Transactions on Control Systems Technology*, 2004, vol. 12, no.1, pp. 211-216.

[2] Grudinin, N. Reactive power optimization using successive quadratic programming method. *IEEE Transactions on Power Systems*, 1998, vol. 13, no. 4, pp. 1219-1225.

[3] Wang, J; Zhang, Y. *Recurrent Neural Networks for Real-Time Computation of Inverse Kinematics of Redundant Manipulators*. Machine Intelligence Quo Vadis?, Singapore: World Scientific, 2004.

[4] Leithead, WE; Zhang, Y. O(N^2)-operation approximation of covariance matrix inverse in Gaussian process regression based on quasi-Newton BFGS method. *Communications in Statistics-Simulation and Computation*, 2007, vol. 36, no. 2, 367-380.

[5] Zhang, Y; Wang, J; Xia, Y. A dual neural network for redundancy resolution of kinematically redundant manipulators subject to joint limits and joint velocity limits. *IEEE Transactions on Neural Networks*, 2003, vol. 14, no. 3, pp. 658-667.

[6] Manherz, RK; Jordan, BW; Hakimi, SL. Analog methods for computation of the generalized inverse. *IEEE Transactions on Automatic Control*, 1968, vol. 13, no. 5, pp. 582-585.

[7] Sturges, Jr RH. Analog matrix inversion (robot kinematics). *IEEE Journal of Robotics and Automation*, 1988, vol. 4, no. 2, pp. 157-162.

[8] Tank, DW; Hopfield, JJ. Simple "neural" optimization networks: an A/D converter, signal decision circuit, and a linear programming circuit. *IEEE Transactions on Circuits and Systems*, 1986, vol. 33, no.5, pp. 533-541.

[9] Costantini, G; Perfetti, R; Todisco, M. Quasi-Lagrangian neural network for convex quadratic optimization. *IEEE Transactions on Neural Networks*, 2008, vol. 19, no. 10, pp. 1804-1809.

[10] Zhang, Y. Revisit the analog computer and gradient-based neural system for matrix inversion. *Proceedings of IEEE International Symposium on Intelligent Control*, 2005, pp. 1411-1416.

[11] Zhang, Y; Jiang, D; Wang, J. A recurrent neural network for solving Sylvester equation with time-varying coefficients. *IEEE Transactions on Neural Networks*, 2002, vol. 13, no. 5, pp. 1053-1063.

[12] Zhang, Y; Ge, SS. Design and analysis of a general recurrent neural network model for time-varying matrix inversion. *IEEE Transactions on Neural Networks*, 2005, vol. 16, no. 6, pp. 1477-1490.

[13] Zhang, Y; Yue, S; Chen, K; Yi, C. MATLAB simulation and comparison of Zhang neural network and gradient neural network for time-varying Lyapunov equation solving. *Lecture Notes in Computer Science, ISNN2008*, Beijing, China, part I, 2008, vol. 5263, pp. 117-127.

[14] Zhang, Y; Chen, K; Li, X; Yi, C; Zhu, H. Simulink modeling and comparison of Zhang neural networks and gradient neural networks for time-varying Lyapunov equation solving. *Proceedings of the Fourth International Conference on Natural Computation*, 2008, pp. 521-525.

[15] Mead, C. *Analog VLSI and Neural Systems*. Reading, MA: Addison-Wesley, 1989.

[16] Tsiotras, P; Corless, M; Rotea, M. Optimal control of rigid body angular velocity with quadratic cost. *Journal of Optimization Theory and Applications*, 1998, vol. 96, no. 3, pp. 507-532.

[17] Zhang, Y. Towards piecewise-linear primal neural networks for optimization and redundant robotics. *Proceedings of IEEE International Conference on Networking, Sensing and Control*, 2006, pp. 374-379.

[18] Davey, K; Ward, MJ. A successive preconditioned conjugate gradient method for the minimization of quadratic and nonlinear functions. *Applied Numerical Mathematics*, 2000, vol. 35, no.2, pp. 129-156.

[19] Zhang, Y; Chen, K. Global exponential convergence and stability of Wang neural network for solving online linear equations. *Electronics Letters*, 2008, vol. 44, no. 2, pp. 145-146.

[20] Boyd, S; Vandenberghe, L. *Convex Optimization*. New York: Cambridge University Press, 2004.

[21] Nocedal, J; Wright, SJ. *Numerical Optimization*. Berlin Heidelberg: Springer-Verlag, 1999.

[22] Wang, J. Electronic realisation of recurrent neural work for solving simultaneous linear equations. *Electronics Letters*, 1992, vol. 28, no. 5, pp. 493-495.

[23] Zhang, Y; Ge, SS; Lee, TH. A unified quadratic-programming-based dynamical system approach to joint torque optimization of physically constrained redundant manipulators. *IEEE Transactions on Systems, Man, and Cybernetics*, Part B, 2004, vol. 34, no. 5, pp. 2126-2132.

References

[24] Zhang, Y. *Dual Neural Networks: Design, Analysis, and Application to Redundant Robotics, in: Progress in Neurocomputing Research.* New York: Nova Science Publishers, 2007.

[25] Hartman, P. *Ordinary Differential Equations.* Boston: Bivkhauser, 1982.

[26] The MathWorks Inc. Using Simulink,
`http://www.mathworks.com/access/helpdesk/help/toolbox/simulink.`

Part II

Pure Matrix Problems

Chapter 4

ZNN for Time-Varying Matrix Inversion and Pseudoinverse-Solving

Abstract

Different from gradient-based neural networks (GNN), another special kind of recurrent neural networks has recently been proposed by Zhang *et al.* for time-varying matrix inversion and pseudoinverse solving. As compared with GNN models, such Zhang neural networks (ZNN) are designed based on a matrix-valued error function, instead of a scalar-valued norm-based energy function. In addition, in this chapter, we simulate and compare ZNN and GNN models for the online solution of time-varying matrix inversion and pseudoinverse using MATLAB-coding and Simulink-modeling techniques as well. Computer-simulation results, including an application to robot kinematic control, substantiate the theoretical analysis and demonstrate the efficacy of ZNN models on linear time-varying inverse and/or pseudoinverse solving, especially when using power-sigmoid activation functions.

4.1. Introduction

The problem of finding the inverse of a matrix online arises in numerous fields of science, engineering, and business. It is usually an essential part of many solutions, e.g., as preliminary steps for optimization, signal-processing, electromagnetic systems, and robot kinematics [1–4]. Since the mid-1980s, efforts have been directed towards computational aspects of fast matrix inversion and many algorithms have been proposed [5–18]. For many numerical methods, the minimal arithmetic operations are usually proportional to the cube of the matrix dimension, and consequently such algorithms performed on digital computers may not be efficient enough in large-scale online applications [9, 10]. In view of this, parallel computation schemes have been investigated for matrix inversion. The dynamical approach is one of the important methods for solving matrix inversion problems [12, 17]. Recently, due to the in-depth research in neural networks, numerous dynamic solvers based on recurrent neural networks have been developed and investigated. The neural approach is now regarded as a powerful alternative for online computation because of its parallel distributed nature and convenience of hardware implementation.

However, almost all the aforementioned numerical, dynamical and neural computation schemes were designed intrinsically for constant matrices rather than time-varying ones. They are in general related to the gradient descent method in optimization, where a scalar-valued cost function is first constructed such that its minimum point is the matrix inverse, and then an algorithm is designed to evolve along a descent direction of this cost function until a minimum is reached. The typical descent direction is the negative gradient [12, 17]. Since the matrix to be inverted online is usually time-varying, such a gradient-based method only works approximately and with appreciable residual errors [17, 19]. Moreover, the gradient-based method also requires much faster convergence in comparison with the time scale of time-varying matrices or imposes very stringent restrictions on design parameters.

In this chapter, different from gradient-based neural networks (GNN), another special kind of recurrent neural networks has recently been proposed by Zhang *et al.* for time-varying matrix inversion and pseudoinverse solving [20, 21]. As compared to GNN models, such Zhang neural networks (ZNN) are designed based on a matrix-valued error function, instead of a scalar-valued norm-based energy function. In addition, ZNN models are depicted in implicit dynamics rather than explicit dynamics. Furthermore, the state matrix of ZNN globally exponentially converges to the exact inverse and/or pseudoinverse of the time-varying matrix. We simulate and compare ZNN and GNN for the online solution of time-varying matrix inversion and pseudoinverse using MATLAB M-File coding and Simulink-modeling techniques as well. Computer-simulation results, including an application to robot kinematic control, substantiate the theoretical analysis and demonstrate the efficacy of ZNN models on the linear time-varying inverse and/or pseudoinverse solving, especially when using a power-sigmoid activation function array.

4.2. Problem Formulation and Neural Solvers

Consider a smoothly time-varying matrix $A(t) \in \mathbb{R}^{n \times n}$. We are to find $X(t) \in \mathbb{R}^{n \times n}$ such that the following matrix equation holds:

$$A(t)X(t) - I = 0, \ t \in [0, +\infty) \tag{4.1}$$

where $I \in \mathbb{R}^{n \times n}$ is the identity matrix. Without loss of generality, $A(t)$ and its time derivative $\dot{A}(t)$ are assumed to be known or measurable. As a basis of discussion, the existence of the inverse $A^{-1}(t)$ at any time instant t is also assumed. To facilitate the convergence and robustness analysis, the following condition is introduced to guarantee the existence of $A^{-1}(t)$:

Invertibility Condition 4.2.1. There exists a positive real number $\alpha > 0$ such that

$$\min_{\forall i \in \{1, \cdots, n\}} |\lambda_i(A(t))| \geq \alpha, \ \forall t \geq 0 \tag{4.2}$$

where $\lambda_i(\cdot)$ denotes the ith eigenvalue of matrix $A \in \mathbb{R}^{n \times n}$.

If the invertibility condition holds, it is clear that there exists a unique solution to equation (4.1). Let $\|A\|_F = \sqrt{\mathrm{trace}(A^T A)}$ denote the Frobenius norm. Then, the invertibility condition leads to the following lemma on the boundedness of $\|A(t)\|_F$ and $\|A^{-1}(t)\|_F$.

Lemma 4.2.1. *If $A(t)$ satisfies the invertibility condition* (4.2) *with its norm uniformly upper bounded by* β *(i.e.,* $\|A(t)\|_F \leq \beta, \forall t \geq 0$*), then* $\|A^{-1}(t)\|_F$ *is uniformly upper bounded, i.e.,*

$$\|A^{-1}(t)\|_F \leq \varphi(\alpha, \beta, n) := \sum_{i=0}^{n-2} C_n^i \beta^{n-i-1}/\alpha^{n-i} + n^{3/2}/\alpha \tag{4.3}$$

for any time $t \geq 0$, where $C_n^i := n!/(i!(n-i)!)$.

Proof. By the Cayley-Hamilton Theorem [22], matrix A satisfies its characteristic polynomial:

$$A^n + \omega_{n-1}A^{n-1} + \cdots + \omega_{n-i}A^{n-i} + \cdots + \omega_0 I = 0 \tag{4.4}$$

where ω_i are coefficients defined below in terms of the eigenvalues of A:

$$\begin{cases} \omega_{n-1} = \sum_{j_1=1}^{n} (-\lambda_{j_1}), \\ \cdots \\ \omega_{n-i} = \sum_{j_i > \cdots > j_2 > j_1 = 1}^{n} (-\lambda_{j_1})(-\lambda_{j_2}) \cdots (-\lambda_{j_i}), \\ \cdots \\ \omega_0 = (-\lambda_1)(-\lambda_2) \cdots (-\lambda_n). \end{cases}$$

Specifically, $\omega_0 \neq 0$ and $|\omega_0| \geq \alpha^n$ due to the invertibility condition (4.2). It follows from (4.4) that,

$$A^{-1} = -(\upsilon_n A^{n-1} + \upsilon_{n-1}A^{n-2} + \cdots + \upsilon_{n-i}A^{n-i-1} + \cdots + \upsilon_1 I) \tag{4.5}$$

where the coefficients υ_j are

$$\begin{cases} \upsilon_n = \frac{1}{\omega_0} = \frac{1}{(-\lambda_1)(-\lambda_2)\cdots(-\lambda_n)}, \\ \upsilon_{n-1} = \frac{\omega_{n-1}}{\omega_0} = \sum_{j_1=1}^{n} \frac{1}{(-\lambda_1)\cdots(-\lambda_{j_1-1})(-\lambda_{j_1+1})\cdots(-\lambda_n)}, \\ \cdots \\ \upsilon_{n-i} = \frac{\omega_{n-i}}{\omega_0} = \sum_{j_i > \cdots > j_2 > j_1 = 1}^{n} \frac{(-\lambda_{j_1})(-\lambda_{j_2})\cdots(-\lambda_{j_i})}{(-\lambda_1)(-\lambda_2)\cdots(-\lambda_n)}, \\ \cdots \\ \upsilon_1 = \frac{\omega_1}{\omega_0} = \sum_{j_1=1}^{n} \frac{1}{(-\lambda_{j_1})}. \end{cases}$$

In view of (4.2), by defining the operator $C_n^i = n!/(i!(n-i)!)$ for any $i \in \{1, \cdots, n\}$, we have

$$\begin{cases} |\upsilon_n| \leq 1/\alpha^n, \\ |\upsilon_{n-1}| \leq n/\alpha^{n-1}, \\ \cdots \\ |\upsilon_{n-i}| \leq C_n^i/\alpha^{n-i}, \\ \cdots \\ |\upsilon_1| \leq n/\alpha. \end{cases}$$

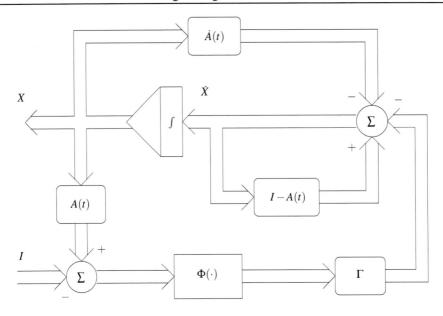

Figure 4.1. Block diagram of ZNN model (4.7) for online time-varying matrix inversion. *Reproduced from Y. Zhang, X. Guo et al., MATLAB Simulink modeling and simulation of Zhang neural network for online time-varying matrix inversion, Figure 1, Proceedings of the 5th IEEE International Conference on Networking, Sensing and Control, pp. 1480-1485. ©[2008] IEEE. Reprinted, with permission.*

It follows from the above inequalities and (4.5) that

$$\|A^{-1}\|_F \leq |\upsilon_n| \, \|A\|_F^{n-1} + |\upsilon_{n-1}| \, \|A\|_F^{n-2} +$$
$$\cdots + |\upsilon_{n-i}| \, \|A\|_F^{n-i-1} + \cdots + |\upsilon_1| \, \|I\|_F$$
$$\leq \beta^{n-1}/\alpha^n + n\beta^{n-2}/\alpha^{n-1} +$$
$$\cdots + C_n^i \beta^{n-i-1}/\alpha^{n-i} + \cdots + n\sqrt{n}/\alpha$$
$$= \sum_{i=0}^{n-2} C_n^i \beta^{n-i-1}/\alpha^{n-i} + n^{3/2}/\alpha,$$

which completes the proof of Lemma 4.2.1. □

It is worth mentioning that the invertibility condition, parameter α in (4.2), and Lemma 4.2.1 are used only for analytic purposes, and that there is no need to know the exact value of α. Instead, in a practical application via the ensuing ZNN model, we could check the invertibility condition by simply monitoring whether or not the value of $\|AX - I\|_F$ is becoming very small (near zero) after network started. For example, in view of the existence of exponential convergence, after $4/\gamma$ seconds, $\|AX - I\|_F$ should be less than $1.85\% (\approx e^{-4})$ of $\zeta \|AX(0) - I\|_F$ [23], where γ is to be defined as the standard convergence rate and also a design parameter of the neural network, and constant $\zeta > 0$. If $\gamma = 10^6$, such an exponential-convergence time is around $4 \sim 6\mu s$. Otherwise, the matrix could be singular.

4.2.1. ZNN Models

In the literature, traditional GNN approaches [2, 8, 11, 12, 15, 16] have been developed to compute the inverse of a time-invariant matrix. For the time-varying case, much faster convergence rate of the gradient-based networks is usually required in comparison with the time scale of time-varying matrices. Otherwise, the gradient-based networks often yield relatively large computational errors. To solve time-varying Sylvester equations online, an elegant recurrent neural network was first proposed in [19] by utilizing the time derivatives of the coefficient matrices. The following general ZNN design method is an important extension of the aforementioned neural network approach [19] to a time-varying one with general nonlinear activation functions and various implementation errors considered.

To monitor the matrix-inversion process, we could firstly define the following matrix-valued error function [21] (instead of the scalar-valued nonnegative energy functions used in gradient-based neural approaches [17]):

$$E(X(t),t) := A(t)X(t) - I \in \mathbb{R}^{n \times n}.$$

The error-function derivative $\dot{E}(X(t),t)$ should be made such that every entry $e_{ij}(t)$, $i, j = 1, \cdots, n$, of $E(X(t),t)$ converges to zero. Specifically, $\dot{E}(X(t),t)$ can be described in the ZNN design-formula [21]:

$$\frac{dE(X(t),t)}{dt} = -\Gamma \Phi(E(X(t),t)) \tag{4.6}$$

where Γ is in general a positive-definite matrix used to scale the convergence rate of the solution with Γ simplified as γI under $\gamma > 0$ (which being a set of reciprocals of capacitance-parameters, should be set as large as hardware permits or set appropriately for simulative/experimental purposes [24]), and $\Phi(\cdot) : \mathbb{R}^{n \times n} \to \mathbb{R}^{n \times n}$ denotes an activation-function processing-array. In addition, each scalar-valued processing-unit $\phi(\cdot)$ of array $\Phi(\cdot)$ should be a monotonically-increasing odd activation function. ZNN design formula (4.6) leads to the following implicit dynamic equation of the generalized neural-network model:

$$A(t)\dot{X}(t) = -\dot{A}(t)X(t) - \Gamma \Phi(A(t)X(t) - I) \tag{4.7}$$

where $X(t)$, starting from an initial condition $X(0) = X_0 \in \mathbb{R}^{n \times n}$, is the activation state matrix corresponding to the theoretical solution $X^*(t) = A^{-1}$ of (4.1). The block diagram realization of the ZNN model (4.7) is shown in Figure 4.1 in the most general form.

For hardware implementation with lower complexity, however, instead of using ZNN (4.7), we could have the following simplified neural-dynamic model by removing the D-circuit [i.e., the term of time derivative $\dot{A}(t)$] from (4.7):

$$A(t)\dot{X}(t) = -\gamma \Phi(A(t)X(t) - I), \tag{4.8}$$

which could solve approximately for time-varying $A^{-1}(t)$.

4.2.2. GNN Models

For comparison, it is worth pointing out here that we can develop a GNN model to solve for a time-varying matrix inverse. However, similar to almost all numerical algorithms and neural-dynamic schemes mentioned before, the GNNs are designed intrinsically for problems with constant coefficient matrices and/or vectors [12]. Now we show the GNN design procedure as the following.

1) Firstly, a scalar-valued norm-based energy function, such as $\|AX(t) - I\|_F^2/2$, is constructed such that its minimum point is the exact solution $A^{-1} \in \mathbb{R}^{n \times n}$.

2) Secondly, an algorithm is designed to evolve along a descent direction of this energy function until the minimum point is reached. The typical descent direction is the negative of the gradient of energy function $\|AX(t) - I\|_F^2/2$, i.e.,

$$-\frac{\partial \|AX(t) - I\|_F^2/2}{\partial X} = -A^T (AX(t) - I).$$

3) Thirdly, by using the above negative gradient to construct and apply the neural network to the time-varying situation, we could have a linear GNN model exploited for time-varying matrix inversion,

$$\dot{X}(t) = -\gamma A^T(t)(A(t)X(t) - I),$$

and a generalized nonlinear GNN model,

$$\dot{X}(t) = -\gamma A^T(t)\Phi(A(t)X(t) - I). \tag{4.9}$$

4.3. Theoretical Results

For the aforementioned ZNN models, we have the following theorems on their convergence and robustness properties.

Theorem 4.3.1. *Given a smoothly time-varying nonsingular matrix $A(t) \in \mathbb{R}^{n \times n}$, if a monotonically-increasing odd activation-function array $\Phi(\cdot)$ is used, then the state matrix $X(t)$ of ZNN model (4.7), starting from any initial state $X(0)$, will always converge to the time-varying theoretical inverse $A^{-1}(t)$. In addition, ZNN (4.7) possesses the following properties.*

1) *If the linear activation function is used, then exponential convergence with the rate γ is achieved for (4.7).*

2) *If the bipolar sigmoid activation-function is used, then the superior convergence can be achieved for (4.7) for error range $[-\varepsilon, \varepsilon]$, as compared to the case of using the linear activation function.*

3) *If the power activation function is used, then the superior convergence can be achieved for (4.7) for error ranges $(-\infty, -1]$ and $[1, +\infty)$, as compared to the linear case.*

4) *If the power-sigmoid activation function is used, then the superior convergence can be achieved for the whole error range $(-\infty, +\infty)$, as compared to the linear case. This is in view of the above mentioned properties.*

Proof. Let $\tilde{X}(t) := X(t) - X^*(t)$ denote the difference between the solution $X(t)$ generated by ZNN model (4.7) and the theoretical solution $X^*(t)$ of (4.1). By using the identity $\dot{A}(t)X^*(t) + A(t)\dot{X}^*(t) = 0$ which is the time derivative of (4.1), it follows that $\tilde{X}(t)$ is the solution to the ensuing dynamics with the initial state $\tilde{X}(0) = X(0) - X^*(0)$:

$$A(t)\dot{\tilde{X}}(t) + \dot{A}(t)\tilde{X}(t) = -\gamma\Phi\left(A(t)\tilde{X}(t)\right). \tag{4.10}$$

Since $E(t) = A(t)\tilde{X}(t)$, (4.10) can be rewritten as $\dot{E}(t) = -\gamma\Phi(E(t))$, which is a compact matrix form of the following set of n^2 equations

$$\dot{e}_{ij}(t) = -\gamma\phi\left(e_{ij}(t)\right), \quad \forall i, j \in \{1, 2, \cdots, n\}. \tag{4.11}$$

Clearly, we can define a Lyapunov function candidate $v_{ij} = e_{ij}^2/2 \geq 0$ for the ijth subsystem (4.11) with its time derivative

$$\frac{dv_{ij}}{dt} = e_{ij}\dot{e}_{ij} = -\gamma e_{ij}\phi(e_{ij}). \tag{4.12}$$

Because monotonically-increasing odd functions are used as activation functions, we have $\phi(-u) = -\phi(u)$, and

$$\phi(u) \begin{cases} > 0, & \text{if } u > 0 \\ = 0, & \text{if } u = 0 \\ < 0, & \text{if } u < 0 \end{cases}$$

which guarantees the negative definiteness of $\dot{v}_{ij}$; i.e., $\dot{v}_{ij} < 0$ for $e_{ij} \neq 0$ and $\dot{v}_{ij} = 0$ for $e_{ij} = 0$. By the Lyapunov stability theory [25], $e_{ij}(t)$ globally converges to zero for any $i, j \in \{1, \cdots, n\}$. Thus, in view of $E(t) = A(t)\tilde{X}(t)$ and (4.2), we have $\tilde{X}(t) \to 0 \in \mathbb{R}^{n \times n}$ as $t \to \infty$, i.e., the neural state $X(t)$ is globally convergent to the theoretical inverse $X^*(t)$. The proof on global convergence is thus complete.

In view of $E = A\tilde{X}$, (4.2) and Lemma 4.2.1, we have $\|X(t) - X^*(t)\|_F \leq \|A^{-1}\|_F\|E\|_F \leq \phi(\sum_i^n \sum_j^n e_{ij}^2)^{1/2} \leq n\phi\max_{1 \leq i,j \leq n}|e_{ij}|$, which implies the computation error and the network convergence can be estimated by those of maximum entry error $e_{ij}(t)$ in (4.11) [23,26]. We now come to prove the additional convergence properties corresponding to specific kinds of activation function $\phi(\cdot)$.

(i) For the simple linear case, $\dot{e}_{ij} = -\gamma e_{ij}$, and the entry error is $e_{ij}(t) = \exp(-\gamma t)e_{ij}(0)$. Thus there exists a constant $\zeta(\alpha, \beta, n, e_{ij}(0)) > 0$ such that $\|X(t) - X^*(t)\|_F \leq \zeta\exp(-\gamma t)$. This means that ZNN model (4.7) possesses the exponential convergence with rate γ when using linear activation function $\phi(u) = u$.

(ii) For the bipolar sigmoid function $\phi(u) = (1 - \exp(-\xi u))/(1 + \exp(-\xi u))$, define the constant $\eta = (\exp(-\xi e_{ij}(0)) - 1)^2/\exp(-\xi e_{ij}(0))$. The solution to (4.11) is given as $e_{ij}(t) = -\ln(1 + z(t))/\xi$ where

$$z(t) = \frac{\eta\exp(-\xi\gamma t)}{2} - \text{sgn}(e_{ij}(0))\sqrt{(1 + \frac{\eta\exp(-\xi\gamma t)}{2})^2 - 1}. \tag{4.13}$$

Using the Taylor series expansion formula $\ln(1+z) = z - z^2/2 + z^3/3 - \cdots$ and the algebraic formula $z + z^2 + z^3 + \cdots = z/(1-z)$ for $|z| < 1$, we have

$$|e_{ij}(t)| \leq |-z/\xi + z^2/(2\xi) - z^3/(3\xi) + \cdots|$$
$$\leq (|z| + |z|^2 + |z|^3 + \cdots)/\xi$$
$$= |z|/(\xi(1 - |z|)).$$

It follows from the formulas' requirement, $|z| < 1$, that the error range is defined as $e_{ij}(t) > -\ln 2/\xi$. From (4.13), we have

$$|z(t)| \leq \frac{\eta \exp(-\xi\gamma t)}{2} + \sqrt{\frac{\eta^2 \exp(-2\xi\gamma t) + 4\eta \exp(-\xi\gamma t)}{4}}$$

$$\leq \frac{\eta}{2} \exp(-\xi\gamma t) + \frac{\sqrt{\eta^2 + 4\eta}}{2} \exp(-\xi\gamma t/2)$$

$$\leq \frac{\eta + \sqrt{\eta^2 + 4\eta}}{2} \exp(-\xi\gamma t/2).$$

In view of the global convergence of $e_{ij}(t)$ and $z(t)$ to zero, there exists $z_0 = |z(t)| < 1$ such that

$$|e_{ij}(t)| \leq \frac{1}{\xi(1-z_0)} |z| \leq \frac{\eta + \sqrt{\eta^2 + 4\eta}}{2\xi(1-z_0)} \exp(-\xi\gamma t/2),$$

which means that the neural network (4.7) possesses the exponential convergence with rate $\xi\gamma/2$ for $e_{ij}(t) > -\ln 2/\xi$ when using the sigmoid activation function.

(iii) For the pth power activation function $\phi(u) = u^p$, (4.11) becomes $\dot{e}_{ij} = -\gamma e_{ij}^p$ and its general solution is

$$e_{ij}(t) = e_{ij}(0) \left\{ (p-1) e_{ij}^{p-1}(0)\gamma t + 1 \right\}^{-\frac{1}{p-1}}.$$

Specifically, when $p = 3$, the entry error $e_{ij}(t) = e_{ij}(0)/\sqrt{2e_{ij}^2(0)\gamma t + 1}$. Clearly, as $t \to \infty$, $e_{ij}(t) \to 0$. Review the Lyapunov function $v_{ij} = e_{ij}^2(t)$ and its time derivative $\dot{v}_{ij} = -\gamma e_{ij}\phi(e_{ij})$ in (4.12). For the error range $|e_{ij}| \gg 1$, we have $e_{ij}^{p+1} \gg e_{ij}^2 \gg |e_{ij}|$. In other words, the deceleration magnitude of power activation function is much greater than those of linear function and sigmoid function. This means when using power function, much faster convergence is achieved by the network for $|e_{ij}(t)| > 1$ in comparison with cases (i) and (ii). $\qquad\square$

Remark 4.3.1. It follows from the above theorem that to achieve superior convergence, a high-performance neural network can be developed by switching power activation function to sigmoid or linear activation function at the switching points $|e_{ij}(t)| = 1$, $i, j \in \{1, \cdots, n\}$. For example, the power-sigmoid activation function (1.7) with suitable design parameters ξ and p is preferable if the hardware permits. One more advantage of using the sigmoid function over the linear function lies in the extra parameter ξ, which is a multiplier of the exponential convergence rate. When there is an upper bound on γ due to hardware implementation, the parameter ξ will be another effective factor expediting the network convergence. For example, the convergence for nonlinear activation functions could be much faster than that for linear functions when using the same level of ξ and p for the power and sigmoid activation function as that of design parameter γ for linear function (e.g., $10 \sim 100$). $\qquad\square$

In the realization of neural networks, there may be some errors involved. Regarding the linear network for solving time-varying Sylvester equation [19], robustness results were given for coefficient matrix perturbation only. For matrix-inversion, if $A(t)$ in (4.7) is perturbed with an additive term $\Delta_A(t)$ where $\|\Delta_A(t)\|_F \leq \varepsilon_1$ for any $t \in [0, \infty)$, then the following lemma can be similarly generalized from [19].

Lemma 4.3.1. *Consider the perturbed ZNN model $\hat{A}\dot{X} = -\dot{\hat{A}}X - \gamma\Phi(\hat{A}X - I)$ with $\hat{A} = A + \Delta_A$. The steady-state residual error $\lim_{t\to\infty}\|X(t) - X^*(t)\|_F$ is uniformly upper bounded by $\varphi(\hat{\alpha}, \beta + \varepsilon_1, n) + \varphi(\alpha, \beta, n)$, provided that the invertibility condition (4.2) still holds with $\hat{\alpha} = (\sqrt{\alpha} - \varepsilon_1)^2$ instead of α for matrix $\hat{A}$.*

Proof. It can be readily obtained by following the proof of Theorem 2 in [19] and taking into account the inequality $\sum_{i=1}^{n}|\lambda_i(A)|^2 \leq \|A\|_F^2$, Lemma 4.2.1, and Theorem 4.3.1 of this chapter. $\square$

Lemma 4.3.1 guarantees that the coefficient perturbation will not derail the neural network if the invertibility condition still holds true. However, such a result does not cover the differentiation error and the model-implementation error, which may appear more frequently than matrix perturbation does in neural network realization [24]. Thus, the following dynamics are considered for the general robustness properties of the proposed neural system (4.7):

$$A(t)\dot{X}(t) = -\left(\dot{A}(t) + \Delta_B(t)\right)X(t) - \gamma\Phi\left(A(t)X(t) - I\right) + \Delta_R(t) \tag{4.14}$$

where $\Delta_B(t) \in \mathbb{R}^{n\times n}$ and $\Delta_R(t) \in \mathbb{R}^{n\times n}$ denote respectively the differentiation error and the model-implementation error, which result from truncating/roundoff errors in digital realization or high-order residual errors of circuit components in analog realization.

Theorem 4.3.2. *Consider the general ZNN model with implementation errors $\Delta_B(t)$ and $\Delta_R(t)$ in (4.14). If $\|\Delta_B(t)\|_F \leq \varepsilon_2$ and $\|\Delta_R(t)\|_F \leq \varepsilon_3$ for any $t \in [0,\infty)$, then the computation error $\|X(t) - X^*(t)\|_F$ is bounded with steady-state residual error as $n\varphi(\varepsilon_3 + \varepsilon_2\varphi)/(\gamma\rho - \varepsilon_2\varphi)$ under the design-parameter requirement $\gamma > \varepsilon_2\varphi/\rho$, where the parameter $\rho > 0$ is defined between $\phi(e_{ij}(0))/e_{ij}(0)$ and $\phi'(0)$. Furthermore, as the design parameter γ tends to positive infinity, the steady-state residual error can be diminished to zero.*

Proof. By defining the error matrix $E(t) = A(t)(X(t) - X^*(t))$ with $X^*(t) = A^{-1}(t)$, (4.14) is finally reformulated as $\dot{E} = -\gamma\Phi(E) - \Delta_B X^* E + (\Delta_R - \Delta_B X^*)$. This is equivalent to the following vector form [22]:

$$\dot{e}(t) = -\gamma\Phi\left(e(t)\right) + B(t)e(t) + c(t) \tag{4.15}$$

where $e := \text{vec}(E) \in \mathbb{R}^{n^2 \times 1}$ denotes a column vector obtained by stacking all column vectors of E together, and the activation-function array Φ is here of dimension $(n^2 \times 1)$ due to the vectorization. In addition, $B := I \otimes (-\Delta_B X^*) \in \mathbb{R}^{n^2 \times n^2}$ and $c := \text{vec}(\Delta_R - \Delta_B X^*) \in \mathbb{R}^{n^2 \times 1}$ with the symbol $\otimes$ denoting the Kronecker product; i.e., $P \otimes Q$ is a large matrix made by replacing the ijth entry p_{ij} of P with the matrix $p_{ij}Q$. For detailed properties of Kronecker product, see [19, 22].

Define the Lyapunov function candidate $v = e^T e/2 = \sum_{i=1}^{n}\sum_{j=1}^{n}e_{ij}^2(t)/2 \geq 0$ for the error dynamics (4.15). The time derivative of v is

$$\begin{aligned}
\frac{dv}{dt} &= e^T(t)\dot{e}(t) \\
&= e^T(t)\left\{-\gamma\Phi\left(e(t)\right) + B(t)e(t) + c(t)\right\} \\
&= -\gamma e^T\Phi(e) + e^T Be + e^T c \\
&= -\gamma e^T\Phi(e) + e^T\frac{B + B^T}{2}e + e^T c.
\end{aligned}$$

It follows from the inequality $\sum_{i=1}^{n}|\lambda_i(A)|^2 \le \|A\|_F^2$ and Lemma 4.2.1 that

$$e^T \frac{B+B^T}{2} e \le e^T e \max_{1\le i\le n^2}\left|\lambda_i(\frac{B+B^T}{2})\right|$$

$$= e^T e \max_{1\le i\le n^2}\left|\lambda_i(\frac{I\otimes(\Delta_B X^* + (\Delta_B X^*)^T)}{2})\right|$$

$$= e^T e \max_{1\le i\le n}\left|\lambda_i(\frac{\Delta_B X^* + (\Delta_B X^*)^T}{2})\right|$$

$$\le e^T e \left\|\frac{\Delta_B X^* + (\Delta_B X^*)^T}{2}\right\|_F$$

$$\le e^T e \|\Delta_B\|_F \|X^*\|_F$$

$$\le e^T e \, \varepsilon_2 \varphi.$$

Similarly, it follows from $\|A\|_F^2 = \sum_{i=1}^{n}\sum_{j=1}^{n}|a_{ij}|^2$ that $|c_i| \le \|\Delta_R - \Delta_B X^*\|_F \le \|\Delta_R\|_F + \|\Delta_B X^*\|_F = \varepsilon_3 + \varepsilon_2\varphi$, $\forall i \in \{1,\cdots,n^2\}$. Thus, $e^T c \le (\varepsilon_3 + \varepsilon_2\varphi)\sum_{i=1}^{n}\sum_{j=1}^{n}|e_{ij}|$. In view of the above facts and the symmetry property of $\phi(\cdot)$, we finally have

$$\frac{dv}{dt} \le -\gamma e^T \Phi(e) + (\varepsilon_2\varphi)e^T e + e^T c \tag{4.16}$$

$$= -\sum_{i=1}^{n}\sum_{j=1}^{n}|e_{ij}|(\gamma\phi(|e_{ij}|) - \varepsilon_2\varphi|e_{ij}| - \varepsilon_3 - \varepsilon_2\varphi).$$

During the time evolution of $e_{ij}(t)$, the above equation falls into two situations: $\gamma\phi(|e_{ij}|) - \varepsilon_2\varphi|e_{ij}| - \varepsilon_3 - \varepsilon_2\varphi \ge 0$, $\forall i,j \in \{1,\cdots,n\}$ or $\gamma\phi(|e_{ij}|) - \varepsilon_2\varphi|e_{ij}| - \varepsilon_3 - \varepsilon_2\varphi < 0$, $\exists i,j \in \{1,\cdots,n\}$.

If in the time interval $[t_0,t_1)$ the trajectory of the system (4.14) is in the first situation, $\dot{v} \le 0$ and (4.16) implies $X(t)$ converges to $X^*(t)$ as time evolves.

For any time t that the trajectory falls into the second situation, the distance between $X(t)$ and $X^*(t)$ may not decrease again. But, even in the worst case, the entry error $|e_{ij}(t)|$ is also upper bounded by the steady-state entry residual error $\bar{e}_{ij} = (\varepsilon_3 + \varepsilon_2\varphi)/(\gamma\rho - \varepsilon_2\varphi)$, where the insensitivity parameter $\rho > 0$ is defined between $\phi(e_{ij}(0))/e_{ij}(0)$ and $\phi'(0)$, and the design parameter γ is required as

$$\gamma > \varepsilon_2\varphi/\rho. \tag{4.17}$$

Thus, it follows that

$$\lim_{t\to\infty} \|X(t) - X^*(t)\|_F \le n\varphi(\varepsilon_3 + \varepsilon_2\varphi)/(\gamma\rho - \varepsilon_2\varphi) \tag{4.18}$$

and evidently, this steady-state residual error caused by implementation errors can be made arbitrarily small as design parameter γ increases. □

Corollary 4.3.1. *In addition to the general robustness results in Theorem 4.3.2, the imprecisely-constructed ZNN model (4.14) possesses the following properties that*

(i) *if linear activation function used, then the entry residual error $\bar{e}_{ij} = (\varepsilon_3 + \varepsilon_2\varphi)/(\gamma - \varepsilon_2\varphi)$ under the requirement $\gamma > \varepsilon_2\varphi$,*

(ii) *if sigmoid activation function used, then the steady-state residual error can be made smaller by increasing γ or ξ, and superior robustness property exists for $\bar{e}_{ij} \leq \varepsilon$, $\exists \varepsilon > 0$, as compared to linear function, and*

(iii) *if power activation function used, then the design-parameter requirement* (4.17) *always holds true for any positive γ and superior robustness property exists for $\bar{e}_{ij} > 1$, as compared to linear function.*

Proof. It follows from (4.16) that $\forall i, j \in \{1, \cdots, n\}$,

$$\gamma\phi(|e_{ij}|) - \varepsilon_2\varphi|e_{ij}| - \varepsilon_3 - \varepsilon_2\varphi \geq 0 \tag{4.19}$$

is a sufficient condition for ensuring $\dot{v} \leq 0$.

(i) For the linear activation function, the insensitivity parameter $\rho = \phi(|e_{ij}|)/|e_{ij}| \equiv 1$ and the design-parameter requirement on γ becomes $\gamma > \varepsilon_2\varphi$. From Theorem 4.3.2 and its proof, we have the steady-state entry residual error $\lim_{t\to\infty} |e_{ij}(t)| \leq (\varepsilon_3 + \varepsilon_2\varphi)/(\gamma - \varepsilon_2\varphi)$.

(ii) For the bipolar sigmoid function, as $|e_{ij}(t)| \to 0$, the value $\phi(e_{ij}(t))/e_{ij}(t)$ changes smoothly from $\phi(e_{ij}(0))/e_{ij}(0)$ to ξ. In view of $\phi(e_{ij}(0))/e_{ij}(0) < 1$ and $\xi > 1$, there exists $t_m(\xi)$ such that $\phi(e_{ij}(t_m))/e_{ij}(t_m) = 1$. Denoting $|e_{ij}(t_m)|$ as $\varepsilon(\xi)$, we have the insensitivity parameter $\rho(\xi) > 1$ for the error range $|e_{ij}(t)| \leq \varepsilon(\xi)$, which, compared to the linear case $\rho \equiv 1$, means easier satisfaction of (4.19) and a smaller steady-state residual error in (4.18), i.e., $\bar{e}_{ij} = (\varepsilon_3 + \varepsilon_2\varphi)/(\gamma\rho(\xi) - \varepsilon_2\varphi)$. In addition, the design-parameter requirement (4.17) is also relaxed by a factor of $\rho(\xi)$. Especially, when the implementation error $\varepsilon_3 + \varepsilon_2\varphi$ is small (e.g., when the steady-state entry residual error is near zero), the insensitivity parameter ρ is evaluated as $f'(0) = \xi$, which means much superior robustness property to the linear/power activation function cases.

(iii) For the power-function case, (4.19) becomes $(\gamma|e_{ij}|^{p-1} - \varepsilon_2\varphi)|e_{ij}| - \varepsilon_3 - \varepsilon_2\varphi \geq 0$. Clearly, there always exists $\zeta > 1$ such that both the previous equation and $\gamma|e_{ij}|^{p-1} - \varepsilon_2\varphi > 0$ hold true for $|e_{ij}(t)| \geq \zeta$. Thus, design-parameter inequality (4.17) always holds and such a requirement can be removed in this case. In the situation that the differentiation/implementation error $\varepsilon_3 + \varepsilon_2\varphi$ is so large that $\bar{e}_{ij} > 1$, insensitivity parameter $\rho = e_{ij}^{p-1} > e_{ij}^2 > 1$. This means easier satisfaction of (4.19) and smaller steady-state residual error (4.18), as compared to linear or sigmoid case under the same design specification. $\square$

Remark 4.3.2. From the above theoretical analysis, for large matrix-inversion error, using the power activation function has much better convergence and robustness than using the linear function. This is because for large entry error (e.g., $|e_{ij}| > 1$), the power activation function could amplify the signal ($e_{ij}^{p-1} > \cdots > e_{ij}^2 > 1$), automatically eliminate the insensitivity condition (4.17), and also expedite the network convergence. On the other hand, for small matrix-inversion error, using a sigmoid activation function has much better convergence and robustness than using the linear function. This is because of the larger slope of the sigmoid function near the origin, which implies the stronger insensitivity/robustness of the sigmoid-based neural model, as compared to the linear ones. It follows from Remark 4.3.1 and the above analysis that for superior convergence and robustness, the power-sigmoid activation function in (1.7) might be a better choice than other functions. $\square$

In the reminder of this section, the following theoretical results are established about the solution-error bound of the simplified ZNN model (4.8) inverting time-varying nonsingular matrix $A(t)$ online.

Theorem 4.3.3. *Consider time-varying nonsingular matrix $A(t) \in \mathbb{R}^{n \times n}$ which satisfies invertibility condition (4.2) and norm condition (4.3). If a monotonically-increasing odd activation function array $\Phi(\cdot)$ is used, then the computational error $\|X(t) - A^{-1}(t)\|_F$ of ZNN (4.8) starting from any initial state $X(0) \in \mathbb{R}^{n \times n}$ is always upper bounded, with its steady-state solution-error no greater than $n\varepsilon\varphi^2/(\gamma\rho - \varepsilon\varphi)$, provided that $\|\dot{A}(t)\|_F \leq \varepsilon$ for any $t \in [0, \infty)$ and design-parameter γ is large enough ($\gamma > \varepsilon\varphi/\rho$), where coefficient*

$$\rho := \min \left(\max_{i,j \in \{1, \cdots, n\}} \left(\phi(|e_{ij}(0)|)/|e_{ij}(0)| \right), \phi'(0) \right), \tag{4.20}$$

with $e_{ij}(0) := [A(0)X(0) - I]_{ij}, i, j \in \{1, 2, \cdots, n\}$.

Proof. We can reformulate ZNN (4.8) as the following [with $\Delta_B(t) := -\dot{A}(t)$ and $\Delta_R(t) := 0 \in \mathbb{R}^{n \times n}$]:

$$\begin{aligned} A(t)\dot{X}(t) = & - \left(\dot{A}(t) + \Delta_B \right) X(t) \\ & - \gamma\Phi\left(A(t)X(t) - I \right) + \Delta_R, \end{aligned} \tag{4.21}$$

which becomes exactly equation (4.14). In view of $\|\Delta_B\|_F = \|-\dot{A}\|_F = \|\dot{A}\|_F \leq \varepsilon$ and $\|\Delta_R\|_F = 0$ for any $t \in [0, \infty)$, we could now reuse the theoretical results of Theorem 4.3.2. That is, the computational error $\|X(t) - A^{-1}(t)\|_F$ of neural-dynamics (4.21) [equivalently, ZNN (4.8)] is always upper bounded. In addition, it follows immediately from Theorem 4.3.2 and equation (4.18) that its steady-state computational error

$$\lim_{t \to \infty} \|X(t) - A^{-1}(t)\|_F \leq n\varepsilon\varphi^2/(\gamma\rho - \varepsilon\varphi).$$

Furthermore, design-parameter γ is required therein to be greater than $\varepsilon\varphi/\rho$. In the original proof of Theorem 4.3.2, coefficient $\rho > 0$ is defined between $\phi(e_{ij}(0))/e_{ij}(0)$ and $\phi'(0)$. Following that proof and considering the worst case or such an error bound, we could determine the value of ρ as in (4.20). Specifically speaking, if the linear activation-function (1.4) is used, then $\rho \equiv 1$; if the bipolar sigmoid activation-function (1.5) is used, then $\rho = \max_{i,j \in \{1,2,\cdots,n\}}(\phi(|e_{ij}(0)|)/|e_{ij}(0)|)$; and, if the power-sigmoid activation-function (1.7) is used, then $\rho \geq 1$ (where the sign of inequality ">" is taken in most situation). The proof is thus complete. □

4.4. MATLAB-Coding Simulation Techniques

In this section, the following MATLAB-coding simulation techniques are investigated to show the characteristics of such ZNN models. Note that, the coding of activation functions could be seen in Subsection 2.4.1..

4.4.1. Kronecker Product and Vectorization

Review ZNN model (4.7) and GNN model (4.9). Their dynamic equations are all described in matrix form, which cannot be directly simulated. Thus, the Kronecker product and vectorization techniques are needed to transform such matrix-form differential equations to vector-form differential equations. Based on the definition in Appendix A.2. about Kronecker product and vectorization techniques, for simulation proposes, let us transform the matrix differential equation (4.7) to a vector differential equation. The following theorem is thus obtained:

Theorem 4.4.1. *Matrix differential equation* (4.7) *can be reformulated as the following vector differential equation:*

$$(I \otimes A)\operatorname{vec}(\dot{X}) = -(I \otimes \dot{A})\operatorname{vec}(X) - (I \otimes \Gamma)\Phi\Big((I \otimes A)\operatorname{vec}(X) - \operatorname{vec}(I)\Big), \qquad (4.22)$$

or, simply put,

$$M\dot{x} = -\dot{M}x - \gamma\Phi\big(Mx - \operatorname{vec}(I)\big),$$

where the activation-function array $\Phi(\cdot)$ in (4.22) *is defined the same as in* (4.7), *except that its dimensions are changed hereafter as $\Phi(\cdot) : \mathbb{R}^{n^2 \times 1} \to \mathbb{R}^{n^2 \times 1}$, the so-called mass matrix $M(t) := I \otimes A(t)$, and the integration vector $x := \operatorname{vec}(X)$.*

Proof. For the reader's convenience, the matrix-form differential equation (4.7) is repeated here: $A(t)\dot{X}(t) = -\dot{A}(t)X(t) - \Gamma\Phi(A(t)X(t) - I)$. By vectorizing equation (4.7) based on the Kronecker product and the above $\operatorname{vec}(\cdot)$ operator, the left hand side of ZNN model (4.7) becomes

$$\operatorname{vec}\big(A(t)\dot{X}(t)\big) = \big(I \otimes A(t)\big)\operatorname{vec}(\dot{X}(t)). \qquad (4.23)$$

The right hand side of ZNN model (4.7) correspondingly becomes

$$\begin{aligned}
&\operatorname{vec}\big(-\dot{A}(t)X(t) - \Gamma\Phi(A(t)X(t) - I)\big) \\
&= \operatorname{vec}\big(-\dot{A}(t)X(t)\big) + \operatorname{vec}\Big(-\Gamma\Phi\big(A(t)X(t) - I\big)\Big) \\
&= -\operatorname{vec}\big(\dot{A}(t)X(t)\big) - \operatorname{vec}\Big(\Gamma\Phi\big(A(t)X(t) - I\big)\Big) \\
&= -\big(I \otimes \dot{A}(t)\big)\operatorname{vec}(X(t)) - (I \otimes \Gamma)\operatorname{vec}\Big(\Phi\big(A(t)X(t) - I\big)\Big).
\end{aligned} \qquad (4.24)$$

Note that, as shown in Subsection 2.4.1., the definition and coding of the activation-function array $\Phi(\cdot)$ are very flexible and could also be those of a vectorized array from $\mathbb{R}^{n^2 \times 1}$ to $\mathbb{R}^{n^2 \times 1}$. We thus have

$$\begin{aligned}
&(I \otimes \Gamma)\operatorname{vec}\Big(\Phi\big(A(t)X(t) - I\big)\Big) \\
&= (I \otimes \Gamma)\Phi\Big(\operatorname{vec}\big(A(t)X(t) - I\big)\Big) \\
&= (I \otimes \Gamma)\Phi\Big(\operatorname{vec}\big(A(t)X(t)\big) + \operatorname{vec}(-I)\Big) \\
&= (I \otimes \Gamma)\Phi\Big(\big(I \otimes A(t)\big)\operatorname{vec}(X(t)) - \operatorname{vec}(I)\Big).
\end{aligned} \qquad (4.25)$$

Combining equations (4.24) and (4.25) yields the vectorization result of the right hand side of matrix differential equation (4.7):

$$\text{vec}\left(-\dot{A}(t)X(t) - \Gamma\Phi\left(A(t)X(t) - I\right)\right)$$
$$= -\left(I \otimes \dot{A}(t)\right)\text{vec}(X(t)) - (I \otimes \Gamma)\Phi\left(\left(I \otimes A(t)\right)\text{vec}(X(t)) - \text{vec}(I)\right). \tag{4.26}$$

Evidently, the vectorization of both sides of matrix differential equation (4.7) should be equal; i.e., equation (4.23) should be equal to equation (4.26). Hence, the presented ZNN model (4.7) can be transformed to the vector form (4.22). In addition, if we define matrix $M(t) := I \otimes A(t)$ and consider $\Gamma = \gamma I$ for the present research, then vector form (4.22) could be further simplified as

$$M\dot{x} = -\dot{M}x - \gamma\Phi\left(Mx - \text{vec}(I)\right).$$

The proof is thus complete. $\qquad\qquad\qquad\qquad\qquad\qquad\qquad\qquad\qquad\qquad\qquad\square$

Remark 4.4.1. The Kronecker product can be generated readily by using routine "kron"; in other words, $A \otimes B$ can be generated by code "kron(A,B)". To generate vec(X), the routine "reshape" can be used. Specifically, if the matrix X has m rows and n columns, then the code of vectorizing X is "reshape(X,m*n,1)" which generates a column vector, $\text{vec}(X) = [x_{11}, \cdots, x_{m1}, x_{12}, \cdots, x_{m2}, \cdots, x_{1n}, \cdots, x_{mn}]^T$. $\qquad\qquad\square$

Based on routines "kron" and "vec", the following code could be used to define a function "ZnnRightHandSide" which evaluates the right hand side of vector-form ZNN model (4.22). In other words, it returns the evaluation of $g(t,x)$ in equation $M\dot{x} = g(t,x)$, where, in our case, $g(t,x) := -\dot{M}x - \gamma\Phi\left(Mx - \text{vec}(I)\right)$.

```
function output=ZnnRightHandSide(t,x,gamma)
if nargin==2, gamma=1; end
n = sqrt(size(x,1));
% The following generates the time derivative of matrix M
syms u; D=diff(MatrixM(u));
u=t; dotM=eval(D);
% The following generates the vectorization of identity matrix I
vecI=reshape(eye(n),n^2,1);
% The following calculates the right hand side of equation (5)
% If using power-sigmoid function (firstly preferred):
output=-dotM*x-gamma*AFMpowersigmoid(MatrixM(t,x)*x-vecI);
% If using sigmoid activation function (secondly preferred):
% output=-dotM*x-gamma*AFMsigmoid(MatrixM(t,x)*x-vecI);
% If using linear activation function (thirdly preferred):
% output=-dotM*x-gamma*AFMlinear(MatrixM(t,x)*x-vecI);
% If using power activation function (fourthly preferred):
% output=-dotM*x-gamma*AFMpower(MatrixM(t,x)*x-vecI);
```

4.4.2. ODE with Mass Matrix

In the simulation of ZNN model (4.7), the routine "ode45" is preferred because "ode45" can solve the initial-value ODE problem with a mass matrix, e.g., $M(t,x)\dot{x} = g(t,x)$, where the nonsingular matrix $M(t,x)$ on the left hand side of such an equation is termed the mass matrix, and $g(t,x) := -\dot{M}x - \gamma\Phi(Mx - \text{vec}(I))$ for our case (see Theorem 4.4.1).

The following code could be used to generate a mass matrix $M(t,x) := I \otimes A(t)$ in equation (4.22), where a two-dimensional time-varying matrix $A(t)$ is defined to be $[\sin t, \cos t; -\cos t, \sin t]$ as an example.

```
function output=MatrixM(t,x)
% M(t,x) for ODE problem M(t,x)*x'=g(t,x)
A=[sin(t) cos(t);-cos(t) sin(t)];
output=kron(eye(size(A)),A);
```

To solve the ODE with a mass matrix, the routine "odeset" should also be used. Its "Mass" property should be assigned to be the function handle "@MatrixM", which returns the value of the mass matrix $M(t,x)$. Note that, if $M(t,x)$ does not depend on the state variable x and if the function "MatrixM" is to be invoked with only one input argument t, then the value of the "MStateDep" property of routine "odeset" could be set as "none". Thus, for example, the following code could be used to solve an initial-value ODE problem with state-independent mass matrix $M(t)$ and random initial state $x(0)$.

```
tspan=[0 10]; x0=4*(rand(4,1)-0.5*ones(4,1));
options=odeset('Mass',@MatrixM,'MStateDep','none');
[t,x]=ode45(@ZnnRightHandSide,tspan,x0,options);
```

4.4.3. Obtaining Matrix Derivative

In the time-varying matrix inversion using ZNN model (4.7), the time derivative $\dot{A}(t)$ is assumed to be known or measurable. This implies that $\dot{A}(t)$ can be given directly in an analytical form or can be estimated via finite difference. Similarly, in the vector-form ZNN model (4.22), the time derivative $\dot{M}(t) = I \otimes \dot{A}(t)$ is also needed (i.e., known or measurable). Thus for the simulation, the routine "diff" has to be introduced, as follows, in order to automatically generate the time derivative $\dot{M}(t)$ and/or $\dot{A}(t)$.

- For a vector or matrix X, "diff(X)" returns the row difference of X. For example, for a matrix $X = [3,7,5;0,9,2]$, "diff(X)" is $[-3,2,-3]$.

- For a symbolic expression X, "diff(X)" differentiates X with respect to its free variable as determined by routine "findsym". For example, the derivative of $\sin x^2$ with respect to variable x can be obtained by using the following commands: "x=sym('x')"; "diff(sin(x^2))". The result of such commands is thus "2*cos(x^2)*x".

As can be seen in the code of the function "ZnnRightHandSide" presented in Subsection 4.4.1., in order to get the time derivative $\dot{M}(t)$ of $M(t)$, a symbolic argument "u" can be constructed and then the command "D=diff(MatrixM(u))" can be used to generate $\dot{M}(t)$. Note that, without using the above symbolic "u", row differences of "MatrixM(t)" are generated by calling the command "diff(MatrixM(t))", which is not the desired time derivative of $M(t)$.

4.4.4. Illustrative Examples

For illustration and simulation purposes, in the first example, both ZNN model (4.7) and GNN model (4.9) are exploited to solve online for a time-varying sinusoidal-matrix inverse. Two other illustrative examples are also presented to further substantiate the efficacy of the ZNN model on time-varying matrix inversion.

Example 1. Let us consider the following time-varying matrix $A(t)$ and its time derivative $\dot{A}(t)$:

$$A(t) = \begin{bmatrix} \sin t & \cos t \\ -\cos t & \sin t \end{bmatrix}, \quad \dot{A}(t) = \begin{bmatrix} \cos t & -\sin t \\ \sin t & \cos t \end{bmatrix}.$$

Simple manipulations can verify that

$$A^{-1}(t) = A^T(t) = \begin{bmatrix} \sin t & -\cos t \\ \cos t & \sin t \end{bmatrix},$$

which is used to compare the correctness of the neural-network solutions. When used to solve for the above $A^{-1}(t)$ in real time, ZNN model (4.7) can take the following specific and simplified form:

$$\begin{bmatrix} \sin t & \cos t \\ -\cos t & \sin t \end{bmatrix} \begin{bmatrix} \dot{x}_{11} & \dot{x}_{12} \\ \dot{x}_{21} & \dot{x}_{22} \end{bmatrix} = - \begin{bmatrix} \cos t & -\sin t \\ \sin t & \cos t \end{bmatrix} \begin{bmatrix} x_{11} & x_{12} \\ x_{21} & x_{22} \end{bmatrix}$$
$$- \gamma \Phi \left(\begin{bmatrix} \sin t & \cos t \\ -\cos t & \sin t \end{bmatrix} \begin{bmatrix} x_{11} & x_{12} \\ x_{21} & x_{22} \end{bmatrix} - \begin{bmatrix} 1 & 0 \\ 0 & 1 \end{bmatrix} \right)$$

where array Φ is constructed by one of the presented four types of aforementioned activation functions in Section 1.5., such as the power-sigmoid functions with $\xi = 4$ and $p = 3$. By using Kronecker product and vectorization techniques, we have

$$M(t) = I \otimes A(t) = \begin{bmatrix} \sin t & \cos t & 0 & 0 \\ -\cos t & \sin t & 0 & 0 \\ 0 & 0 & \sin t & \cos t \\ 0 & 0 & -\cos t & \sin t \end{bmatrix}, \quad \dot{M}(t) = I \otimes \dot{A}(t),$$

$x = \text{vec}(X) = [x_{11}, x_{21}, x_{12}, x_{22}]^T$, and $\text{vec}(I) = [1, 0, 0, 1]^T$. The above matrix-form differential equation can thus be rewritten as the following vector-form differential equation, which can be solved directly with "ode45":

$$\begin{bmatrix} \sin t & \cos t & 0 & 0 \\ -\cos t & \sin t & 0 & 0 \\ 0 & 0 & \sin t & \cos t \\ 0 & 0 & -\cos t & \sin t \end{bmatrix} \begin{bmatrix} \dot{x}_{11} \\ \dot{x}_{21} \\ \dot{x}_{12} \\ \dot{x}_{22} \end{bmatrix} = - \begin{bmatrix} \cos t & -\sin t & 0 & 0 \\ \sin t & \cos t & 0 & 0 \\ 0 & 0 & \cos t & -\sin t \\ 0 & 0 & \sin t & \cos t \end{bmatrix} \begin{bmatrix} x_{11} \\ x_{21} \\ x_{12} \\ x_{22} \end{bmatrix}$$
$$- \gamma \Phi \left(\begin{bmatrix} \sin t & \cos t & 0 & 0 \\ -\cos t & \sin t & 0 & 0 \\ 0 & 0 & \sin t & \cos t \\ 0 & 0 & -\cos t & \sin t \end{bmatrix} \begin{bmatrix} x_{11} \\ x_{21} \\ x_{12} \\ x_{22} \end{bmatrix} - \begin{bmatrix} 1 \\ 0 \\ 0 \\ 1 \end{bmatrix} \right).$$

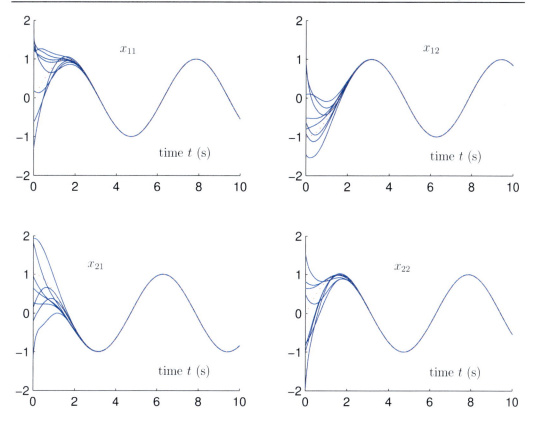

Figure 4.2. Online inversion of time-varying sinusoidal matrix $A(t)$ by ZNN model (4.7).

4.4.4.1. Simulation of Convergence

The following code could be used to simulate ZNN model (4.7), starting from eight random initial states. Figure 4.2, which shows the global exponential convergence of such a neural network, is then generated. In the figure, the theoretical inverse $A^{-1}(t)$ is denoted by dotted red lines, and the neural-network solutions $X(t)$ are denoted by solid blue lines.

```
gamma=1; %just for illustrative purposes
tspan=[0 10];
options=odeset('Mass',@MatrixM,'MStateDep','none');
for iter=1:8
    x0=4*(rand(4,1)-0.5*ones(4,1));
    [t,x]=ode45(@ZnnRightHandSide,tspan,x0,options,gamma);
    subplot(2,2,1);plot(t,x(:,1));hold on
    subplot(2,2,3);plot(t,x(:,2));hold on
    subplot(2,2,2);plot(t,x(:,3));hold on
    subplot(2,2,4);plot(t,x(:,4));hold on
end
subplot(2,2,1);plot(t,sin(t),'r:'); %compare with A^{-1}(t)
subplot(2,2,3);plot(t,cos(t),'r:');
subplot(2,2,2);plot(t,-cos(t),'r:');
subplot(2,2,4);plot(t,sin(t),'r:');
```

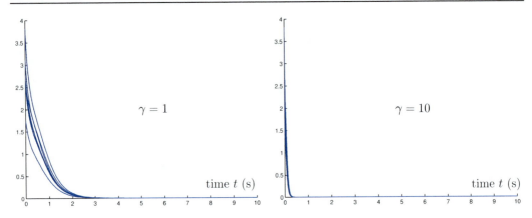

Figure 4.3. Convergence of computational error $\|X(t)-A^{-1}(t)\|$ by ZNN model (4.7).

The norm of the solution error, $\|X(t)-A^{-1}(t)\|$, can also be used and shown to monitor the network convergence. The related codes are shown as follows, i.e., the user-defined functions "NormError" and "ZnnNormError".

```
function NormError(x0,gamma)
tspan=[0 10];
options=odeset('Mass',@MatrixM,'MStateDep','none');
[t,x]=ode45(@ZnnRightHandSide,tspan,x0,options,gamma);
Ainv=[sin(t) cos(t) -cos(t) sin(t)]';
err=x'-Ainv; total=length(t);
for i=1:total,
    nerr(i) = norm(err(:,i));
end
plot(t,nerr); hold on
```

```
function ZnnNormError(gamma)
% usage: figure; ZnnNormError(1);
if nargin<1, gamma = 1; end
for i=1:8
    x0=4*(rand(4,1)-0.5*ones(4,1));
    NormError(x0,gamma);
end
```

By calling "ZnnNormError" twice with different γ values, Figure 4.3 can be generated. It shows that, starting from eight initial states randomly selected in $[-2,2]$, the state matrices of the presented neural network (4.7) all converge to the theoretical inverse $A^{-1}(t)$, where the computational errors $\|X(t)-A^{-1}(t)\|$ all converge correspondingly to zero. Such a convergence can be expedited by increasing γ. For example, if γ is increased to 10^3, the convergence time is within 4 milliseconds; and, if γ is increased to 10^6, the convergence time is within 4 microseconds.

Note that, for comparison, by applying the above-presented simulation techniques, GNN model (4.9) can be used to invert the same time-varying matrix $A(t)$ as ZNN model (4.7) does. Figures 4.4 and 4.5 illustrate the performance of GNN model (4.9) under the same simulating conditions. It follows from the GNN-related figures that the GNN state

ZNN for Time-Varying Matrix Inversion and Pseudoinverse-Solving 87

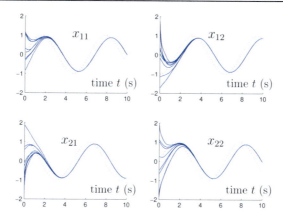

Figure 4.4. Inversion of time-varying sinusoidal matrix $A(t)$ by GNN (4.9).

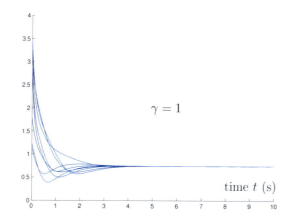

Figure 4.5. Convergence of solution error $\|X(t) - A^{-1}(t)\|$ as of GNN model (4.9).

$X(t)$ (denoted by solid blue curves) does not fit well with the theoretical inverse $A^{-1}(t)$ (denoted by dotted red curves). In other words, appreciable computational errors always exist between the GNN solution and the theoretical inverse $A^{-1}(t)$. In addition, it can be seen from Figure 4.5 that, by using GNN (4.9) for the online time-varying matrix inversion, its computational error $\|X(t) - A^{-1}(t)\|$ is quite large. This possibly occurs because GNN (4.9) does not make full use of the time-derivative information of matrix $A(t)$.

4.4.4.2. Robustness Simulation

Similar to the transformation of matrix-form differential equation (4.7) to vector-form differential equation (4.22), the perturbed ZNN model (4.14) can be rewritten as the following vector-form differential equation:

$$(I \otimes A)\operatorname{vec}(\dot{X}) = -(I \otimes \dot{A} + I \otimes \Delta_B)\operatorname{vec}(X) \\ - \gamma \Phi\big((I \otimes A)\operatorname{vec}(X) - \operatorname{vec}(I)\big) + \operatorname{vec}(\Delta_R) \quad (4.27)$$

or, simply put,

$$M\dot{x} = -(\dot{M} + I \otimes \Delta_B)x - \gamma \Phi(Mx - \operatorname{vec}(I)) + \operatorname{vec}(\Delta_R),$$

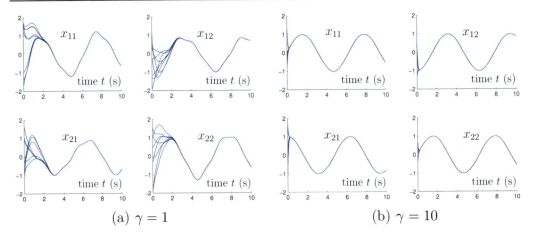

Figure 4.6. Online inversion of sinusoidal matrix $A(t)$ by perturbed ZNN model (4.14).

where the activation-function array $\Phi(\cdot): \mathbb{R}^{n^2 \times 1} \to \mathbb{R}^{n^2 \times 1}$, mass matrix $M(t) := I \otimes A(t)$, and integration vector $x := \text{vec}(X)$.

To show the robustness characteristics of ZNN model (4.7), the following differentiation error $\Delta_B(t)$ and model-implementation error $\Delta_R(t)$ are added in a higher-frequency sinusoidal form (with $\varepsilon_2 = \varepsilon_3 = 0.5$):

$$\Delta_B(t) = \varepsilon_2 \begin{bmatrix} \cos 3t & -\sin 3t \\ \sin 3t & \cos 3t \end{bmatrix}, \quad \Delta_R(t) = \varepsilon_3 \begin{bmatrix} \sin 3t & 0 \\ 0 & \cos 3t \end{bmatrix}.$$

We could define the following function "ZnnRightHandSideImprecise" for ODE solvers, which returns the value of the right-hand side of perturbed vector-form differential equation (4.27) [or equivalently, perturbed matrix-form differential equation (4.14)].

```
function output=ZnnRightHandSideImprecise(t,x,gamma)
if nargin==2, gamma=1; end
e2=.5; e3=.5;
deltaB=e2*[cos(3*t) -sin(3*t); sin(3*t) cos(3*t)];
kronB=kron(eye(2), deltaB);
deltaR=e3*[sin(3*t) 0; 0 cos(3*t)];
vecR=reshape(deltaR,4,1);
vecI=reshape(eye(2),4,1);
syms u; D=diff(MatrixM(u)); u=t; dotM=eval(D);
output=-(dotM+kronB)*x-gamma*...
    AFMpowersigmoid(MatrixM(t)*x-vecI)+vecR;
```

ZNN for Time-Varying Matrix Inversion and Pseudoinverse-Solving

```
function ZnnRobust(gamma)
tspan=[0 10];
options=odeset('Mass',@MatrixM,'MStateDep','none');
for iter=1:8
  x0=4*(rand(4,1)-0.5*ones(4,1));
  [t,x]=ode45(@ZnnRightHandSideImprecise,tspan,x0,options,gamma);
  subplot(2,2,1);plot(t,x(:,1));hold on
  subplot(2,2,3);plot(t,x(:,2));hold on
  subplot(2,2,2);plot(t,x(:,3));hold on
  subplot(2,2,4);plot(t,x(:,4));hold on
end
subplot(2,2,1);plot(t,sin(t),'r:');
subplot(2,2,3);plot(t,cos(t),'r:');
subplot(2,2,2);plot(t,-cos(t),'r:');
subplot(2,2,4);plot(t,sin(t),'r:');
```

Based on the function "ZnnRightHandSideImprecise" and the user-defined function "ZnnRobust", Figure 4.6 can be generated using $\gamma = 1$ and $\gamma = 10$. As shown in Figure 4.6(a), with a large differentiation error $\Delta_B(t)$ and model-implementation error $\Delta_R(t)$, when $\gamma = 1$, the neural network solution is not very close to the theoretical inverse $A^{-1}(t)$. But with a larger value of design parameter γ (e.g., 10), as shown in Figure 4.6(b), the neural-network solution again becomes very close to the theoretical inverse $A^{-1}(t)$.

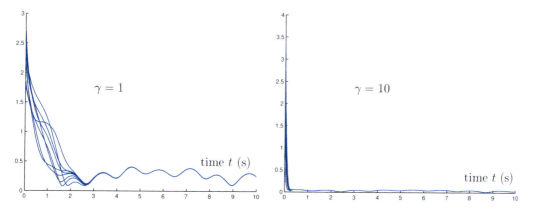

Figure 4.7. Convergence of computational error $\|X(t) - A^{-1}(t)\|$ of perturbed ZNN (4.14).

Similarly, the computational error $\|X(t) - A^{-1}(t)\|_F$ of the perturbed ZNN model (4.14) can be shown in the presence of large differentiation and model-implementation errors. See Figure 4.7. Even with large realization errors, the computational error $\|X(t) - A^{-1}(t)\|_F$ synthesized by the perturbed neural network (4.14) is still bounded and very small. Moreover, as the design parameter γ increases from 1 to 100, the convergence is expedited and the steady-state computational error is decreased. It is observed from other simulation data that when using power-sigmoid activation functions, the maximum steady-state computational error is 6×10^{-3} and 6×10^{-4}, respectively, for $\gamma = 100$ and $\gamma = 1000$. These simulation results have substantiated the theoretical results presented in previous sections and in [21].

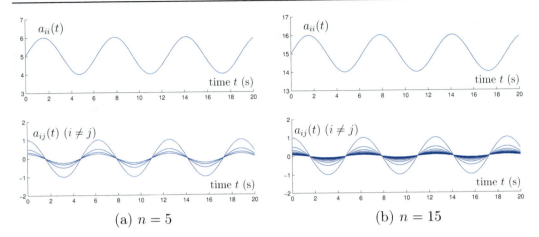

Figure 4.8. Entry trajectories of time-varying Toeplitz matrix $A(t) = [a_{ij}(t)] \in \mathbb{R}^{n \times n}$.

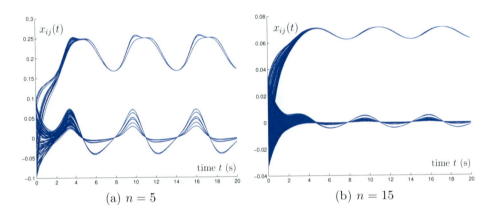

Figure 4.9. Online inversion of time-varying Toeplitz matrix $A(t)$ by ZNN model (4.7).

Example 2. Consider the following time-varying Toeplitz matrix:

$$A(t) = \begin{bmatrix} a_1(t) & a_2(t) & a_3(t) & \cdots & a_n(t) \\ a_2(t) & a_1(t) & a_2(t) & \cdots & a_{n-1}(t) \\ a_3(t) & a_2(t) & a_1(t) & \cdots & \vdots \\ \vdots & \vdots & \ddots & \ddots & a_2(t) \\ a_n(t) & a_{n-1}(t) & \cdots & a_2(t) & a_1(t) \end{bmatrix} \in \mathbb{R}^{n \times n},$$

where $a(t) := [a_1(t), a_2(t), a_3(t), \cdots, a_n(t)]^T$ denotes the first column vector of matrix $A(t) = [a_{ij}(t)] \in \mathbb{R}^{n \times n}$ ($i, j = 1, 2, \cdots, n$). Let $a_1(t) = n + \sin(t)$ and $a_k(t) = \cos(t)/(k-1)$ ($k = 2, 3, \cdots, n$). The entry trajectories of matrix $A(t) = [a_{ij}(t)] \in \mathbb{R}^{n \times n}$ are shown in Figure 4.8 for the situation of $n = 5$ and $n = 15$. Evidently, matrix $A(t)$ is strictly diagonally-dominant for any time instant $t \geq 0$, and is thus invertible.

Similarly, by applying the above-presented simulation techniques, ZNN model (4.7) can be exploited to invert the time-varying Toeplitz matrix $A(t)$. When design parameter $\gamma = 1$ and power-sigmoid activation functions (with $\xi = 4$ and $p = 3$) are used, the inversion

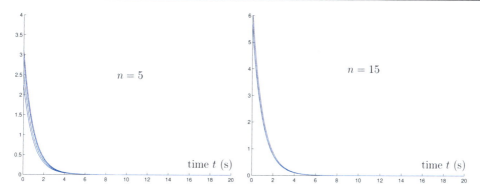

Figure 4.10. Convergence of residual error $\|A(t)X(t)-I\|$ by ZNN (4.7) with $\gamma = 1$.

performance of ZNN model (4.7) can be seen in Figures 4.9 and 4.10. As shown in Figure 4.9, starting from four randomly-generated initial states within $[-1/(2n), 1/(2n)]$, the state matrix $X(t) = [x_{ij}(t)] \in \mathbb{R}^{n \times n}$ ($i, j = 1, 2, \cdots, n$) of ZNN model (4.7) always converges to the theoretical time-varying inverse $A^{-1}(t)$ (which corresponds to the convergence of residual error $\|A(t)X(t) - I\|$ to zero, as depicted in Figure 4.10).

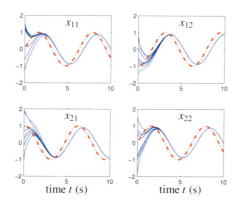

Figure 4.11. Inversion of time-varying matrix $A(t)$ by model (4.8) using power-sigmoid activation function array and $\gamma = 1$, where dashed-dotted curves denote the theoretical time-varying inverse $A^{-1}(t)$. *Reproduced from Y. Zhang, Z. Chen et al., Zhang neural network without using time-derivative information for constant and time-varying matrix inversion, Figure 4, Proceedings of 2008 International Joint Conference on Neural Networks, pp. 142-146. ©[2008] IEEE. Reprinted, with permission.*

Example 3. For illustration and verification of Theorem 4.3.3, let us consider the same time-varying matrix as that of Example 1; i.e., $A(t) = [\sin t, \cos t; -\cos t, \sin t]$.

Figures 4.11 and 4.12 could thus be generated to show the performance of ZNN (4.8). According to Figures 4.11 and 4.12, starting from initial states randomly selected in $[-2, 2]^{2 \times 2}$, the state matrices of the presented ZNN model (4.8) could not converge to the theoretical inverse $A^{-1}(t)$ exactly. Instead, it could only approach to an approximate solution of $A^{-1}(t)$. In addition, as shown in Figure 4.12, when we increase design-parameter γ from 1 to 10, the steady-state computational error $\lim_{t \to +\infty} \|X(t) - A^{-1}(t)\|_F$ decreases rapidly. However, there always exists a steady-state solution-error which could not van-

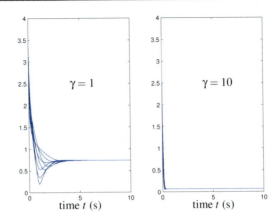

Figure 4.12. Computational error $\|X(t) - A^{-1}(t)\|_F$ by model (4.8) using power-sigmoid activation-function array and different values of parameter γ. *Reproduced from Y. Zhang, Z. Chen et al., Zhang neural network without using time-derivative information for constant and time-varying matrix inversion, Figure 5, Proceedings of 2008 International Joint Conference on Neural Networks, pp. 142-146.* ©*[2008] IEEE. Reprinted, with permission.*

ish to zero. These computer-simulation results have substantiated the theoretical results presented in Section 4.3..

For comparison between the simplified ZNN model (4.8) and the original ZNN model (4.7) [which has the time-derivative term $\dot{A}(t)X(t)$], by applying ZNN model (4.7), we have the same figure as Figure 4.2. It shows the performance of the original ZNN model (4.7) for the time-varying matrix inversion under the same design parameters. Seeing and comparing Figures 4.2 and 4.11, we know that the time derivative information $\dot{A}(t)$ plays an important role on the convergence of ZNN models for time-varying matrix inversion. On the other hand, without exploiting the $\dot{A}(t)$ information, the simplified ZNN model only approaches to the approximate inverse (instead of the exact one). Figures 4.2, 4.11 and 4.12 have demonstrated the importance of the time derivative information $\dot{A}(t)$, affecting the performance of the simplified ZNN model for online time-varying matrix inversion.

4.5. Simulink-Modeling Approach

In this section, we will investigate the MATLAB Simulink modeling techniques and then build up the neural-network system [27–33]. To do so, we could firstly transform ZNN (4.7) into the following equation to lay a basis of modeling, simulation and realization:

$$\dot{X}(t) = -\dot{A}(t)X(t) + \Big(I - A(t)\Big)\dot{X}(t) \\ - \gamma\Phi\Big(A(t)X(t) - I\Big). \qquad (4.28)$$

From (4.28), the ijth-neuron dynamic-equation of ZNN (4.7) is

$$\dot{x}_{ij} = -\sum_{k=1}^{n}\dot{a}_{ik}x_{kj} - \gamma\phi\left(\sum_{k=1}^{n}(a_{ik}x_{kj} - \delta_{ij})\right) + \sum_{k=1}^{n}v_{ik}\dot{x}_{kj},$$

ZNN for Time-Varying Matrix Inversion and Pseudoinverse-Solving 93

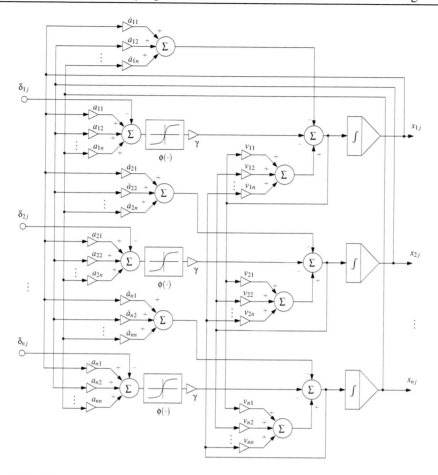

Figure 4.13. Circuit schematics of the jth neuron column used in ZNN (4.7). *Reproduced from Y. Zhang, X. Guo et al., MATLAB Simulink modeling and simulation of Zhang neural network for online time-varying matrix inversion, Figure 2, Proceedings of the 5th IEEE International Conference on Networking, Sensing and Control, pp. 1480-1485. ©[2008] IEEE. Reprinted, with permission.*

where

- x_{ij} denotes the ijth neuron of ZNN (4.7) corresponding to the ijth entry of state matrix $X(t)$, $i, j = 1, 2, ..., n$;

- time-varying weights a_{ij} and $\dot{a}_{ij}$ are defined respectively as the ijth entries of matrix $A(t)$ and its time-derivative measurement $\dot{A}(t)$;

- δ_{ij} is the Kronecker delta (defined here as the ijth entry of identity matrix I), and $v_{ij} = \delta_{ij} - a_{ij}$.

The jth-column circuit schematics of ZNN model (4.7) are thus shown in Figure 4.13.

4.5.1. ZNN-Convergence Simulation

In this subsection, we are simulating ZNN (4.7) for online time-varying matrix inversion based on MATLAB Simulink modeling techniques. For illustrative and comparative pur-

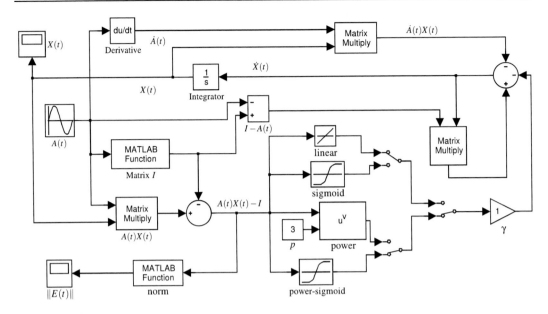

Figure 4.14. Overall Simulink modeling of ZNN (4.7) for time-varying matrix inversion. *Reproduced from Y. Zhang, X. Guo et al., MATLAB Simulink modeling and simulation of Zhang neural network for online time-varying matrix inversion, Figure 3, Proceedings of the 5th IEEE International Conference on Networking, Sensing and Control, pp. 1480-1485. ©[2008] IEEE. Reprinted, with permission.*

poses, let us consider the time-varying matrix $A(t)$ illustrated in Example 1 of Subsection 4.4.4.. Since $A(t)$ is a unitary matrix, we could readily write out its time-varying inverse as

$$X^*(t) := A^{-1}(t) = A^T(t) = \begin{bmatrix} \sin t & -\cos t \\ \cos t & \sin t \end{bmatrix},$$

which can be used to verify the solution correctness of ZNN model and/or other computational models.

The overall Simulink model of ZNN (4.7) is depicted in Figure 4.14, where $p = 3$ and $\xi = 4$ are the default parameter-values used in the activation-function-arrays. The detailed consideration is given below for constructing and simulating such a neural model.

1) Generating Matrix $A(t)$

If nonsingular matrix A is a constant matrix, we can simply use the *Constant* block to generate it by directly specifying the matrix in the "Constant value" parameter of such a block. Note that the default option of "Interpret vector parameters as 1-D" should be canceled.

To generate a time-varying matrix $A(t)$, the *MATLAB Function* block (with the *Clock* block as its input) could be exploited. We could simply specify the matrix in the "MATLAB Function" parameter and deselect the default option "Collapse 2-D results to 1-D". Note that the input argument "t" should be changed as "u" when we specify such a matrix, e.g., [sin(u) cos(u); -cos(u) sin(u)] for our example. Moreover, as the entries of matrix $A(t)$ we are going to invert here are sine and cosine functions, we could use the *Sine Wave* block to generate the matrix, where its "Phase" parameter is defined as [0 pi/2; -pi/2 0] and its default option "Interpret vector parameters as 1-D" should be deselected. In addition, four types

ZNN for Time-Varying Matrix Inversion and Pseudoinverse-Solving 95

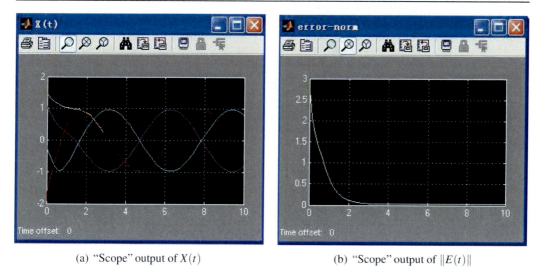

(a) "Scope" output of $X(t)$

(b) "Scope" output of $\|E(t)\|$

Figure 4.15. "Scope" outputs of ZNN (4.7) with design parameter $\gamma = 1$. *Reproduced from Y. Zhang, X. Guo et al., MATLAB Simulink modeling and simulation of Zhang neural network for online time-varying matrix inversion, Figure 5, Proceedings of the 5th IEEE International Conference on Networking, Sensing and Control, pp. 1480-1485. ©[2008] IEEE. Reprinted, with permission.*

of activation functions and their arrays are investigated (for more details, see Subsection 2.5.2.).

2) Parameters and Options Setting

Other important parameters and options related to the blocks used could be specified as follows.

- The default option "Element-wise" of three *Product* blocks have to be changed to "Matrix" so as to perform the standard matrix multiplication.

- To generate different initial state X_0, we could set the "Initial condition" parameter of the *Integrator* block to be 4*(rand(2,2)-0.5*ones(2,2).

As for the Simulink model of ZNN (4.7) established, prior to running it, we have to config the work environment by opening the "Configuration Parameters" dialog box and setting the options as follows:

- Solver: "ode45".

- Max step size: 0.2; Min step size: auto.

- Algebraic loop: none.

We could change the start and stop time as well to meet our simulation requirements.

3) Results Illustration

After running the Simulink model of ZNN (4.7) established, the "scope" output of $X(t)$ is depicted in Figure 4.15(a). As seen from Figure 4.15(a), started from a randomly-generated initial state $X_0 \in [-2,2]^{2\times 2}$, $X(t)$ converges to the theoretical inverse $X^*(t) =$

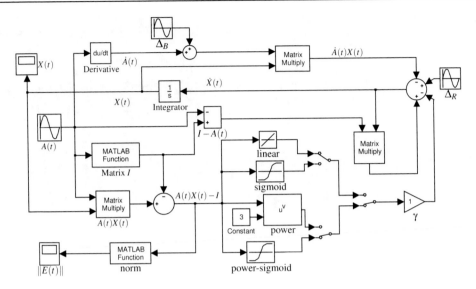

Figure 4.16. Overall Simulink modeling of imprecisely-constructed ZNN (4.14) used in online time-varying matrix inversion. *Reproduced from Y. Zhang, X. Guo et al., MATLAB Simulink modeling and simulation of Zhang neural network for online time-varying matrix inversion, Figure 6, Proceedings of the 5th IEEE International Conference on Networking, Sensing and Control, pp. 1480-1485. ©[2008] IEEE. Reprinted, with permission.*

$[\sin t, -\cos t; \cos t, \sin t]$. To monitor the neural-network convergence, we could also show the error-norm profile (i.e., $\|E(t)\|$ over time t), which is in Figure 4.15(b). Evidently, as shown in Figure 4.15(b), the computational error converges to zero (without appreciable error) in 2.74 seconds. The convergence will become much faster if we increase the design parameter γ. The observation is that, if design parameter γ is increased to 10^2, the convergence time is within 0.03 seconds; if γ is increased to 10^3, the convergence time is within 3 milliseconds; and if γ is increased to 10^4, the convergence time is within 0.3 milliseconds.

4.5.2. ZNN-Robustness Simulation

As mentioned in Section 4.3., some errors may appear in the circuits realization of ZNN model. For this reason, we are simulating the imprecisely-constructed ZNN model in this subsection to show the robustness of such a neural-network system. The corresponding Simulink modeling diagram is shown in Figure 4.16, where the differentiation error $\Delta_B(t)$ and model-implementation error $\Delta_R(t)$ are added the same as those in Example 1 of Subsubsection 4.4.4.2..

Given large implementation errors $\Delta_B(t)$ and $\Delta_R(t)$, the "scope" output of the computational error $\|E(t)\|$ of the imprecisely-constructed ZNN model (4.14) is shown in Figure 4.17. Evidently, the residual error is upper bounded by a small positive scalar, and with the increase of design parameter γ, the residual error could become smaller [i.e., $\lim_{t \to +\infty} X(t)$ of ZNN (4.14) becomes much closer to the theoretical inverse $X^*(t) := A^{-1}(t)$, as γ increases]. In addition, if the value of γ is increased, the convergence is expedited simultaneously. Compared to the case of using linear or pure power activation-function array, superior convergence and robustness performance could be achieved for the ZNN model by using

ZNN for Time-Varying Matrix Inversion and Pseudoinverse-Solving

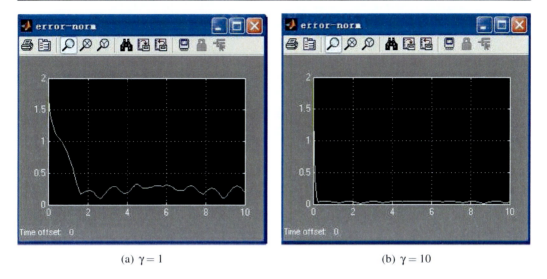

(a) $\gamma = 1$ (b) $\gamma = 10$

Figure 4.17. "Scope" outputs of computational error $\|E(t)\|$ synthesized by the imprecisely-constructed ZNN (4.14). *Reproduced from Y. Zhang, X. Guo et al., MATLAB Simulink modeling and simulation of Zhang neural network for online time-varying matrix inversion, Figure 7, Proceedings of the 5th IEEE International Conference on Networking, Sensing and Control, pp. 1480-1485. ©[2008] IEEE. Reprinted, with permission.*

power-sigmoid or sigmoid activation-function array under the same design specifications.

4.6. Extension to Time-Varying Matrix Pseudoinverse Solving

The solution and application of matrix pseudoinverse (e.g., Moore-Penrose inverse) [34–36] are widely encountered in various science and engineering fields, such as, robotics [37], signal processing [38], and applied mathematics [34]. Since numerical algorithms performed on digital computers may not be efficient enough in large-scale online applications, parallel computation schemes have been investigated intensively for solving online linear matrix equations. In view of these, Zhang neural networks have been extended for such time-varying matrix pseudoinverse solving.

4.6.1. Problem Formulation and ZNN Solver

Let us consider a smoothly time-varying matrix $A(t) \in \mathbb{R}^{m \times n}$ with $m \neq n$. If matrix $A(t)$ has full rank, i.e., $\mathrm{rank}(A(t)) = \min\{m,n\}$, $A^T(t)A(t)$ or $A(t)A^T(t)$ is nonsingular [where $A^T(t)$ denotes the transpose of $A(t)$]. Evidently, there exists a unique pseudoinverse $A^+(t) \in \mathbb{R}^{n \times m}$ for matrix $A(t)$, which could be generalized as [35, 36]:

$$A^+(t) := \begin{cases} A^T(t)\left(A(t)A^T(t)\right)^{-1}, & \text{if } m < n, \quad (4.29\text{a}) \\ \left(A^T(t)A(t)\right)^{-1} A^T(t), & \text{if } m > n, \quad (4.29\text{b}) \end{cases}$$

where (4.29a) and (4.29b) correspond to the right and left pseudoinverses, respectively. In this section, as an illustrative example, the right pseudoinverse (4.29a) is considered hereafter.

To solve online time-varying matrix pseudoinverse (4.29a), we could now follow the design method proposed by Zhang *et al* [21] to develop a recurrent neural network below.

Firstly, as the pseudoinverse of $A(t)$, $A^+(t)$ satisfies $A^+(t)A(t)A^T(t) = A^T(t)$, we can define the following matrix-valued error function:

$$E(X(t),t) := X(t)A(t)A^T(t) - A^T(t), \qquad (4.30)$$

with $X(t)$ corresponding to the pseudoinverse we are to solve for. To make $E(X(t),t)$ converge to zero (specifically, every entry $e_{ij}(t)$ converges to zero, $i = 1,2,\cdots,n$; $j = 1,2,\cdots,m$), its time-derivative $\dot{E}(X(t),t)$ can be

$$\dot{E}(X(t),t) = -\gamma\Phi\big(E(X(t),t)\big), \qquad (4.31)$$

where design parameter $\gamma > 0$ should be set as large as the hardware permits or selected appropriately for modeling purposes. $\Phi(\cdot) : \mathbb{R}^{n\times m} \to \mathbb{R}^{n\times m}$ denotes an activation-function array. Generally speaking, any monotonically-increasing odd activation-functions $\phi(\cdot)$, being the ijth element of $\Phi(\cdot)$, can be used. In this section, the linear activation function (1.4) and power-sigmoid activation function (1.7) are investigated for ZNN construction.

Secondly, it follows from (4.30) that

$$\dot{E}(X(t),t) = \dot{X}(t)A(t)A^T(t) + X(t)\Big(\dot{A}(t)A^T(t) + A(t)\dot{A}^T(t)\Big) - \dot{A}^T(t). \qquad (4.32)$$

Substituting (4.30) and (4.32) into (4.31), we can obtain the following ZNN model which solves for time-varying $A^+(t)$:

$$\begin{aligned}
\dot{X}(t)A(t)A^T(t) = &-\gamma\Phi\Big(X(t)A(t)A^T(t) - A^T(t)\Big) \\
&- X(t)\Big(\dot{A}(t)A^T(t) + A(t)\dot{A}^T(t)\Big) + \dot{A}^T(t).
\end{aligned} \qquad (4.33)$$

Thirdly, we could have the following theoretical results on global exponential convergence of ZNN (4.33) [17, 19–21].

Proposition 4.6.1. *Consider a smoothly time-varying full-rank matrix $A(t) \in \mathbb{R}^{m\times n}$. If monotonically increasing odd activation functions $\phi(\cdot)$ are used, state matrix $X(t)$ of ZNN (4.33) starting from any initial state $X(0) \in \mathbb{R}^{n\times m}$ converges to the time-varying theoretical pseudoinverse $A^+(t)$. In addition, if linear activation functions are used, global exponential convergence with rate γ can be achieved for ZNN (4.33). As compared to the linear situation, superior convergence can be achieved for ZNN (4.33) by using power-sigmoid activation-functions.* □

4.6.2. ZNN Simulink Modeling

While main theoretical results are presented in Subsection 4.6.1., we investigate the MATLAB Simulink modeling techniques to build up such a dynamic system in this subsection.

ZNN for Time-Varying Matrix Inversion and Pseudoinverse-Solving

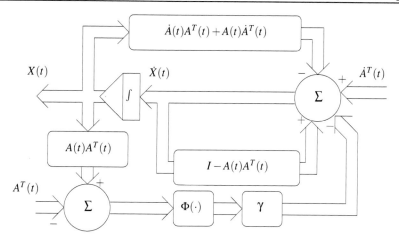

Figure 4.18. ZNN block diagram for online time-varying matrix pseudoinverse solving. *Reproduced from Y. Zhang, N. Tan et al., MATLAB Simulink modeling of Zhang neural network solving for time-varying pseudoinverse in comparison with gradient neural network, Figure 1, Proceedings of 2008 International Symposium on Intelligent Information Technology Application, pp. 39-43. ©[2008] IEEE. Reprinted, with permission.*

Before doing it, we may need to transform ZNN (4.33) into the following (with its block diagram depicted in Figure 4.18):

$$\dot{X}(t) = \dot{X}(t)\left(I - A(t)A^T(t)\right) - \gamma\Phi\left(X(t)A(t)A^T(t) - A^T(t)\right)$$
$$- X(t)\left(\dot{A}(t)A^T(t) + A(t)\dot{A}^T(t)\right) + \dot{A}^T(t).$$

Now, we come to simulate the ZNN model for online matrix pseudoinverse solving by using MATLAB Simulink. For illustrative purposes, the following specified time-varying matrix $A(t)$ is employed as an example:

$$A(t) = \begin{bmatrix} 0.5\sin t & -\cos t & 0.5\sin t \\ 0.5\cos t & \sin t & 0.5\cos t \end{bmatrix},$$

and evidently, its theoretical pseudoinverse of $A(t)$ is

$$A^+(t) = \begin{bmatrix} \sin t & \cos t \\ -\cos t & \sin t \\ \sin t & \cos t \end{bmatrix},$$

being used to verify the correctness of the neural network solution. The overall Simulink model of ZNN (4.33) is depicted in Figure 4.19, where parameters $p = 3$ and $\xi = 4$. Detailed description on the model construction is in the next subsections.

4.6.3. Generating $A(t)$

To generate a time-varying matrix $A(t)$, the *MATLAB Function* block could be employed by using the *Clock* block as its input. We need specify the matrix in the "MATLAB Function" parameter and deselect the default option "Collapse 2-D results to 1-D". It is worth

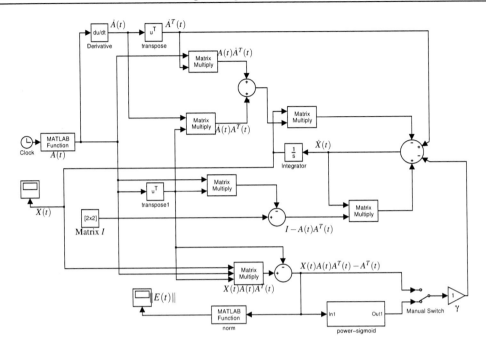

Figure 4.19. Overall ZNN Simulink model which solves online for the time-varying matrix pseudoinverse. *Reproduced from Y. Zhang, N. Tan et al., MATLAB Simulink modeling of Zhang neural network solving for time-varying pseudoinverse in comparison with gradient neural network, Figure 2, Proceedings of 2008 International Symposium on Intelligent Information Technology Application, pp. 39-43.* ©*[2008] IEEE. Reprinted, with permission.*

mentioning that the input argument should be set as "u" when we specify the matrix, e.g., [0.5*sin(u) -cos(u) 0.5*sin(u); 0.5*cos(u) sin(u) 0.5*cos(u)] for our example. Specially, all entries of matrix $A(t)$ used here are sine or cosine functions, and thus we can also use the *Sine Wave* block to generate the matrix directly, where "Amplitude" and "Phase" parameters are set as [0.5 1 0.5; 0.5 1 0.5] and [0 -pi/2 0; pi/2 0 pi/2], respectively. Besides, the default option, "Interpret vector parameters as 1-D", should be deselected when we use the *Sine Wave* block.

4.6.4. Other Parameters Setting

Some important parameters in the blocks used for ZNN modeling have been set as follows.

1) For all *MATLAB Function* blocks, the option "Collapse 2-D results to 1-D" should be deselected.

2) The default operation of the six *Product* blocks must be altered from "Element-wise" to "Matrix" multiplication.

3) To generate different initial state $X(0)$, we can set the "Initial condition" parameter of the *Integrator* block as "4*(rand(3,2)-0.5*ones(3,2))".

After the ZNN Simulink model depicted in Figure 4.19 has been built up, we also need to change some of the default simulation options. To do this, we firstly open the "Configuration Parameters" dialog box, and then set the simulation options as follows: 1)

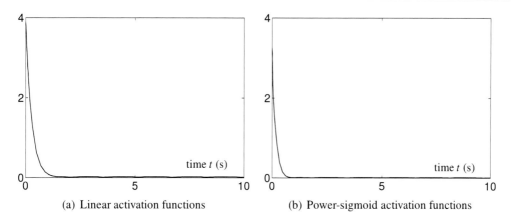

Figure 4.20. Solution-error $\|X(t) - A^+(t)\|$ of ZNN (4.33) solving for time-varying matrix pseudoinverse (4.29a) using different activation functions and with $\gamma = 4$. *Reproduced from Y. Zhang, N. Tan et al., MATLAB Simulink modeling of Zhang neural network solving for time-varying pseudoinverse in comparison with gradient neural network, Figure 3, Proceedings of 2008 International Symposium on Intelligent Information Technology Application, pp. 39-43. ©[2008] IEEE. Reprinted, with permission.*

Solver: "ode45"; 2) Max step size: "0.2"; 3) Min step size: "auto"; 4) Algebraic loop: "none". In addition, to generate $\Phi(\cdot)$, interested readers could see Subsection 2.5.2..

4.6.5. Modeling Results

The Simulink modeling results of Zhang neural network (i.e., via Figure 4.19) are shown in Figures 4.20 and 4.21. We can see that the ZNN state can always converge to the theoretical pseudoinverse of time-varying matrix $A(t)$. In addition, superior convergence can be achieved by using power-sigmoid activation functions (as compared to using linear activation functions) and/or by increasing design-parameter γ.

4.6.6. Comparisons with GNN

To show our method's superiority, we compare in this subsection the ZNN model with the conventional GNN model for the same online time-varying pseudoinverse-solving task. In view of [12], we could have the following GNN model for time-varying pseudoinverse solving:

$$\dot{X}(t) = -\gamma A^T(t) \Phi(A(t)X(t) - I). \qquad (4.34)$$

The simulation and comparison results are shown in Figure 4.22, where ZNN state $X(t)$ could always converge to $A^+(t)$, whereas GNN state $X(t)$ does not fit well with $A^+(t)$ (i.e., always lagging behind the theoretical time-varying pseudoinverse with appreciably large errors).

Evidently, the ZNN superiority comes from the fact that ZNN (4.33) exploits the time-derivative information of matrix $A(t)$ during the real-time inverting process, which ensures that ZNN (4.33) could globally exponentially converge to the exact solution of the time-varying pseudoinverse problem. In contrast, GNN (4.34) has not exploited such important information, and thus it may not be effective on solving such a time-varying pseudoinverse

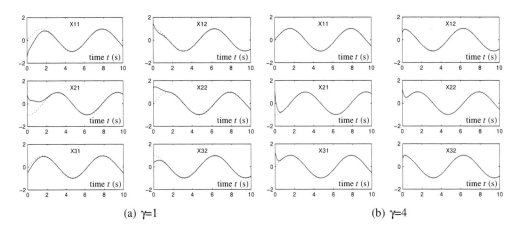

Figure 4.21. State trajectories of ZNN (4.33) using power-sigmoid activation functions, where dotted curves correspond to theoretical time-varying pseudoinverse $A^+(t)$. *Reproduced from Y. Zhang, N. Tan et al., MATLAB Simulink modeling of Zhang neural network solving for time-varying pseudoinverse in comparison with gradient neural network, Figure 4, Proceedings of 2008 International Symposium on Intelligent Information Technology Application, pp. 39-43. ©[2008] IEEE. Reprinted, with permission.*

problem. In addition, ZNN (4.33) is designed based on the elimination of every entry of matrix-valued error function $E(X)$, whereas GNN (4.34) is based on the elimination of norm-based scalar-valued error function $\|AX - I\|_F^2/2$ or its lower-bounded variants. Moreover, ZNN (4.33) is depicted in an implicit dynamics, whereas GNN (4.34) is depicted in an explicit dynamics.

4.7. Application to Robot Kinematic Control

This section presents the application of the ZNN model (4.7) to kinematic control of redundant manipulators via online solution of time-varying pseudoinverse.

4.7.1. Preliminaries on Inverse Kinematics

Consider a redundant manipulator of which the end-effector position/orientation vector $r(t) \in \mathbb{R}^m$ in Cartesian space is related to the joint-space vector $\theta(t) \in \mathbb{R}^n$ through the following forward kinematic equation

$$r(t) = f(\theta(t)) \tag{4.35}$$

where $f(\cdot)$ is a continuous nonlinear mapping function with a known structure and parameters for a given manipulator. The inverse kinematics problem is to find the joint variable $\theta(t)$ for any given $r(t)$ through the inverse mapping of (4.35), i.e., $\theta(t) = f^{-1}(r(t))$.

Unfortunately, it is usually impossible to find an analytic solution of f^{-1} due to the serious nonlinearity. The inverse kinematics problem is thus usually solved at the velocity level. Differentiating (4.35) with respect to time t yields a linear relation between velocities

ZNN for Time-Varying Matrix Inversion and Pseudoinverse-Solving

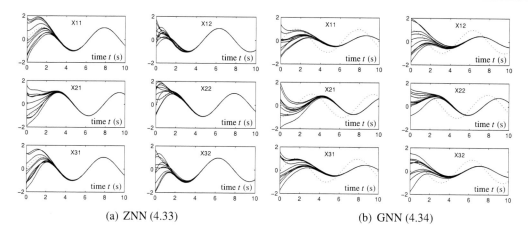

Figure 4.22. Comparison of ZNN and GNN models for online time-varying pseudoinverse solving using linear activation functions with design parameter $\gamma=1$, where dotted curves correspond to theoretical time-varying pseudoinverse $A^+(t)$. *Reproduced from Y. Zhang, N. Tan et al., MATLAB Simulink modeling of Zhang neural network solving for time-varying pseudoinverse in comparison with gradient neural network, Figure 5, Proceedings of 2008 International Symposium on Intelligent Information Technology Application, pp. 39-43. ©[2008] IEEE. Reprinted, with permission.*

$\dot{r}$ and $\dot{\theta}$:

$$J(\theta(t))\dot{\theta}(t) = \dot{r}(t) \quad (4.36)$$

where $J(\theta) \in \mathbb{R}^{m \times n}$ is the Jacobian matrix defined as $J(\theta) = \partial f(\theta)/\partial \theta$. Since $m < n$ in a redundant manipulator, (4.36) is underdetermined and may admit an infinite number of solutions.

The pseudoinverse/nullspace-type solution to (4.36), widely used by most of the current researchers, is generally formulated as a minimum-norm particular solution plus a homogeneous solution [25, 39]:

$$\dot{\theta}(t) = J^+(t)\dot{r}(t) + (I - J^+(t)J(t))z(t) \quad (4.37)$$

where $J^+(t) \in \mathbb{R}^{n \times m}$ denotes the pseudoinverse of $J(t)$. The vector $z(t) \in \mathbb{R}^n$ is arbitrary and can be chosen as the negative gradient of a performance index to be minimized, e.g., the optimization criteria of avoiding joint limits, singularity, and/or obstacles.

In the sense of Moore-Penrose generalized inverse, J^+ is defined as $J^T(JJ^T)^{-1}$ if J is of full row rank [22]. However, like matrix inversion, the usual numerical ways for solving $J^+(t)$ are in general computationally intensive and with large relative computational error because of no information of $\dot{J}(t)$ considered. In addition, with multiple tasks and constraints included by (4.37) via z, the heavy-burden J^+-computing procedure may hinder on-line applications, especially in high-DOF (degrees-of-freedom) sensor-based robotic systems [39, 40].

Defining $A(t) = J(t)J^T(t)$, we could re-exploit ZNN (4.7) to solve $J^+(t)$ in a parallel manner so as to expedite the computation process and achieve better control precision. That is, $J^+(t) = J^T(t)X(t)$ where $X(t)$ is the state matrix generated online by ZNN (4.7).

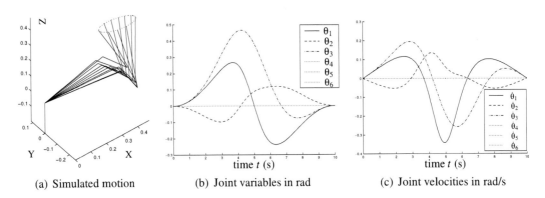

Figure 4.23. Motion trajectories of PUMA560 manipulator synthesized by ZNN (4.7). *Reproduced from Y. Zhang and S. S. Ge, Design and analysis of a general recurrent neural network model for time-varying matrix inversion, Figure 11, IEEE Transactions on Neural Networks, pp. 1477-1490. ©[2005] IEEE. Reprinted, with permission.*

4.7.2. Simulation Based on PUMA560 Robot Arm

The Unimation PUMA560 robot arm has six joints [41]. When we consider only the positioning of the end-effector, the PUMA560 becomes a redundant manipulator with the dimensionality of J being 3×6 [42]. The ZNN model (4.7) is first applied to the PUMA560 robot arm with the design parameter $\gamma = 10^3$ and $\dot{A}(t) = \dot{J}(t)J^T(t) + J(t)\dot{J}^T(t)$. The desired motion of the end-effector is a circle of radius $r = 10$cm and with the revolute angle about X-axis being $-\pi/6$. The motion duration is 10 seconds, and the initial joint variables $\theta(0) = [0,0,0,0,0,0]^T$ in radians. Figure 4.23(a) illustrates the simulated motion of the

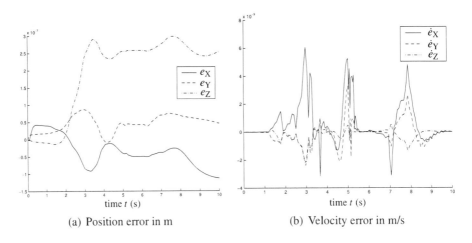

Figure 4.24. End-effector errors of PUMA560 manipulator tracking a 10cm-radius circle. *Reproduced from Y. Zhang and S. S. Ge, Design and analysis of a general recurrent neural network model for time-varying matrix inversion, Figure 12, IEEE Transactions on Neural Networks, pp. 1477-1490. ©[2005] IEEE. Reprinted, with permission.*

robot arm in the 3-dimensional work space, which is sufficiently close to the desired one. Specifically, as shown in Figure 4.24, the maximal Cartesian position and velocity tracking errors at the end-effector are respectively less than 3×10^{-7}m and 7×10^{-9}m/s. The corresponding joint variables and joint velocities are depicted in the subplots (b) and (c) of Figure 4.23. The network states $X(t)$ of ZNN model (4.7) is shown in Figure 4.25.

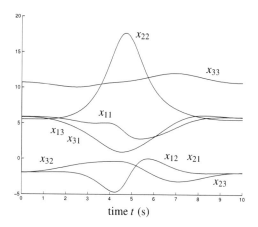

Figure 4.25. States of ZNN (4.7) for kinematic control of PUMA560, where $x_{ij} = x_{ji}$ shows the symmetry. *Reproduced from Y. Zhang and S. S. Ge, Design and analysis of a general recurrent neural network model for time-varying matrix inversion, Figure 13, IEEE Transactions on Neural Networks, pp. 1477-1490. ©[2005] IEEE. Reprinted, with permission.*

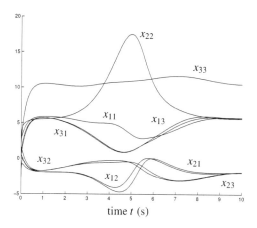

Figure 4.26. States of the traditional GNN model for kinematic control of PUMA560. *Reproduced from Y. Zhang and S. S. Ge, Design and analysis of a general recurrent neural network model for time-varying matrix inversion, Figure 14, IEEE Transactions on Neural Networks, pp. 1477-1490. ©[2005] IEEE. Reprinted, with permission.*

For comparison, the traditional GNN is also simulated for solving $J^+(t)$ online. Under the same design condition $\gamma = 10^3$, the end-effector position and velocity errors in this case are considerably large (i.e., $\geq$ 1.0cm and 1.5cm/s, respectively). In addition, to achieve the similar precision of the proposed neural model, the traditional gradient-based neural

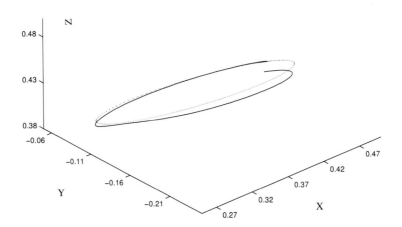

Figure 4.27. End-effector motion trajectory of PUMA560 synthesized by the traditional GNN model, where the dotted curve denotes the desired circular path. *Reproduced from Y. Zhang and S. S. Ge, Design and analysis of a general recurrent neural network model for time-varying matrix inversion, Figure 15, IEEE Transactions on Neural Networks, pp. 1477-1490.* ©*[2005] IEEE. Reprinted, with permission.*

network requires $\gamma \geq 10^9$, which is a very stringent restriction on system design, either analogue or digital. To facilitate the comparison with the proposed model, the states $X(t)$ of the traditional gradient-based neural network are also depicted, i.e., in Figure 4.26. Clearly, in the solution generated by traditional GNN, the state x_{ij} does not equals x_{ji} violating the symmetric property of $(JJ^T)^{-1}$, and the settling time-period from initial states to steady states is relatively long (about 0.6s). These lead to the considerably larger Cartesian positioning error at the PUMA560 end-effector as in Figure 4.27. The reason is also that the $\dot{A}(t)$ information has not been utilized in traditional gradient-based approaches.

4.7.3. Conclusion

A special type of recurrent neural networks (i.e., ZNN) with implicit dynamics has been presented in this chapter for solving the inverse/pseudoinverse of time-varying matrix in real time. Different from the traditional GNN models used in time-varying cases, the proposed ZNN approach fully and methodically utilizes the time-derivative information of the matrix to be inverted. It is thus able to guarantee the global exponential convergence to the exact solution of such time-varying problems. Moreover, it has been shown that superior convergence and robustness can be achieved by using sigmoid or power-sigmoid activation functions. In addition, the MATLAB Simulink modeling of such ZNN models has been investigated for the purposes of its final FPGA/ASIC circuits implementation. We build up such neural systems by simply connecting Simulink function blocks instead of writing lengthy M-files. Programming efforts could thus be reduced drastically. Simulation results including kinematic control of the PUMA560 manipulator have demonstrated the effectiveness and efficiency of the proposed recurrent neural network model. Further efforts are to be directed at the design and analysis of discrete-time neural networks, numerical algorithms, and electronic circuits for solving the inverse of time-varying matrix.

References

[1] Zhang, Y. Towards piecewise-linear primal neural networks for optimization and redundant robotics. *Proceedings of IEEE International Conference on Networking, Sensing and Control*, 2006, pp. 374-379.

[2] Steriti, RJ; Fiddy, MA. Regularized image reconstruction using SVD and a neural network method for matrix inversion. *IEEE Transactions on Signal Processing*, 1993, vol. 41, no. 10, pp. 3074-3077.

[3] Yeung, KS; Kumbi, F. Symbolic matrix inversion with application to electronic circuits. *IEEE Transactions on Circuits and Systems*, 1988, vol. 35, no. 2, pp. 235-238.

[4] Sturges, Jr RH. Analog matrix inversion (robot kinematics). *IEEE Journal of Robotics and Automation*, 1988, vol. 4, no. 2, pp. 157-162.

[5] Neagoe, VE. Inversion of the Van der Monde matrix. *IEEE Signal Processing Letters,* 1996, vol. 3, no. 4, pp. 119-120.

[6] Wang, YQ; Gooi, HB. New ordering methods for space matrix inversion via diagonaliztion. *IEEE Transactions on Power Systems,* 1997, vol. 12, no. 3, pp. 1298-1305.

[7] Koc, CK; Chen, G. Inversion of all principal submatrices of a matrix. *IEEE Transactions on Aerospace Electrical Systems,* 1994, vol. 30, no. 1, pp. 280-281.

[8] Manherz, RK; Jordan, BW; Hakimi, SL. Analog methods for computation of the generalized inverse. *IEEE Transactions on Automatic Control,* 1968, vol. 13, no. 5, pp. 582-585.

[9] Zhang, Y; Leithead, WE; Leith, DJ. Time-series Gaussian process regression based on Toeplitz computation of $O(N^2)$ operations and $O(N)$-level storage. *Proceedings of the 44th IEEE Conference on Decision and Control*, 2005, pp. 3711-3716.

[10] Leithead, WE; Zhang, Y. $O(N^2)$-operation approximation of covariance matrix inverse in Gaussian process regression based on quasi-Newton BFGS methods. *Communications in Statistics - Simulation and Computation*, 2007, vol. 36, no. 2, pp. 367-380.

[11] Jang, J; Lee, S; Shin, S. *An optimization network for matrix inversion, in: Anderson, DZ (Ed.).* New York: Neural Information Processing Systems, American Institute of Physics, 1988, pp. 397-401.

[12] Wang, J. A recurrent neural network for real-time matrix inversion. *Applied Mathematics and Computation*, 1993, vol. 55, no. 1, pp. 89-100.

[13] El-Amawy, A. A systolic architecture for fast dense matrix inversion. *IEEE Transactions on Computers*, 1989, vol. 38, no. 3, pp. 449-455.

[14] Carneiro, NCF; Caloba, LP. A new algorithm for analog matrix inversion. *Proceedings of the 38th Midwest Symposium on Circuits and Systems*, 1995, vol. 1, pp. 401-404.

[15] Luo, FL; Zheng, B. Neural network approach to computing matrix inversion. *Applied Mathematics and Computation,* 1992, vol. 47, no. 2-3, pp. 109-120.

[16] Song, J; Yam, Y. Complex recurrent neural network for computing the inverse and pseudo-inverse of the complex matrix. *Applied Mathematics and Computation,* 1998, vol. 93, no. 2-3, pp. 195-205.

[17] Zhang, Y. Revisit the analog computer and gradient-based neural system for matrix inversion. *Proceedings of IEEE International Symposium on Intelligent Control*, 2005, pp. 1411-1416.

[18] Zhang, Y; Leithead, WE. Exploiting Hessian matrix and trust-region algorithm in hyperparameters estimation of Gaussian process. *Applied Mathematics and Computation,* 2005, vol. 171, no. 2, pp. 1264-1281.

[19] Zhang, Y; Jiang, D; Wang, J. A recurrent neural network for solving Sylvester equation with time-varying coefficients. *IEEE Transactions on Neural Networks*, 2002, vol. 13, no. 5, pp. 1053-1063.

[20] Zhang, Y; Ge, SS. A general recurrent neural network model for time-varying matrix inversion. *Proceedings of the 42nd IEEE Conference on Decision and Control*, 2003, pp. 6169-6174.

[21] Zhang, Y; Ge, SS. Design and analysis of a general recurrent neural network model for time-varying matrix inversion. *IEEE Transactions on Neural Networks*, 2005, vol. 16, no. 6, pp. 1477-1490.

[22] Horn, RA; Johnson, CR. *Topics in Matrix Analysis.* Cambridge: Cambridge University Press, 1991.

[23] Zhang, Y; Wang, J. Global exponential stability of recurrent neural networks for synthesizing linear feedback control systems via pole assignment. *IEEE Transactions on Neural Networks,* 2002, vol. 13, no. 3, pp. 633-644.

[24] Mead, C. *Analog VLSI and Neural Systems.* Reading, MA: Addison-Wesley, 1989.

[25] Ge, SS; Lee, TH; Harris, CJ. *Adaptive Neural Network Control of Robotic Manipulators.* London: World Scientific, 1998.

[26] Zhang, Y; Heng, PA; Fu, AWC. Estimate of exponential convergence rate and exponential stability for neural networks. *IEEE Transactions on Neural Networks,* 1999, vol. 10, no. 6, pp. 1487-1493.

References

[27] Shanblatt, MA; Foulds, B. A Simulink-to-FPGA implementation tool for enhanced design flow. *Proceedings of IEEE International Conference on Microelectronic Systems Education*, 2005, pp. 89-90.

[28] Grout, IA. Modeling, simulation and synthesis: from Simulink to VHDL generated hardware. *Proceedings of the 5th World Multi-Conference on Systemics, Cybernetics and Informatic*, 2001, vol. 15, pp. 443-448.

[29] Ma, W; Zhang, Y; Wang, J. MATLAB Simulink modeling and simulation of Zhang neural networks for online time-varying Sylvester equation solving. *Proceedings of International Joint Conference on Neural Networks*, 2008, pp. 286-290.

[30] Grout, IA; Keane, K. A MATLAB to VHDL conversion toolbox for digital control. *Proceedings of IFAC Symposium on Computer Aided Control Systems Design*, 2000, pp. 164-169.

[31] Shi, KL; Chan, TF; Wong, YK. Modeling of the three-phase induction motor using Simulink. *Proceedings of IEEE International Electric Machines and Drives Conference*, 1997, pp. WB3/6.1-WB3/6.3.

[32] Tank, D; Hopfield, JJ. Simple neural optimization networks: an A/D converter, signal decision circuit, and a linear programming circuit. *IEEE Transactions on Circuits and Systems*, 1986, vol. 33, no. 5, pp. 533-541.

[33] Anderson, JA; Rosenfeld, E. *Neurocomputing: Foundations of Research.* Cambridge, MA: The MIT Press, 1988.

[34] Wei, Y. Recurrent neural networks for computing weighted Moore-Penrose inverse. *Applied Mathematics and Computation*, 2000, vol. 116, no. 3, pp. 279-287.

[35] Wang, J. Recurrent neural networks for computing pseudoinverses of rank-deficient matrices. *SIAM Journal on Scientific Computing*, 2007, vol. 18, no. 5, pp. 1479-1493.

[36] Parks-Gornet, J; Imam, IN. Using rank factorization in calculating the Moore-Penrose generalized inverse. *Proceedings of IEEE Energy and Information Technologies in the Southeast*, 1989, vol. 2, pp. 427-431.

[37] Klein, CA; Kee, KB. The nature of drift in pseudoinverse control of kinematically redundant manipulators. *IEEE Transactions on Robotics and Automation,* 1989, vol. 5, no. 2, pp. 231-234.

[38] Yahagi, T. A deterministic approach to optimal linear digital equalizers. *IEEE Transactions on Acoustics, Speech, and Signal Processing,* 1983, vol. ASSP-31, no. 2, pp. 491-500.

[39] Sciavicco, L; Siciliano, B. *Modeling and control of robot manipulators.* London, U.K.: Springer-Verlag, 2000.

[40] Zhang, Y; Wang, J; Xu, Y. A dual neural network for bi-criteria kinematic control of redundant manipulators, *IEEE Transactions on Robotics and Automation,* 2002, vol. 18, no. 6, pp. 923-931.

[41] Corke, PI; Armstrong-Helouvry, B. A search for consensus among model parameters reported for the PUMA 560 robot. *Proceedings of IEEE International Conference Robotics and Automation,* 1994, vol. 2, pp. 1608-1613.

[42] Zhang, Y; Wang, J. A dual neural network for constrained joint torque optimization of kinematically redundant manipulators. *IEEE Transactions on Systems, Man, and Cybernectics,* Part B, 2002, vol. 32, no. 5, pp. 654-662.

Chapter 5

ZNN for Linear Time-Varying Matrix Equation Solving

Abstract

For solving online the linear matrix equation (e.g., Sylvester equation, Lyapunov equation, $AXB = C$) with time-varying coefficients, this chapter presents a special kind of recurrent neural networks by using the design method recently proposed by Zhang *et al*. Compared with gradient neural networks (abbreviated as GNN, or termed as gradient-based neural networks), the resultant Zhang neural networks (termed as such and abbreviated as ZNN for presentation convenience) are designed based on matrix-valued error functions, instead of scalar-valued energy functions. The ZNN model is deliberately developed in the way that its trajectory could be guaranteed to globally exponentially converge to the time-varying theoretical solution of a given linear matrix equation. For comparison, we develop and simulate the GNN as well, which is exploited to solve online the same time-varying linear matrix equation. In addition, towards the final purpose of hardware realization, this chapter highlights the model building and convergence illustration of the ZNN model in comparison with GNN. Computer-simulation results substantiate the theoretical efficacy and superior performance of ZNN for the online solution of time-varying linear matrix equation, especially when using a power-sigmoid activation function array.

5.1. Introduction

The problem of linear matrix equation solving (including matrix-inverse as a sub-topic) is considered to be a very fundamental problem widely encountered in science and engineering. It could usually be an essential part of many solutions; e.g., in control system design [1,2] and image-processing [3]. There are two general types of solutions to the problem of linear matrix equations. One is the numerical algorithms performed on digital computers (i.e., on today's computers). Usually, such numerical algorithms are of serial-processing nature and may not be efficient enough for large-scale online or real-time applications. Being the second general type of solution, many parallel-processing computational methods have been developed, analyzed, and implemented on specific architectures [3–13].

The dynamic-system approach is one of such important parallel-processing methods for solving linear matrix equations. Recently, due to the in-depth research, neural-dynamic approaches have been widely considered as a powerful alternative to solving online linear matrix equations and related issues [3,6,14–16]. Compared to numerical algorithms, neural-dynamic methods mainly in the form of recurrent neural networks (or to say, Hopfield-type neural networks) appear to be more suitable for large-scale online application owing to their intrinsically-parallel-distributed processing nature and circuit-implementation convenience [7, 10, 17, 18]. Different from GNN for constant problems solving [2, 3, 6, 10, 17, 19, 20], a special kind of recurrent neural network designed for online solution of time-varying problems has been proposed and established by Zhang *et al.* [11–13]. The resultant ZNN could converge (globally exponentially) to theoretical solutions of time-varying problems exactly, while gradient-based neural networks may approximate the solution only. The design method of Zhang neural network is completely different from that of gradient neural networks. In this chapter, we generalize such a design method to solving online the time-varying linear matrix equation. Theoretical and simulative results both demonstrate the efficacy of the proposed ZNN neural approach.

For the purpose of field programmable gate array (FPGA) and application-specific integrated circuit (ASIC) realization, we investigate the blocks modeling and verification of ZNN model for time-varying linear matrix equations solving as well as GNN model for comparison. Simulink [21] is a graphical-design based modeling tool which could exploit existing function blocks by using click-and-drag operations to construct mathematical and logical models as well as process flow instead of time-consuming coding. In addition to mathematical analysis, by our understanding, blocks modeling is an important (and key) step towards the final hardware implementation of artificial neural networks with interconnecting structures, which could be viewed as a virtual implementation of a real system satisfying a set of engineering requirements.

5.2. ZNN for Time-Varying Sylvester Equation Solving

In this section, we exploit the ZNN models for time-varying Sylvester equation solving, with the main theoretical results presented. MATLAB-based simulation results substantiate the theoretical analysis and demonstrate the efficacy of ZNN on time-varying problem solving, especially when using a power-sigmoid activation function array.

5.2.1. Problem Formulation and Solvers

Consider a smoothly time-varying Sylvester equation:

$$A(t)X(t) - X(t)B(t) + C(t) = 0, \quad 0 \le t < +\infty, \tag{5.1}$$

where $A(t) \in \mathbb{R}^{m \times m}$, $B(t) \in \mathbb{R}^{n \times n}$, and $C(t) \in \mathbb{R}^{m \times n}$ are smoothly time-varying coefficients; and, $X(t) \in \mathbb{R}^{m \times n}$ is the unknown matrix to be obtained. We are going to find $X(t)$ in real time and in an error-free manner such that the above equation holds true. Without loss of generality, $A(t)$, $B(t)$ and $C(t)$ are assumed to be known, while their time derivatives $\dot{A}(t)$, $\dot{B}(t)$ and $\dot{C}(t)$ are assumed at least to be measurable (if unknown). The existence of the theoretical solution $X^*(t) \in \mathbb{R}^{m \times n}$ at any time instant t is also assumed for comparative purposes.

Here, it is worth mentioning that simplifications of Sylvester equation may result in various kinds of new problems, such as, the matrix-inverse problem and linear-equation problem. Specifically speaking,

- if $n = m$, $B(t) = 0$ and $C(t) = -I \in \mathbb{R}^{n \times n}$, then the above time-varying Sylvester equation reduces to the matrix-inverse problem, $A(t)X(t) = I$; and

- if $n = 1$, $B(t) = 0$ and $C(t) = -c(t) \in \mathbb{R}^m$, then $X(t) \in \mathbb{R}^{m \times n}$ reduces to $x(t) \in \mathbb{R}^m$ and the above time-varying Sylvester equation reduces to a set of linear time-varying equations, $A(t)x(t) = c(t)$.

The results achieved for time-varying Sylvester equation, for time-varying matrix inversion, and/or for linear time-varying equations could thus be tailored and used for one another.

5.2.1.1. Zhang Neural Networks

For the time-varying problem solving, a special kind of recurrent neural network was proposed by Zhang *et al.* [11–13]. By following such a design method, we can generalize a recurrent neural network to solving (5.1) in real time t as follows.

Step 1. To monitor the equation-solving procedure, the following matrix-valued error function is defined, instead of scalar-valued norm-based energy functions.

$$E(t) := A(t)X(t) - X(t)B(t) + C(t) \in \mathbb{R}^{m \times n}.$$

Step 2. The error-function time-derivative $\dot{E}(t) \in \mathbb{R}^{m \times n}$ should be made such that every element $e_{ij}(t) \in \mathbb{R}$ of $E(t) \in \mathbb{R}^{m \times n}$ converges to zero, $\forall\, i = 1, \cdots, m$ and $j = 1, \cdots, n$. In mathematics, we need choose $\dot{E}(t)$ and/or $\dot{e}_{ij}(t)$ such that $\lim_{t \to +\infty} e_{ij}(t) = 0$, $\forall\, i, j$.

Step 3. A general form of $\dot{E}(t)$ could thus be

$$\frac{\mathrm{d}E(X(t),t)}{\mathrm{d}t} = -\Gamma \Phi\left(E(X(t),t)\right) \tag{5.2}$$

where design parameter Γ (which could simply be γI) and activation-function array $\Phi(\cdot)$ are described and discussed the same as in the preceding chapters.

Step 4. Expanding the design formula (5.2) leads to the following implicit dynamic equation of ZNN model for solving online the time-varying Sylvester equation, (5.1).

$$\begin{aligned} A(t)\dot{X}(t) - \dot{X}(t)B(t) = &- \dot{A}(t)X(t) + X(t)\dot{B}(t) - \dot{C}(t) \\ &- \gamma \Phi\left(A(t)X(t) - X(t)B(t) + C(t)\right) \end{aligned} \tag{5.3}$$

where

- $X(t)$, starting from an initial condition $X(0) = X_0 \in \mathbb{R}^{m \times n}$, is the activation state matrix corresponding to the theoretical solution of (5.1); and

- $\dot{A}(t)$, $\dot{B}(t)$ and $\dot{C}(t)$ denote the measurements or known analytical forms of the time derivatives of matrices $A(t)$, $B(t)$ and $C(t)$ [simply, $\dot{A}(t) := \mathrm{d}A(t)/\mathrm{d}t$, and so on].

It is worth mentioning that when using the linear activation function array $\Phi(E) = E$, ZNN model (5.3) reduces to the following classic linear one:

$$A(t)\dot{X}(t) - \dot{X}(t)B(t) = -\left(\gamma A(t) + \dot{A}(t)\right)X(t) + X(t)\left(\gamma B(t) + \dot{B}(t)\right)$$
$$- \left(\gamma C(t) + \dot{C}(t)\right).$$

5.2.1.2. Gradient-Based Neural Networks

For comparison, it is also worth mentioning that almost all the numerical algorithms and neural-dynamic computational schemes were designed intrinsically for constant coefficient matrices A, B and C, rather than time-varying matrices $A(t)$, $B(t)$ and $C(t)$. The aforementioned neural-dynamic computational schemes [6, 10, 15, 17, 22] are in general related to the gradient-descent method in optimization. Its design procedure is as follows.

- A scalar-valued energy function, such as $\|AX - XB + C\|^2/2$, is firstly constructed such that its minimum point is the solution of Sylvester equation $AX - XB + C = 0$.

- Then, the typical descent direction is the negative gradient of $\|AX - XB + C\|^2/2$; i.e.,

$$-\frac{\partial\left(\|AX - XB + C\|^2/2\right)}{\partial X} = -A^T\left(AX - XB + C\right) + \left(AX - XB + C\right)B^T.$$

By using the above design method of gradient-based neural networks [6, 10, 15, 17, 22], we could have the classic linear gradient-based neural network

$$\dot{X}(t) = -\gamma A^T\left(AX - XB + C\right) + \gamma\left(AX - XB + C\right)B^T$$

and the general nonlinear form of gradient-based neural network

$$\dot{X}(t) = -\gamma A^T\Phi\left(AX - XB + C\right) + \gamma\Phi\left(AX - XB + C\right)B^T. \tag{5.4}$$

5.2.1.3. Method and Model Comparisons

As compared to the above GNN model (5.4), the difference and novelty of ZNN model (5.3) for solving online the time-varying Sylvester equation lie in the following facts.

- Firstly, the above ZNN model (5.3) is designed based on the elimination of every entry of the matrix-valued error function $E(t) = A(t)X(t) - X(t)B(t) + C(t)$. In contrast, the GNN model (5.4) was designed based on the elimination of the scalar-valued energy function $\mathcal{E}(t) = \|AX(t) - X(t)B + C\|^2/2$ (note that here A, B and C could only be constant in the design of the GNN model).

- Secondly, ZNN model (5.3) methodically and systematically exploits the time derivatives of coefficient matrices $A(t)$, $B(t)$ and $C(t)$ during the problem-solving procedure. This is the reason why ZNN model (5.3) globally exponentially converges to the exact solution of a time-varying problem. In contrast, the GNN model (5.4) has not exploited such important information.

- Thirdly, ZNN model (5.3) is depicted in an implicit dynamics, or to say, in the concisely arranged form $M(t)\dot{y}(t) = P(t)y(t) + q(t)$ where $M(t)$, $P(t)$, $q(t)$ and $y(t)$ can be given by using Kronecker product and vectorization technique [12, 13, 22]. In contrast, the GNN model (5.4) is depicted in an explicit dynamics, or to say, in the concisely arranged form $\dot{y}(t) = \bar{P}y(t) + \bar{q}$.

It is worth pointing out that the implicit dynamic equations (or to say, implicit systems) frequently arise in analog electronic circuits and systems due to Kirchhoff's rules. Furthermore, implicit systems have higher abilities in representing dynamic systems, as compared to explicit systems. The implicit dynamic equations could preserve physical parameters in the coefficient/mass matrix, i.e., $M(t)$ in $M(t)\dot{y}(t) = P(t)y(t) + q(t)$. They could describe the usual and unusual parts of a dynamic system in the same form. In this sense, implicit systems are much superior to the systems represented by explicit dynamics. In addition, the implicit dynamic equations (or implicit systems) could mathematically be transformed to explicit dynamic equations (or explicit systems) if needed.

5.2.2. Theoretical Results

From analysis results of [10–13], the following propositions on global (exponential) convergence of ZNN (5.3) are obtained finally.

Proposition 5.2.1. *Given the smoothly time-varying matrices $A(t) \in \mathbb{R}^{m \times m}$, $B(t) \in \mathbb{R}^{n \times n}$ and $C(t) \in \mathbb{R}^{m \times n}$, if a monotonically-increasing odd activation function array $\Phi(\cdot)$ is used, then the state matrix $X(t)$ of ZNN model (5.3) starting from any state $X_0 \in \mathbb{R}^{m \times n}$ will always converge to the theoretical solution $X^*(t)$ of Sylvester equation (5.1).* □

In the realization of neural network models, there are always some errors involved. For the differentiation errors and model-implementation errors, the following dynamics of Zhang neural network is considered, compared to the original dynamic equation (5.3).

$$
\begin{aligned}
A(t)\dot{X}(t) - \dot{X}(t)B(t) = &- \big(\dot{A}(t) + \Delta_{\dot{A}}(t)\big)X(t) + X(t)\big(\dot{B}(t) + \Delta_{\dot{B}}(t)\big) \\
&- \dot{C}(t) - \gamma\Phi\big(A(t)X(t) - X(t)B(t) + C(t)\big) + \Delta_R(t),
\end{aligned}
\tag{5.5}
$$

where $\Delta_{\dot{A}}(t)$ and $\Delta_{\dot{B}}(t)$ denote respectively the differentiation errors of matrices $A(t)$ and $B(t)$, $\Delta_R(t)$ denotes the model-implementation error (including the differentiation error of matrix C as a part).

Proposition 5.2.2. *Consider the ZNN model with implementation errors $\Delta_{\dot{A}}(t)$, $\Delta_{\dot{B}}(t)$ and $\Delta_R(t)$ finally depicted in (5.5). If $\|\Delta_{\dot{A}}(t)\|$, $\|\Delta_{\dot{B}}(t)\|$ and $\|\Delta_R(t)\|$ are uniformly upper bounded by some positive scalars, then the computational error $\|X(t) - X^*(t)\|$ over time t and the steady-state residual error $\lim_{t \to \infty} \|X(t) - X^*(t)\|$ are uniformly upper bounded by some positive scalars, provided that the design parameter γ is large enough. More importantly, as γ tends to positive infinity, the steady-state residual error vanishes to zero.* □

Proposition 5.2.3. *If the power-sigmoid activation function is used in ZNN models, the design parameter γ could be any positive value. In addition, superior convergence/robustness properties could be obtained by using the power-sigmoid activation function in ZNN models (5.3) and (5.5), compared to the case of using the linear activation function.* □

5.2.3. MATLAB-Coding Simulation Techniques

While the above subsections present ZNN models and theoretical results, we will investigate the related MATLAB simulation techniques in this subsection. Note that, the definition and coding of activation functions could be seen in Subsection 2.4.1..

5.2.3.1. Kronecker Product and Vectorization

For the proposes of MATLAB simulation, we have to transform the matrix-form differential equation (5.3) into a vector-form differential equation via the following theorem (with proof similar to Theorem 4.4.1 in Subsection 4.4.1. and thus omitted here). With symbol $\otimes$ denoting the Kronecker product, $B \otimes A$ is a large matrix made by replacing the ijth entry b_{ij} of B with the matrix $b_{ij}A$. Operator $\text{vec}(X) \in \mathbb{R}^{nm}$ denotes a column vector obtained by stacking all column vectors of $X \in \mathbb{R}^{m \times n}$ together.

Theorem 5.2.1. *The matrix-form ZNN model* (5.3) *can be transformed to the following vector-form differential equation:*

$$
\begin{aligned}
M(t)\dot{x}(t) = &-\gamma\Phi\big(M(t)x(t) + \text{vec}(C(t))\big) \\
&-\dot{M}(t)x(t) - \text{vec}(\dot{C}(t)),
\end{aligned}
\tag{5.6}
$$

where mass matrix $M(t) := I \otimes A(t) - B^T(t) \otimes I$, integration vector $x(t) := \text{vec}(X(t))$, and activation-function array $\Phi(\cdot)$ is defined the same as in (5.3) *except that its dimensions are changed hereafter as $\Phi(\cdot) : \mathbb{R}^{mn \times 1} \to \mathbb{R}^{mn \times 1}$.* □

The above Kronecker product could be generated easily by using MATLAB routine "kron". For example, the following MATLAB code is used to generate the mass matrix $M = I \otimes A - B^T \otimes I$, where matrices A and B are defined respectively as $[\sin t, \cos t; -\cos t, \sin t]$ and $[0.1 \sin t, 0; 0, 0.2 \cos t]$.

```
function output=M(t,x)
A=[sin(t) cos(t); -cos(t) sin(t)];
B=[0.1*sin(t) 0; 0 0.2*cos(t)];
output=kron(eye(size(B)),A)-kron(B.',eye(size(A)));
```

Based on MATLAB routine "reshape", the vectorization of a matrix could be achieved easily. For example, the following MATLAB code is used to evaluate the right-hand side of differential equation (5.11), where user-defined MATLAB functions "DiffC" and "DiffM" will be introduced later for evaluating time derivatives of matrices C and M.

```
function output=ZnnRHS(t,x,gamma)
if nargin==2, gamma=1; end
C=MatrixC(t); [m,n]=size(C); dotC=DiffC(t,x);
vecC=reshape(C,m*n,1);
vecDotC=reshape(dotC,m*n,1);
M=M(t,x); dotM=DiffM(t,x);
output=-gamma*AFMpowersigmoid(M*x+vecC)-vecDotC-dotM*x;
```

5.2.3.2. ODE with Mass Matrix

In the simulation of ZNN models (5.3) and (5.6), the MATLAB routine "ode45" is preferred. This is because "ode45" can solve initial-value ODE problems with nonsingular mass matrices, e.g., $M(t,x)\dot{x} = g(t,x)$, $x(0) = x_0$, where matrix $M(t,x)$ on the left-hand side of such an equation is termed the mass matrix, and $g(t,x) := -\dot{M}x - \gamma\Phi(Mx - \text{vec}(C)) - \text{vec}(\dot{C})$ for our case.

To solve an ODE problem with a mass matrix, the MATLAB routine "odeset" should also be used. Its "Mass" property should be assigned to be the function handle "@MatrixM", which returns the evaluation of mass matrix $M(t,x)$. Note that, if 1) $M(t,x)$ does not depend on state variable x and 2) the function "M" is to be invoked with only one input argument t, then the value of the "MStateDep" property of "odeset" should be set to "none". For example, the following MATLAB code can be used to solve an initial-value ODE problem with state-independent mass matrix $M(t)$ and random initial state x_0.

```
tspan=[0 10]; x0=4*(rand(4,1)-0.5*ones(4,1));
options=odeset('Mass',@M,'MStateDep','none');
[t,x]=ode45(@ZnnRHS,tspan,x0,options);
```

5.2.3.3. Obtaining Matrix Derivatives

While matrix $M(t)$ can be generated by the user-defined function "M" shown in Subsubsection 5.2.3.1., its time derivative $\dot{M}(t)$ required by (5.6) should also be obtained. Without loss of generality, we can use MATLAB routine "diff" to generate $\dot{M}(t)$ and $\dot{C}(t)$, of which the latter is also required in ZNN model (5.3).

There are several overloaded functions "diff" in MATLAB environment. In order to get the time derivative of $M(t)$, a symbolic object "u" is constructed firstly, and then the MATLAB code "DM=diff(M(u))" is used to generate the analytical form of $\dot{M}$. Finally, evaluating such an analytical form of $\dot{M}$ with a numerical t will generate the required $\dot{M}(t)$. The MATLAB code for generating $\dot{M}(t)$ is as follows. Similarly, MATLAB function "DiffC" could be coded to generate $\dot{C}(t)$.

```
syms u; DM=diff(M(u)); u=t; output=eval(DM);
```

5.2.4. Illustrative Examples by MATLAB Code

For simulation and comparison purposes, let us consider Sylvester equation (5.1) with the following time-varying coefficient matrices $A(t)$, $B(t)$, and $C(t)$.

$$A(t) = \begin{bmatrix} \sin t & \cos t \\ -\cos t & \sin t \end{bmatrix}, B(t) = \begin{bmatrix} 0.1\sin t & 0 \\ 0 & 0.2\cos t \end{bmatrix},$$

$$C(t) = \begin{bmatrix} 0.1\sin^2 t - 1 & -0.2\cos^2 t \\ 0.1\sin t \cos t & 0.2\sin t \cos t - 1 \end{bmatrix}.$$

Note that, for comparative proposes, the theoretical solution $X^*(t)$ of Sylvester equation (5.1) could be verified to be $X^*(t) = [\sin t, -\cos t; \cos t, \sin t]$.

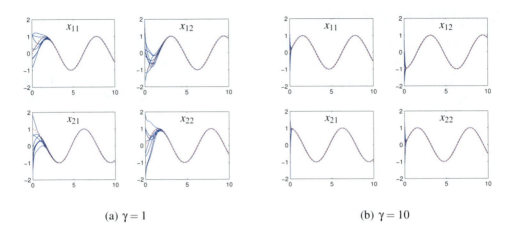

Figure 5.1. Online solution of Sylvester equation (5.1) by ZNN model (5.3).

5.2.4.1. Simulation of Convergence

To solve Sylvester equation (5.1) with the above coefficient matrices and to show the convergence characteristics of ZNN model (5.3), the following MATLAB code is used.

```
function ZnnConvergence(gamma)
if nargin < 1, gamma=1; end
tspan=[0 10]; [m,n]=size(MatrixC(0));
options=odeset('Mass',@M,'MStateDep','none');
for iter=1:8
  x0=4*(rand(4,1)-0.5*ones(4,1));
  [t,x]=ode45(@ZnnRHS,tspan,x0,options,gamma);
  subplot(2,2,1);plot(t,x(:,1));hold on
  subplot(2,2,3);plot(t,x(:,2));hold on
  subplot(2,2,2);plot(t,x(:,3));hold on
  subplot(2,2,4);plot(t,x(:,4));hold on
end
subplot(2,2,1);plot(t,sin(t),'r:');
subplot(2,2,3);plot(t,cos(t),'r:');
subplot(2,2,2);plot(t,-cos(t),'r:');
subplot(2,2,4);plot(t,sin(t),'r:');
```

By using the above function "ZnnConvergence" with different input arguments $\gamma = 1$ and $\gamma = 10$, we can generate Figure 5.1. As seen from Figure 5.1, the design parameter γ has remarkable effectiveness on the convergence rate of ZNN model (5.3).

The computational error $\|X(t) - X^*(t)\|$ could also be used to monitor the network convergence. The MATLAB code to show the convergence properties of $\|X(t) - X^*(t)\|$ are given below, i.e., the user-defined functions "NormError" and "ZnnNormError".

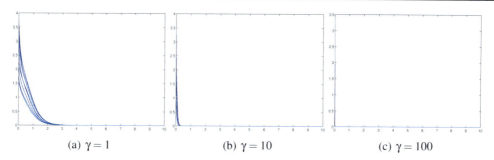

Figure 5.2. Convergence of computational error $\|X(t)-X^*(t)\|$ of ZNN model (5.3).

```
function NormError(x0,gamma)
tspan=[0 10];
options=odeset('Mass',@M,'MStateDep','none');
[t,x]=ode45(@ZnnRHS,tspan,x0,options,gamma);
Xstar=[sin(t) cos(t) -cos(t) sin(t)]';
err=x'-Xstar; total=length(t);
for i=1:total, nerr(i)=norm(err(:,i)); end
plot(t,nerr); hold on
```

```
function ZnnNormError(gamma)
if nargin<1, gamma=1; end
for i=1:8
    x0=4*(rand(4,1)-0.5*ones(4,1));
    NormError(x0,gamma);
end
```

By calling "ZnnNormError" with different design-parameter values (e.g., $\gamma = 1$, 10 and 100), we can generate Figure 5.2. It shows that starting from any initial state randomly selected in $[-2,2]$, the state matrices of neural network (5.3) all converge to the theoretical solution $X^*(t)$, while the computational errors $\|X(t)-X^*(t)\|$ all converge to zero. Moreover, such a convergence can be expedited by increasing γ. The larger the γ is, the faster the neural network (5.3) converges to the theoretical solution $X^*(t)$.

5.2.4.2. Simulation of Robustness

To show the robustness characteristics, the following matrix implementation error $\Delta_C(t)$ and the model-implementation error $\Delta_R(t)$ are added to ZNN (5.3) in a higher-frequency sinusoidal form (with $\varepsilon_1 = 0.05$ and $\varepsilon_2 = 0.5$); i.e., the resultant perturbed model (5.5).

$$\Delta_C(t) = \varepsilon_1 \begin{bmatrix} \cos 2t & 0 \\ 0 & \sin 2t \end{bmatrix}, \quad \Delta_R(t) = \varepsilon_2 \begin{bmatrix} \sin 2t & 0 \\ 0 & \cos 2t \end{bmatrix}.$$

As shown in Figure 5.3, even with imprecise matrix implementation $C(t) + \Delta_C(t)$ and large model-implementation error $\Delta_R(t)$, ZNN still performs well. Note that, in Figure 5.3, the theoretical solution $X^*(t)$ is denoted by dotted lines in red, while the neural network solution $X(t)$ is denoted by solid lines in blue. Similarly, we can show the computational error $\|X(t)-X^*(t)\|$ of the perturbed ZNN model (5.5). As seen from Figure 5.4, even with

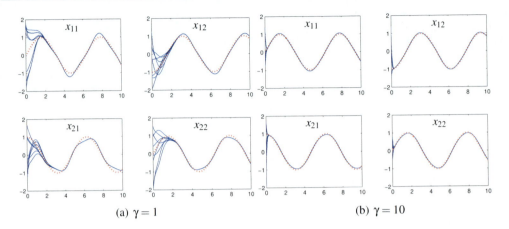

Figure 5.3. Online solution of Sylvester equation (5.1) by perturbed ZNN model (5.5) with large differentiation and implementation errors.

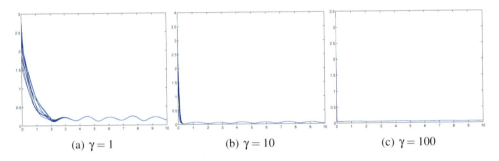

Figure 5.4. Convergence of $\|X(t) - X^*(t)\|$ of perturbed ZNN model (5.5).

large realization errors, the computational error $\|X(t) - X^*(t)\|$ synthesized by the perturbed ZNN model (5.5) is still bounded and very small. Moreover, as the design parameter γ increases from 1 to 100, the convergence is expedited and the steady-state computational error is decreased. It is worth mentioning that using power-sigmoid or sigmoid activation functions has a smaller steady-state residual error than using linear or pure power activation functions. Compared to the case of using linear or pure power activation functions, superior performance can be achieved by using power-sigmoid functions under the same design specification.

5.2.5. Simulink-Modeling Studies

While the above subsection presents the related MATLAB-coding simulation techniques, we now show an illustrative example by Simulink-modeling techniques in this subsection.

5.2.5.1. Illustrative Example

Let us consider the same example presented in Subsection 5.2.4.. Before simulating ZNN model (5.3), we may need to transform it into the following equation to lay a basis for the

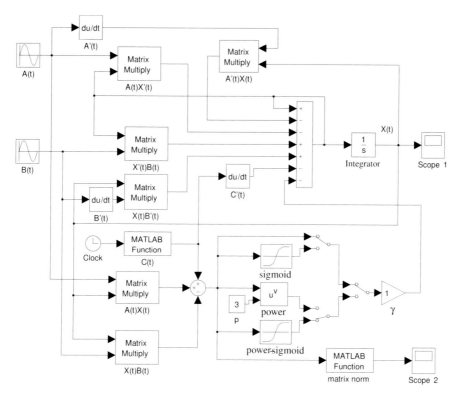

Figure 5.5. Overall Simulink model of ZNN (5.3) for online time-varying Sylvester equation solving. *Reproduced from W. Ma, Y. Zhang et al., MATLAB Simulink modeling and simulation of Zhang neural networks for online time-varying Sylvester equation solving, Figure 1, Proceedings of 2008 International Joint Conference on Neural Networks, pp. 286-290. ©[2008] IEEE. Reprinted, with permission.*

purpose of modeling and networking:

$$\dot{X}(t) = -A(t)\dot{X}(t) + \dot{X}(t)B(t) - \dot{A}(t)X(t) + X(t)\dot{B}(t) - \dot{C}(t) \\ - \gamma\Phi\big(A(t)X(t) - X(t)B(t) + C(t)\big) + \dot{X}(t).$$

According to the above transformation, the overall Simulink model of ZNN (5.3) could be established and shown in Figure 5.5.

5.2.5.2. Generating Coefficient Matrices

If A is a constant matrix, we can simply use the *Constant* block to generate it by specifying the matrix directly in the "Constant value" parameter of such a block. Note that the default option of "Interpret vector parameters as 1-D" should be deselected. However, ZNN model (5.3) is exploited for solving time-varying Sylvester equation, so other methods for constructing time-varying matrices $A(t)$, $B(t)$ and $C(t)$ in Simulink are introduced here as well. The *MATLAB Function* block together with the *Clock* block as its input could be used to generate a time-varying matrix. To do so, we just need to specify the matrix in the "MATALB Function" parameter; e.g., [sin(u) cos(u); -cos(u) sin(u)] for generating

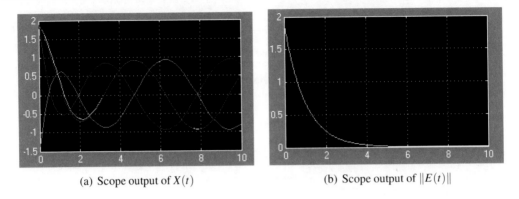

(a) Scope output of $X(t)$ (b) Scope output of $\|E(t)\|$

Figure 5.6. Scope outputs of ZNN (5.3) when parameter $\gamma = 1$. *Reproduced from W. Ma, Y. Zhang et al., MATLAB Simulink modeling and simulation of Zhang neural networks for online time-varying Sylvester equation solving, Figure 3, Proceedings of 2008 International Joint Conference on Neural Networks, pp. 286-290. ©[2008] IEEE. Reprinted, with permission.*

$A(t)$, and [0.1*sin(u)*sin(u)-1 -0.2*cos(u)*cos(u); 0.1*sin(u)*cos(u) 0.2*sin(u)*cos(u)-1] for generating $C(t)$.

It is worth mentioning specially that all entries of $A(t)$ and $B(t)$ we used here are sine/cosine functions, so we can also use the *Sine Wave* block to generate $A(t)$ by altering the "Phase" parameter as [0 pi/2; -pi/2 0]. To generate $B(t)$, the "Phase" parameter is set to be [0 0; 0 pi/2] and the "Amplitude" parameter is set to be [0.1 0; 0 0.2]. Other parameters could be set to be their default values. In addition, generating activation function $\Phi(\cdot)$ could be seen in Subsection 2.5.2..

5.2.5.3. Parameter Settings

Some important parameter settings of the blocks used in Figure 5.5 are specified as follows.

- For most *MATLAB Function* blocks, the default option "Collapse 2-D results to 1-D" should be deselected.

- The default options of the *Product* blocks should be altered from "Element-wise" to "Matrix" multiplication.

- To get different initial state X_0, we can set the "Initial condition" parameter of *Integrator* block to be random, e.g., "$4*(rand(2,2) - 1/2)$".

After the overall Simulink model depicted in Figure 5.5 has been built up, we need to change some of the default simulation options. To do so, we firstly open the "Configuration Parameters" dialog box, and then set the options as:

- Solver: "ode45".

- Max step size: 0.2; min step size: auto.

- Algebraic loop: none.

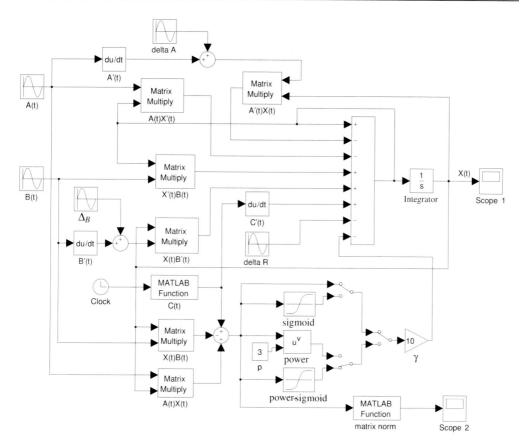

Figure 5.7. Overall Simulink model of imprecisely-constructed ZNN (5.5) for online time-varying Sylvester equation solving. *Reproduced from W. Ma, Y. Zhang et al., MATLAB Simulink modeling and simulation of Zhang neural networks for online time-varying Sylvester equation solving, Figure 4, Proceedings of 2008 International Joint Conference on Neural Networks, pp. 286-290. ⓒ[2008] IEEE. Reprinted, with permission.*

Other simulation options remain as their default settings. We could also change the start and stop time to meet our need.

After the execution of ZNN Simulink model in Figure 5.5, the scope output of $X(t)$ is depicted in Figure 5.6(a). As seen from Figure 5.6(a), starting from an initial state X_0, the state $X(t)$ converges to theoretical solution $X^*(t) = [\sin t, -\cos t; \cos t, \sin t]$. To monitor the neural-network convergence, the computational error $\|E(t)\|$ is also depicted in Figure 5.6(b), which converges to zero approximately in 5 seconds. The convergence could be expedited if we increase the design parameter γ; e.g., converging in 5ms if $\gamma = 10^3$.

It is observed from other simulation results that, compared to the linear or pure-power case, superior convergence can be achieved by using power-sigmoid activation function under the same parameter-specification. Specially saying, the convergence of state matrix $X(t)$ in the pure-power case is only asymptotic and thus appreciably slower than other activation-function cases.

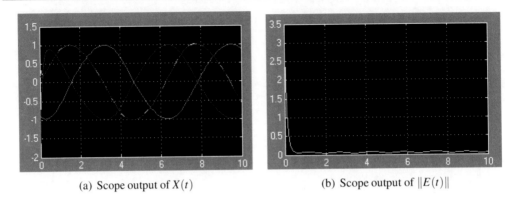

(a) Scope output of $X(t)$ (b) Scope output of $\|E(t)\|$

Figure 5.8. Scope outputs of imprecisely-constructed ZNN (5.5) when $\gamma = 10$. *Reproduced from W. Ma, Y. Zhang et al., MATLAB Simulink modeling and simulation of Zhang neural networks for online time-varying Sylvester equation solving, Figure 5, Proceedings of 2008 International Joint Conference on Neural Networks, pp. 286-290. ©[2008] IEEE. Reprinted, with permission.*

5.2.5.4. Modeling of Imprecisely-Constructed ZNN

As mentioned in Subsection 5.2.2., some errors may appear in the realization of ZNN (5.3). For this reason, we are to simulate the imprecisely-constructed ZNN (5.5) to show the robustness. The corresponding Simulink model is shown in Figure 5.7, in which the differentiation errors $\Delta_{\dot{A}}(t)$, $\Delta_{\dot{B}}(t)$ and model-implementation error $\Delta_R(t)$ are added in a higher-frequency sinusoidal form (with $\varepsilon_1 = 0.05$ and $\varepsilon_2 = 0.5$):

$$\Delta_{\dot{A}}(t) = \Delta_{\dot{B}}(t) = \varepsilon_1 \begin{bmatrix} \cos 3t & -\sin 3t \\ \sin 3t & \cos 3t \end{bmatrix}, \quad \Delta_R(t) = \varepsilon_2 \begin{bmatrix} \sin 2t & 0 \\ 0 & \cos 2t \end{bmatrix}.$$

The scope output of computational error $\|E(t)\|$ is shown in Figure 5.8(b), which is generated by imprecisely-constructed ZNN (5.5). It is natural that the state $X(t)$ of imprecisely-constructed ZNN (5.5) could only converge to an approximate solution nearby $X^*(t)$ of (5.1). In addition, the solution error is upper bounded by some small positive scalar. With the increase of design parameter γ, the solution error could be suppressed smaller. Simply put, for the imprecisely-implemented ZNN model (5.5), its neural state $X(t)$ becomes much closer to the theoretical solution $X^*(t)$ of (5.1). In summary, if γ is increased, the convergence is expedited and the neural-system is more robust in the sense that its steady-state computational error $\|E(t)\|$ is decreased. Compared to linear or pure-power case, superior robustness can be achieved as well by using power-sigmoid or sigmoid activation functions under the same design specification. The simulation results have substantiated the theoretical results presented in Subsection 5.2.2..

5.3. ZNN for Time-Varying Lyapunov Equation Solving

In this section, ZNN and GNN are developed for comparative purposes to solve online the Lyapunov equation with time-varying coefficient matrices.

5.3.1. Problem Formulation and Solvers

Lyapunov methods could be applied effectively to analyze various kinds of nonlinear and linear control systems, which result in the well-known Lyapunov matrix equation, $A^T P + PA = -C$. For its solution with constant coefficients, many numerical algorithms have been developed [23,24]. Neural approaches such as GNN could be employed as well [2,10,19]. However, as far as we know, there is little work on the time-varying Lyapunov equation solving:

$$A^T(t)P(t) + P(t)A(t) + C(t) = 0, \quad 0 \le t < +\infty, \tag{5.7}$$

where $A(t) \in \mathbb{R}^{n \times n}$ and $C(t) \in \mathbb{R}^{n \times n}$ are smoothly time-varying coefficient matrices, while $P(t)$ is the unknown matrix to be obtained. In this section, the objective is to solve the above time-varying Lyapunov equation (5.7) in real time t and in an error-free manner.

To solve time-varying Lyapunov equation (5.7), its coefficient matrices $A(t)$ and $C(t)$, together with their time-derivatives $\dot{A}(t)$ and $\dot{C}(t)$, are assumed to be known or could be estimated. Moreover, before solving Lyapunov equation (5.7), we provide the following solution-uniqueness condition and its related lemma.

Solution-Uniqueness Condition. There exists $\alpha > 0 \in \mathbb{R}$ such that, $\forall t \ge 0$,

$$\min \sigma \Big(I \otimes \big(A^T(t)A(t) \big) + A(t) \otimes A^T(t) + A^T(t) \otimes A(t) + \big(A^T(t)A(t) \big) \otimes I \Big) \ge \alpha,$$

where symbol $\sigma(\cdot)$ denotes the eigenvalues set of an input matrix, and I denotes an approximately-dimensioned identity-matrix. Symbol $\otimes$ denotes the Kronecker product of two matrices [13,25]; e.g., $P \otimes Q$ is a larger matrix made by replacing the ijth entry p_{ij} of P with matrix $p_{ij}Q$. □

Lemma 5.3.1. *Time-varying Lyapunov equation* (5.7) *satisfies the solution-uniqueness condition if and only if there are no two eigenvalues of bounded matrix* $A(t) \in \mathbb{R}^{n \times n}$, $\lambda_i(A(t))$ *and* $\lambda_j(A(t))$ *(with* $i, j = 1, 2, \cdots, n$*), adding up to zero at any time instant* t *[26]. Furthermore, the solution-uniqueness condition means that the minimal sum of any two eigenvalues of* $A(t)$ *are uniformly no less than* $\sqrt{\alpha}$. □

5.3.1.1. Zhang Neural Networks

Conventional GNN approaches [2,10,19] have been developed to compute exactly the time-invariant Lyapunov equation $A^T P + PA = -C$, of which the coefficients are constant matrices. Compared to the gradient-based method, by following Zhang *et al*'s approach [12,13], we could firstly define a matrix-valued error function associated to the Lyapunov equation (5.7):

$$E(t) := A^T(t)P(t) + P(t)A(t) + C(t) \in \mathbb{R}^{n \times n}, \tag{5.8}$$

where, if error function $E(t)$ equals zero, then $P(t)$ becomes the theoretical solution $P^*(t)$ of the time-varying Lyapunov equation (5.7).

Secondly, the error-function derivative $\dot{E}(t) \in \mathbb{R}^{n \times n}$ should be made such that every entry $e_{ij}(t) \in \mathbb{R}$ of $E(t) \in \mathbb{R}^{n \times n}$ converges to zero, $i, j = 1, 2, \cdots, n$. ZNN design-formula (5.2) could be adopted again.

Thirdly, expanding the design formula (5.2) could give the following implicit dynamics of ZNN, which solves Lyapunov equation (5.7):

$$A^T(t)\dot{P}(t) + \dot{P}(t)A(t) = -\dot{A}^T(t)P(t) - P(t)\dot{A}(t) - \dot{C}(t)$$
$$- \gamma\Phi\big(A^T(t)P(t) + P(t)A(t) + C(t)\big), \tag{5.9}$$

where $P(t)$, starting from any initial condition $P(0) = P_0 \in \mathbb{R}^{n \times n}$, is the activation state matrix corresponding to the theoretical solution $P^*(t)$ of (5.7).

In addition, ZNN model (5.9) possesses the following convergence properties [12, 13].

Proposition 5.3.1. *Consider smoothly time-varying coefficient-matrices $A(t) \in \mathbb{R}^{n \times n}$ and $C(t) \in \mathbb{R}^{n \times n}$ of time-varying Lyapunov equation (5.7), which satisfy the solution-uniqueness condition. If a monotonically-increasing odd activation-function-array $\Phi(\cdot)$ is used, then the neural-state matrix $P(t)$ of ZNN (5.9) starting from any initial state $P_0 \in \mathbb{R}^{n \times n}$ could converge to the theoretical solution $P^*(t)$ of time-varying Lyapunov equation (5.7).* $\square$

Proposition 5.3.2. *In addition to Proposition 5.3.1, if a linear activation-function-array $\Phi(\cdot)$ is used, then the neural-state matrix $P(t)$ of ZNN (5.9) starting from any initial state $P_0 \in \mathbb{R}^{n \times n}$ could exponentially converge to the theoretical solution $P^*(t)$ of (5.7) with rate γ. As compared to the situation of using linear activation functions, if the array $\Phi(\cdot)$ of power-sigmoid activation functions (1.7) is used, then superior global exponential convergence can be achieved for (5.9).* $\square$

5.3.1.2. Gradient Neural Networks

For comparison with the ZNN model presented in Subsubsection 5.3.1.1., we would like to develop the following GNN model, which handles intrinsically the situation of constant $A \in \mathbb{R}^{n \times n}$ and $C \in \mathbb{R}^{n \times n}$:

$$\dot{P}(t) = -\gamma\Phi\big(A(A^T P + PA + C) + (A^T P + PA + C)A^T\big), \tag{5.10}$$

where processing-array $\Phi(\cdot)$ can be defined the same as in Subsubsection 5.3.1.1..

Comparing ZNN (5.9) and GNN (5.10), we have the following important remarks.

1) ZNN (5.9) is designed based on the elimination of every entry of the matrix-valued error function $E(t) = A^T(t)P(t) + P(t)A(t) + C(t)$. In contrast, GNN (5.10) is designed through the elimination of scalar-valued norm-based error function $\|A^T P + PA + C\|_F^2/2$, where coefficient-matrices A and C could only be constant.

2) ZNN (5.9) methodically and systematically exploits the time-derivative information of coefficient matrices $A(t)$ and $C(t)$ during its real-time solving process. This is the reason why ZNN (5.9) could globally exponentially converge to the exact solution of a time-varying problem. On the other hand, GNN (5.10) has not exploited such important time-derivative information, and thus may not be effective on solving online such time-varying problems.

3) ZNN (5.9) is depicted in an implicit dynamics, i.e., $A^T(t)\dot{P}(t) + A(t)\dot{P}(t) = \cdots$. In comparison, GNN (5.10) is depicted in an explicit dynamics (without any mass matrix), i.e., $\dot{P}(t) = \cdots$, which is normally associated with classical Hopfield-type recurrent neural networks.

5.3.2. MATLAB-Coding Simulation Techniques

While Subsection 5.3.1. presents the ZNN and GNN models together with related analysis results, we would like in this subsection to investigate the MATLAB-coding simulation techniques for such two models. To simulate the implicit-dynamic neural-network system (5.9), several important simulation techniques have to be employed, e.g., MATLAB routines "ode45" and "diff".

5.3.2.1. Kronecker Product and Vectorization

For the proposes of MATLAB-coding simulation, we have to transform the matrix-form differential equation (5.9) to a vector-form differential equation.

Theorem 5.3.1. *The matrix-form ZNN model* (5.9) *can be transformed to the following vector-form differential equation:*

$$M(t)\dot{p}(t) = -\dot{M}(t)p(t) - \text{vec}(\dot{C}(t)) - \gamma\Phi\big(M(t)p(t) + \text{vec}(C(t))\big), \qquad (5.11)$$

where mass matrix $M(t) := A^T(t) \otimes I + I \otimes A^T(t)$. Operator $\text{vec}(C(t)) \in \mathbb{R}^{n^2 \times 1}$ generates a column vector obtained by stacking all column vectors of $C(t)$ together. Integration vector $p(t) := \text{vec}(P(t)) \in \mathbb{R}^{n^2 \times 1}$. Activation-function-array $\Phi(\cdot)$ is defined the same as in (5.9) except for its dimensions as $\mathbb{R}^{n^2 \times 1}$.

Given matrix A, to generate the mass matrix $M = A^T \otimes I + I \otimes A^T$, we could exploit MATLAB routine "kron" in the following code (as an example):

```
function M=MatrixM(t,x)
A=[-1-1/2*cos(2*t) 1/2*sin(2*t); 1/2*sin(2*t) -1+1/2*cos(2*t)];
M=kron(A.',eye(size(A)))+kron(eye(size(A)),A.');
```

5.3.2.2. Obtaining Matrix Derivatives

To obtain the time derivative $\dot{M}(t)$ required in vector-form ZNN model (5.11), without loss of generality, we can use MATLAB routine "diff" for such $\dot{M}(t)$ and $\dot{C}(t)$ evaluation. However, to do so, a symbolic object "u" has to be constructed firstly, and then the MATLAB command "`DM=diff(MatrixM(u))`" generates the analytical form of $\dot{M}$. Finally, evaluating such an analytical form of $\dot{M}$ with a numerical value of t will give us the required value of $\dot{M}(t)$. MATLAB code for evaluating such a $\dot{M}(t)$ is shown as follows, and we can generate $\dot{C}(t)$ similarly.

```
function output=DiffM(t,x)
syms u; DM=diff(MatrixM(u));
u=t; output=eval(DM);
```

5.3.2.3. ZNN Right-Hand Side

By using MATLAB routine "reshape", the vectorization of a matrix could be done readily. For example, the following MATLAB code is used to evaluate the right-hand side of vector-form ZNN model (5.11).

```
function output=ZnnRightHandSide(t,x,gamma)
if nargin==2, gamma=1; end
C=MatrixC(t); [m,n]=size(C); dotC=DiffC(t,x);
vecC=reshape(C,m*n,1); vecDotC=reshape(dotC,m*n,1);
M=MatrixM(t,x); dotM=DiffM(t,x);
output=-dotM*x-gamma*AFMpowersigmoid(M*x+vecC)-vecDotC;
```

In a similar way, we could have the MATLAB code which evaluates the vectorized right-hand side of GNN model (5.10).

5.3.2.4. ODE with Mass Matrix

For the simulation of ZNN (5.9) [equivalently (5.11)], the MATLAB routine "ode45" is preferred. This is because "ode45" can solve initial-value ordinary differential equation (ODE) problems with nonsingular mass matrices, i.e., $M(t,p)\dot{p} = g(t,p)$, $p(0) = p_0$, where matrix $M(t,p)$ on the left-hand side of such an equation is the so-called mass matrix, and the right-hand side $g(t,p) := -\dot{M}p - \gamma\Phi(Mp + \text{vec}(C)) - \text{vec}(\dot{C})$ in this section.

To solve the ODE problem with a mass matrix, the MATLAB routine "odeset" is also involved. Its "Mass" property should be assigned to be the function handle "@MatrixM", which returns the evaluation of mass matrix $M(t,p)$. Note that, if 1) $M(t,p)$ does not depend on state variable p and 2) the function "MatrixM" is to be invoked with only one input argument t, then the "MStateDep" property of "odeset" should be set "none". The following MATLAB code solves such an initial-value ODE problem with state-independent mass matrix $M(t)$ and from a random initial state p_0; i.e., ZNN (5.11) [equivalently, (5.9)] handling (5.7).

```
tspan=[0 10]; x0=4*(rand(4,1)-0.5*ones(4,1));
options=odeset('Mass',@MatrixM,'MStateDep','none');
[t,x]=ode45(@ZnnRightHandSide,tspan,x0,options);
```

5.3.3. Illustrative Examples by MATLAB Code

In this subsection, for illustrative purposes, a computer-simulation example is presented to show the performance and efficacy of ZNN model (5.9) solving (5.7). For comparison, conventional GNN (5.10) is simulated as well. Let us consider Lyapunov equation (5.7) with the following time-varying coefficient matrices $A(t)$ and $C(t)$:

$$A(t) = \begin{bmatrix} -1-0.5\cos 2t & 0.5\sin 2t \\ 0.5\sin 2t & -1+0.5\cos 2t \end{bmatrix}, \ C(t) = \begin{bmatrix} \sin 2t & \cos 2t \\ -\cos 2t & \sin 2t \end{bmatrix}.$$

Simple manipulations can verify that theoretical solution $P^*(t)$ could be

$$P^*(t) = \begin{bmatrix} -(\sin 2t)(-2+\cos 2t)/3 & -(2\cos 2t-1)(2+\cos 2t)/6 \\ -(2\cos 2t+1)(-2+\cos 2t)/6 & (2+\cos 2t)(\sin 2t)/3 \end{bmatrix}, \quad (5.12)$$

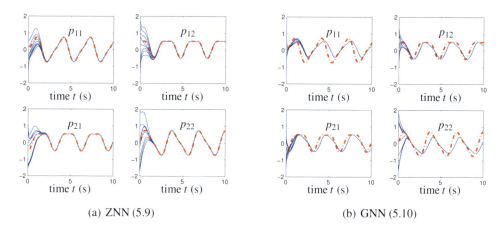

Figure 5.9. Online ZNN and GNN solution of time-varying Lyapunov equation (5.7), where dash-dotted red lines correspond to theoretical solution $P^*(t)$, and solid blue lines correspond to neural-solution $P(t)$ with $\gamma = 1$. *Reproduced from Y. Zhang, S. Yue et al., MATLAB simulation and comparison of Zhang neural network and gradient neural network for time-varying Lyapunov equation solving, Figure 1, F. Sun et al. (Eds.): ISNN 2008, Part I, LNCS 5263, pp. 117-127, 2008. ©Springer-Verlag Berlin Heidelberg 2008. With kind permission of Springer Science+Business Media.*

which is used to check the correctness of the neural-network solutions.

5.3.3.1. Neural-State Simulation

To solve Lyapunov equation (5.7) with the above time-varying coefficients via ZNN (5.9), the following MATLAB code could be used. Figure 5.9(a) is then generated.

```
function ZnnConvergence(gamma)
tspan=[0 10];
options=odeset('Mass',@MatrixM,'MStateDep','none');
for iter=1:8
    x0=4*(rand(4,1)-0.5*ones(4,1));
    [t,x]=ode45(@ZnnRightHandSide,tspan,x0,options,gamma);
    subplot(2,2,1);plot(t,x(:,1));hold on
    subplot(2,2,3);plot(t,x(:,2));hold on
    subplot(2,2,2);plot(t,x(:,3));hold on
    subplot(2,2,4);plot(t,x(:,4));hold on
end
```

As illustrated in Figure 5.9(a), starting from any random initial state in the range of [-2,2], ZNN could always converge to the theoretical solution (5.12) rapidly where the design parameter γ is set to be 1. For comparison, GNN could also be employed to solve Lyapunov equation (5.7) with the same time-varying coefficients and under the same design-parameter value by using MATLAB code below.

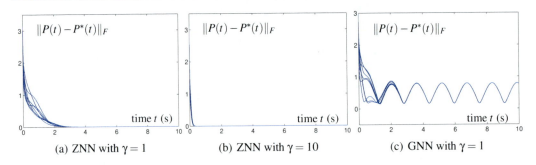

Figure 5.10. Convergence performance of ZNN and GNN solution errors $\|P(t) - P^*(t)\|_F$. *Reproduced from Y. Zhang, S. Yue et al., MATLAB simulation and comparison of Zhang neural network and gradient neural network for time-varying Lyapunov equation solving, Figure 2, F. Sun et al. (Eds.): ISNN 2008, Part I, LNCS 5263, pp. 117-127, 2008. ©Springer-Verlag Berlin Heidelberg 2008. With kind permission of Springer Science+Business Media.*

```
function GnnConvergence(gamma)
tspan=[0 10]; options=odeset();
for iter=1:8
    x0=4*(rand(4,1)-0.5*ones(4,1));
    [t,x]=ode45(@GnnRightHandSide,tspan,x0,options,gamma);
    subplot(2,2,1);plot(t,x(:,1));hold on
    subplot(2,2,3);plot(t,x(:,2));hold on
    subplot(2,2,2);plot(t,x(:,3));hold on
    subplot(2,2,4);plot(t,x(:,4));hold on
end
```

The GNN-convergence is, then, shown in Figure 5.9(b). It follows from Figure 5.9(b) that the GNN solution (denoted by solid blue curves) does not fit well with the theoretical time-varying solution $P^*(t)$ (denoted by dash-dotted red curves). Simply put, there always exists a residual error between the GNN solution and the theoretical solution $P^*(t)$ to time-varying Lyapunov equation (5.7). This is evidently because the time-derivative information of $A(t)$ and $C(t)$ has not been utilized in these gradient-descent computational schemes.

5.3.3.2. Computation-Error Simulation

The computational error $\|P(t) - P^*(t)\|_F$ could also be used to monitor the neural-network convergence. The following MATLAB code shows the ZNN convergence of error $\|P(t) - P^*(t)\|_F$. User-defined MATLAB function "ZnnNormError" depends on another user-defined function "NormError", which could finally generate Figure 5.10(a) and (b). The MATLAB code of applying GNN to generating Figure 5.10(c) is similar but omitted here.

```
function NormError(x0,gamma)
tspan=[0 10]; options=odeset('Mass',@MatrixM,...
 'MStateDep','none','RelTol',1e-6,'AbsTol',1e-10);
[t,x]=ode45(@ZnnRightHandSide,tspan,x0,options,gamma);
xs=[-1/3*sin(2*t).*(-2+cos(2*t))...
    -1/6*(2*cos(2*t)-1).*(2+cos(2*t))...
    -1/6*(2*cos(2*t)+1).*(-2+cos(2*t))...
    1/3*(2+cos(2*t)).*sin(2*t)]';
err=x'-xs; total=length(t);
for i=1:total,
    nerr(i)=norm(err(:,i));
end
plot(t,nerr); hold on
```

```
function ZnnNormError(gamma)
if nargin<1, gamma = 10; end
for i=1:8
    x0=4*(rand(4,1)-0.5*ones(4,1)); NormError(x0,gamma);
end
```

Figure 5.10 illustrates the computational errors of ZNN (5.9) and GNN (5.10) starting from random initial states. It can be seen from the left two graphs of the figure and other simulation data that, by using ZNN (5.9) to solve online the time-varying Lyapunov equation (5.7), the steady-state solution errors of $\|P(t) - P^*(t)\|_F$ are less than 3×10^{-13} (computed as of $t = 100$s), 2×10^{-14} and 1.4×10^{-14}, which correspond respectively to the usage of design parameter $\gamma = 1$, 10 and 100. In contrast, it is seen from the rightmost graph of Figure 5.10 and simulation data that, by using GNN (5.10) to solve online the time-varying Lyapunov equation (5.7) under the same conditions, the computational error $\|P(t) - P^*(t)\|_F$ is considerably large, with steady-state solution errors being around 0.7947, 0.0971 and 0.00965 respectively for $\gamma = 1$, 10 and 100. Furthermore, as seen from the figure, the neural-network convergence can be expedited effectively by increasing the value of design parameter γ, which substantiates the aforementioned exponential-convergence property. For example, the convergence time with no appreciable solution-error is less than 4μs when design parameter $\gamma = 10^6$.

In summary, Figures 5.9 and 5.10 show that ZNN could globally exponentially converge to the theoretical solution $P^*(t)$ of time-varying Lyapunov matrix equation (5.12), whereas the GNN solution-error $\|P(t) - P^*(t)\|_F$ is much larger.

5.3.4. Simulink Modeling and Verification

As a graphical-design based modeling tool, MATLAB Simulink exploits existing function blocks to construct mathematical and logical models as well as process flow. A Simulink model is a representation of the design or implementation of a system satisfying a set of requirements. Following Subsection 5.3.1. about ZNN and GNN models, the techniques for modeling and verifying them are investigated in terms of MATLAB Simulink environment.

5.3.4.1. Simulink Blocks

MATLAB Simulink contains a comprehensive block library including sinks, sources, linear and nonlinear components, as well as connectors. The blocks generally used to construct ZNN (5.9) and GNN (5.10) are discussed, for readers' convenience, as follows.

- The *Constant* block generates a constant scalar or matrix specified by its parameter "Constant value".

- The *Gain* block could be used to scale the neural-network convergence, e.g., as a scaling parameter γ to scale the convergence rate of neural networks.

- The *Product* block provides two types of multiplication, either element-wise or matrix-wise product.

- The *Subsystem* block is used to construct the sigmoid and power-sigmoid activation-function array, which makes the whole system more readable.

- The *MATLAB Fcn* block can be used to 1) compute the matrix norm, or 2) generate time-varying matrices $A(t)$ and $C(t)$ with the output-results of *Clock* blocks as their input-signals.

- The *Math Function* block could perform various general-purpose mathematical operations, and is employed in this section for transposing a matrix.

- The *Switch* block is used in constructing the power-sigmoid subsystem in this section. "u2>=Threshold" is chosen for the option "Criteria for passing first input", and the value of "Threshold" is set to 1. Simply put, if the 2nd input of the *Switch* block is not smaller than 1, then the 1st input of the *Switch* block (i.e., the power activation-function value) will be passed to the output. Otherwise, the 3rd input of the *Switch* block (i.e., the sigmoid function value) will be passed to the output.

- The *Integrator* block makes continuous-time integration on the input signals. In this section, we could set its "Initial condition" as "$4*(rand(2,2)-1/2)$" to generate $P_0 \in \mathbb{R}^{2 \times 2}$ randomly between -2 and 2.

Other Simulink blocks involved are to be introduced in the ensuring subsubsection one by one due to their importance. By interconnecting these basic function blocks and setting appropriate block-parameters, the overall modeling of ZNN (5.9) and GNN (5.10) can then be built up.

5.3.4.2. Parameter Settings

Some important parameter settings of those blocks used in Figures 5.11 and 5.12 have to (or could) be specified as follows.

- For most *MATLAB Function* blocks, the default option "Collapse 2-D results to 1-D" should be deselected.

ZNN for Linear Time-Varying Matrix Equation Solving

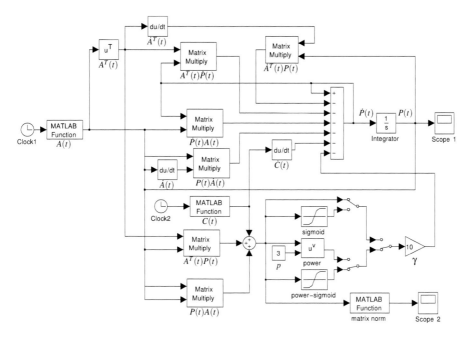

Figure 5.11. Overall Simulink model of ZNN (5.9) for online time-varying Lyapunov equation solving. *Reproduced from Y. Zhang, K. Chen et al., Simulink modeling and comparison of Zhang neural networks and gradient neural networks for time-varying Lyapunov equation solving, Figure 2, Proceedings of the Fourth International Conference on Natural Computation, pp. 521-525. ©[2008] IEEE. Reprinted, with permission.*

- The default option "Element-wise" of the *Product* blocks should be altered to "Matrix" multiplication.

- To generate different initial states P_0, we could set the "Initial condition" parameter of *Integrator* block to be the random one, e.g., using "$4*(rand(2,2) - 1/2)$".

After the overall Simulink models depicted in Figures 5.11 and 5.12 have been built up, we may have to change some of the default simulation options by firstly opening the "Configuration Parameters" dialog box and then setting the options:

- Solver: "ode45 (Dormand-Prince)";

- Absolute tolerance: "1e-6" (i.e., 10^{-6});

- Relative tolerance: "1e-6" (i.e., 10^{-6}).

- Algebraic loop: "none" (or "warning").

5.3.5. An Illustrative Example by Simulink-Modeling Approach

In this subsection, for illustrative purposes, a computer-verification example is presented to show the performance and efficacy of ZNN model (5.9) solving time-varying Lyapunov

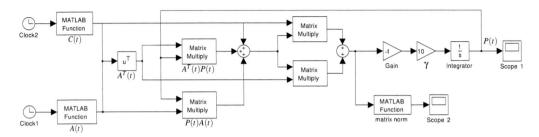

Figure 5.12. Overall Simulink model of GNN (5.10) for online time-varying Lyapunov equation solving. *Reproduced from Y. Zhang, K. Chen et al., Simulink modeling and comparison of Zhang neural networks and gradient neural networks for time-varying Lyapunov equation solving, Figure 3, Proceedings of the Fourth International Conference on Natural Computation, pp. 521-525. ©[2008] IEEE. Reprinted, with permission.*

equation (5.7). For comparison, GNN (5.10) is verified as well. Let us consider Lyapunov matrix equation (5.7) with the following time-varying coefficients $A(t)$ and $C(t)$:

$$A(t) = \begin{bmatrix} -1+1.5c^2 & 1-1.5s \times c \\ -1-1.5s \times c & -1+1.5s^2 \end{bmatrix},$$

$$C(t) = \begin{bmatrix} 2s-3s \times c^2 & -0.5c(1-6s^2); \\ 0.5c(4-3c^2+3s^2) & 0.5s(1-3s^2+3c^2) \end{bmatrix},$$

where s and c denote, for the moment, $\sin(t)$ and $\cos(t)$, respectively. Simple algebraic manipulations could show that the theoretical solution $P^*(t)$ to the time-varying Lyapunov equation (5.7) with the above coefficients is

$$P^*(t) = \begin{bmatrix} \sin(t) & -\cos(t) \\ \cos(t) & \sin(t) \end{bmatrix},$$

which is given just for checking the correctness of neural-network solutions.

5.3.5.1. Convergence Verification

As the Simulink models of ZNN and GNN have already been established, the performance of such two presented neural networks could be simulated and verified readily. As illustrated in Figure 5.13(a), starting from a randomly-generated initial state in the range of [-2,2], ZNN (5.9) could rapidly converge to the theoretical solution $P^*(t)$, where design parameter $\gamma = 10$. For comparison, GNN (5.10) is employed as well to solve Lyapunov matrix equation (5.7) with the same time-varying coefficients and γ-value as those in the ZNN situation. As seen from Figure 5.13(b), the gradient neural network could not track the theoretical solution $P^*(t)$ exactly, because such a neural model makes no use of the time-derivative information of $A(t)$ and $C(t)$.

5.3.5.2. Solution-Error Verification

Solution error $\|P(t) - P^*(t)\|_F$ could be used in this example to monitor the neural-network convergence performance. Figure 5.14 illustrates the solution errors of both ZNN (5.9) and

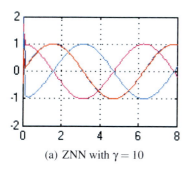
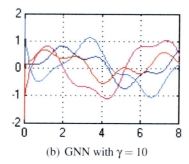

Figure 5.13. Scope outputs of $P(t)$. *Reproduced from Y. Zhang, K. Chen et al., Simulink modeling and comparison of Zhang neural networks and gradient neural networks for time-varying Lyapunov equation solving, Figure 4, Proceedings of the Fourth International Conference on Natural Computation, pp. 521-525. ©[2008] IEEE. Reprinted, with permission.*

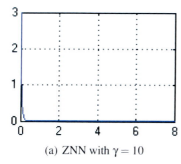
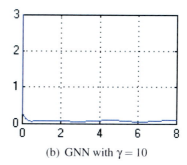

Figure 5.14. Scope outputs of $\|P(t) - P^*(t)\|_F$. *Reproduced from Y. Zhang, K. Chen et al., Simulink modeling and comparison of Zhang neural networks and gradient neural networks for time-varying Lyapunov equation solving, Figure 5, Proceedings of the Fourth International Conference on Natural Computation, pp. 521-525. ©[2008] IEEE. Reprinted, with permission.*

GNN (5.10) starting from random initial states, where the solution error $\|P(t) - P^*(t)\|_F$ by GNN model (5.10) is much larger in comparison with the one by ZNN model (5.9) when both solving online time-varying Lyapunov equation (5.7) under the same conditions. Therefore, from the above and from other verification results we may summarize that the resultant ZNN model performs much better on solving the time-varying matrix (and/or vector) problems, as compared to gradient-based neural networks.

5.4. ZNN for Time-Varying Linear Matrix Equation $AXB = C$ Solving

In this section, ZNN model is designed and exploited to solve online the linear time-varying matrix equation, $A(t)X(t)B(t) - C(t) = 0$, in real time t, of which the solution is compared with conventional GNN model's.

5.4.1. Problem Formulation and Solvers

Consider a time-varying linear matrix equation,

$$A(t)X(t)B(t) - C(t) = 0, \ 0 \leq t < +\infty \tag{5.13}$$

where $A(t) \in \mathbb{R}^{m \times m}$, $B(t) \in \mathbb{R}^{n \times n}$ and $C(t) \in \mathbb{R}^{m \times n}$ are smoothly time-varying coefficient matrices, which, together with their derivatives, are assumed to be estimated accurately. $X(t) \in \mathbb{R}^{m \times n}$ is the unknown matrix to be obtained.

To lay a basis for further discussion, the unique-solution condition is assumed below.

Unique-Solution Condition. Linear matrix equation (5.13) is uniquely solvable, if all eigenvalues of matrices $A(t) \in \mathbb{R}^{m \times m}$ and $B(t) \in \mathbb{R}^{n \times n}$ are nonzero at any time instant $t \in [0, +\infty)$. $\qquad\square$

It follows from Kronecker-product and vectorization techniques [13, 25] that time-varying linear matrix equation (5.13) is equivalent to

$$\left(B^T(t) \otimes A(t)\right)\text{vec}\left(X(t)\right) = \text{vec}\left(C(t)\right). \tag{5.14}$$

Symbol $\otimes$ denotes the Kronecker product; i.e., $P \otimes Q$ is a large matrix made by replacing the ijth entry p_{ij} of P with the matrix $p_{ij}Q$. Operator $\text{vec}(X) \in \mathbb{R}^{mn}$ generates a column vector obtained by stacking all column vectors of X together.

If the unique-solution condition holds true, then it is evident that, at any time instant t, the mn-dimensional square matrix $B^T(t) \otimes A(t)$ is nonsingular with each eigenvalue being nonzero [25]. Thus, a unique solution to (5.13) exists. Moreover, the unique-solution condition equals that a positive real number $\alpha > 0$ exists such that

$$\left(B^T \otimes A\right)^T \left(B^T \otimes A\right) \geq \alpha I, \ \forall t \geq 0, \tag{5.15}$$

where I denotes the identity matrix (with its dimension being mn in the context of this equation).

5.4.1.1. Zhang Neural Network

To solve online the time-varying linear matrix equation (5.13), we could develop a recurrent neural network by using the following design method by Zhang *et al* [10–12].

To monitor the equation-solving process, a matrix-valued error function $E(t)$ is firstly defined below, instead of a scalar-valued norm-based energy function usually associated with gradient-based neural network.

$$E(t) := A(t)X(t)B(t) - C(t) \in \mathbb{R}^{m \times n}. \tag{5.16}$$

Then, combining the design formula (5.2) and equation (5.16) could lead to the following implicit dynamics of the resultant ZNN model for the online solution of time-varying linear matrix equation (5.13):

$$A\dot{X}B = -\dot{A}XB - AX\dot{B} + \dot{C} - \gamma\Phi\left(AXB - C\right), \tag{5.17}$$

where $X(t)$, starting from an initial state $X(0) = X_0 \in \mathbb{R}^{m \times n}$, is the activation state matrix corresponding to theoretical solution $X^*(t)$ of (5.13). We use $\dot{A} \in \mathbb{R}^{m \times m}$, $\dot{B} \in \mathbb{R}^{n \times n}$ and $\dot{C} \in \mathbb{R}^{m \times n}$ to denote the known analytical forms or measurements of the time derivatives of matrices A, B and C, respectively. In addition, when using the linear activation-function array $\Phi(E) = E$, ZNN model (5.17) reduces to the following linear one:

$$A\dot{X}B = -\dot{A}XB - AX\dot{B} - \gamma AXB + \left(\dot{C} + \gamma C\right). \tag{5.18}$$

where argument t is dropped for presentation convenience, and so is in (5.17) and equations afterwards in this section.

5.4.1.2. Gradient Neural Network

For comparison, it is worth mentioning here that we can develop a GNN model to solve online linear matrix equation (5.13). However, similar to almost all the numerical algorithms and neural-dynamic computational schemes [2,3,6,10,17,19,20], the GNN is designed intrinsically for constant coefficient matrices A, B and C. It generally belongs to the gradient descent method in optimization [19,20,27] and can be designed by the following procedure.

- Firstly, a scalar-valued energy function, such as $\|AXB - C\|_F^2 / 2$ with Frobenius norm $\|C\|_F := \sqrt{\operatorname{trace}(C^T C)}$, is constructed such that its minimum point is the solution of equation (5.13) with constant coefficients.

- Secondly, an algorithm is designed to evolve along a descent direction of this error function until the minimum is reached. The typical descent direction is the negative gradient of $\|AXB - C\|_F^2 / 2$, i.e.,

$$-\frac{\partial \left(\|AXB - C\|_F^2 / 2\right)}{\partial X} = -A^T (AXB - C) B^T.$$

- Thirdly, by using the above negative gradient to construct the neural network, we could have a linear GNN model, $\dot{X} = -\gamma A^T (AXB - C) B^T$, and a general nonlinear GNN model as the following:

$$\dot{X}(t) = -\gamma A^T \Phi (AXB - C) B^T, \tag{5.19}$$

where convergence results could be achieved only for the situation of using constant A, B and C [2,10,19].

5.4.1.3. Method and Model Comparisons

In this subsubsection, we would like to compare the two design methods and models of ZNN (5.17) and GNN (5.19), which are exploited for the online solution of time-varying linear matrix equation (5.13). The differences lie in the following facts.

Firstly, ZNN model (5.17) is designed based on the elimination of every entry of the matrix-valued error function $E(t) = A(t)X(t)B(t) - C(t)$. In contrast, GNN model (5.19) is designed based on the elimination of the norm-based scalar-valued energy function $\|AXB - C\|_F^2/2$ (note that here A, B and C could only be constant in the design and analysis of GNN models).

Secondly, ZNN (5.17) methodically and systematically exploits the time-derivative information of coefficient matrices $A(t)$, $B(t)$ and $C(t)$ during its real-time solving process. This is the reason why ZNN (5.17) could globally exponentially converge to the exact solution of a time-varying problem. In contrast, GNN (5.19) has not exploited such important information, and thus may not be effective on solving such a time-varying problem.

Thirdly, ZNN (5.17) is depicted in an implicit dynamics, i.e., $A(t)\dot{X}(t)B(t) = \cdots$. In contrast, GNN (5.19) is depicted in an explicit dynamics, i.e., $\dot{X}(t) = \cdots$, which is usually associated with classic Hopfield-type recurrent neural networks.

5.4.1.4. ZNN Convergence

While the above subsubsections present a general description of Zhang neural network models (5.17) and (5.18), detailed design consideration and main theoretical results about their global exponential convergence are given in this subsubsection. The following propositions might be important.

Proposition 5.4.1. *Consider smoothly time-varying coefficient matrices $A(t) \in \mathbb{R}^{m \times m}$, $B(t) \in \mathbb{R}^{n \times n}$ and $C(t) \in \mathbb{R}^{m \times n}$ of linear matrix equation (5.13), with the unique-solution condition satisfied. If a monotonically-increasing odd activation function processing-array $\Phi(\cdot)$ is used, then the state matrix $X(t)$ of ZNN model (5.17), starting from any initial state $X_0 \in \mathbb{R}^{m \times n}$, converges to the theoretical solution $X^*(t)$ of linear time-varying matrix equation (5.13).* $\square$

Proposition 5.4.2. *In addition to Proposition 5.4.1, ZNN (5.17) possesses the following properties. If linear activation function $\phi(e_{ij}) = e_{ij}$ is used, then global exponential convergence [in terms of error $E(t)$] could be achieved for (5.17) with rate γ. If power-sigmoid function (1.7) is used, then superior convergence for error range $(-\infty, +\infty)$ could be achieved for (5.17), as compared to the linear case.* $\square$

5.4.2. MATLAB-Coding Simulation Techniques

In this subsection, we investigate the MATLAB simulation techniques for ZNN (5.17).

5.4.2.1. Kronecker Product and Vectorization

For the proposes of MATLAB simulation, we have to transform the matrix-form differential equation (5.17) [including (5.18) as a special case] into a vector-form differential equation.

Proposition 5.4.3. *The matrix-form ZNN model (5.17) can be transformed to the vector-form differential equation:*

$$M(t)\dot{x}(t) = -\dot{M}(t)x(t) + \text{vec}(\dot{C}(t)) - \gamma\Phi(M(t)x(t) - \text{vec}(C(t))), \qquad (5.20)$$

where mass matrix $M(t) := B^T(t) \otimes A(t)$, integration vector $x(t) := \text{vec}(X(t))$, and activation-function array $\Phi(\cdot)$ is defined the same as in (5.17) except that its dimensions are flexibly changed hereafter as $\Phi(\cdot) : \mathbb{R}^{mn \times 1} \to \mathbb{R}^{mn \times 1}$. $\qquad\qquad\qquad\qquad\qquad\square$

To generate the mass matrix $M = B^T \otimes A$ from matrices A and B, we could simply use MATLAB routine "kron" as follows (to be an example):

```
function output=MatrixM(t,x)
A=[sin(t) cos(t); -cos(t) sin(t)];
B=[2 cos(t)+sin(t) sin(t);-cos(t) 2 cos(t);...
   sin(t)-2 2-sin(t) 2-sin(t)];
output=kron(B',A);
```

Based on MATLAB routine "reshape", the vectorization of a matrix could be achieved readily as well. For example, the following MATLAB code is used to evaluate the right-hand side of differential equation (5.20) [equivalently, (5.17)], where "DiffA", "DiffB", "DiffC" and "DiffM" denote the MATLAB functions which evaluate the time derivatives of coefficients $A(t), B(t), C(t)$ and mass matrix $M(t)$.

```
function y=ZNNRightHandSide(t,x,gamma)
if nargin==2, gamma=1; end
A=LTVEmatrixA(t,x); B=LTVEmatrixB(t,x); C=LTVEmatrixC(t,x);
dotA=DiffA(t,x);  dotB=DiffB(t,x);  dotC=DiffC(t,x);
[m,n]=size(C); vecC=reshape(C,m*n,1);
vecDotC=reshape(dotC,m*n,1);
M1=kron(B',dotA); M2=kron(dotB',A); M3=kron(B',A);
y=-M1*x-M2*x+vecDotC-gamma*AFMpowersigmoid(M3*x-vecC);
```

Similar to Proposition 5.4.3, the matrix-form GNN (5.19) can be transformed to the following vector-form differential equation:

$$\dot{x}(t) = -\gamma \left(B(t) \otimes A^T(t) \right) \Phi\left(\left(B^T(t) \otimes A(t) \right) \text{vec}(X(t)) - \text{vec}(C(t)) \right). \tag{5.21}$$

The MATLAB code for simulating such a GNN is presented below for comparison. It returns the evaluation of the right-hand side of GNN model (5.21) [equivalently, (5.19)].

```
function y=GNNRightHandSide(t,x,gamma)
if nargin==2, gamma=1; end
A=LTVEmatrixA(t,x); B=LTVEmatrixB(t,x);
C=LTVEmatrixC(t,x);
[m,n]=size(C); vecC=reshape(C,m*n,1);
M3=kron(B,A'); M4=kron(B',A);
y=-gamma*M3*AFMpowersigmoid(M4*x-vecC);
```

5.4.2.2. ODE with Mass Matrix

For the simulation of ZNN model (5.17) [equivalently, (5.20)], the MATLAB routine "ode45" is preferred. This is because "ode45" can solve initial-value ordinary differential equation (ODE) problems with nonsingular mass matrices, e.g., $M(t,x)\dot{x} = g(t,x)$, $x(0) = x_0$, where matrix $M(t,x)$ on the left-hand side of such an equation is termed the mass matrix, and the right-hand side $g(t,x) := -\dot{M}x + \mathrm{vec}(\dot{C}) - \gamma\Phi(Mx - \mathrm{vec}(C))$ for our case here.

To solve an ODE problem with a mass matrix, the MATLAB routine "odeset" should also be used. Its "Mass" property should be assigned to be the function handle "@MatrixM", which returns the evaluation of mass matrix $M(t,x)$. Note that, if 1) $M(t,x)$ does not depend on state variable x and 2) the function "MatrixM" is to be invoked with only one input argument t, then the "MStateDep" property of "odeset" should be "none". For example, the following MATLAB code can be used to solve an initial-value ODE problem with state-independent mass matrix $M(t)$ and starting from a random initial state x_0.

```
tspan=[0 10];
x0=4*(rand(6,1)-0.5*ones(6,1));
options=odeset('Mass',@MatrixM,'MStateDep','none');
[t,x]=ode45(@ZNNRightHandSide,tspan,x0,options,gamma);
```

5.4.3. Illustrative Examples by MATLAB Code

For simulation and comparison purposes, let us consider linear matrix equation (5.13) with the following time-varying coefficient matrices $A(t)$, $B(t)$ and $C(t)$:

$$A(t) = \begin{bmatrix} \sin t & \cos t \\ -\cos t & \sin t \end{bmatrix}, \quad B(t) = \begin{bmatrix} 2 & \sin t + \cos t & \sin t \\ -\cos t & 2 & \cos t \\ \sin t - 2 & 2 - \sin t & 2 - \sin t \end{bmatrix},$$

and $C(t) = [2 + \sin t \cos t - 2\cos t, 3\cos t + \sin t - \sin t \cos t, \sin t + 2\cos t - \sin t \cos t; -\cos t + \sin^2 t - 2\sin t, 2 + 2\sin t - \sin^2 t, \cos t + 2\sin t - \sin^2 t]$. For checking the correctness of the neural solutions, the unique time-varying theoretical-solution $X^*(t)$ could be given for this specific problem as $[\sin t, -\cos t, 0; \cos t, \sin t, 1]$.

To solve linear matrix equation (5.13) with the above coefficients, both GNN model (5.19) and ZNN model (5.17) are employed with the following MATLAB codes.

```
function GNNConvergence(gamma)
for iter=1:5
  x0=4*(rand(6,1)-0.5*ones(6,1)); tspan=[0 10];
  [t,x]=ode45(@GNNRightHandSide,tspan,x0,gamma);
  xStar=[sin(t) cos(t) -cos(t) sin(t) ...
         zeros(length(t),1) ones(length(t),1)];
  for k=1:6
    j=[1 4 2 5 3 6]; subplot(2,3,j(k));
    plot(t,x(:,k)); hold on;
    plot(t,xStar(:,k),'r:'); hold on; end
end
```

ZNN for Linear Time-Varying Matrix Equation Solving

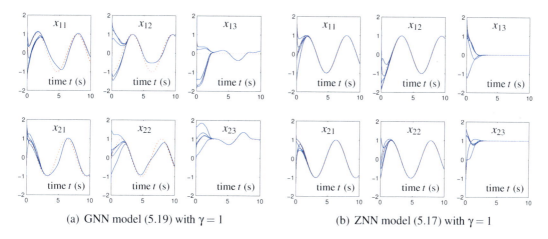

Figure 5.15. Online solution of linear time-varying matrix equation (5.13) by GNN (5.19) and ZNN (5.17) with $\gamma = 1$ and using power-sigmoid functions. *Reproduced from K. Chen, S. Yue and Y. Zhang, MATLAB simulation and comparison of Zhang neural network and gradient neural network for online solution of linear time-varying matrix equation $AXB - C = 0$, Figure 1, D.-S. Huang et al. (Eds.): ICIC 2008, LNAI 5227, pp. 68-75, 2008. ©Springer-Verlag Berlin Heidelberg 2008. With kind permission of Springer Science+Business Media.*

```
function ZNNConvergence(gamma)
tspan=[0 10];
options=odeset('Mass',@MatrixM,'MStateDep','none');
for iter=1:5
  x0=4*(rand(6,1)-0.5*ones(6,1));
  [t,x]=ode45(@ZNNRightHandSide,tspan,x0,options,gamma);
  xStar=[sin(t) cos(t) -cos(t) sin(t) ...
        zeros(length(t),1) ones(length(t),1)];
  for k=1:6
    j=[1 4 2 5 3 6]; subplot(2,3,j(k));
    plot(t,x(:,k)); hold on;
    plot(t,xStar(:,k),'r:'); hold on; end
end
```

By using the above user-defined MATLAB functions "GNNConvergence" and "ZNNConvergennce" with input arguments $\gamma = 1$, we can generate Figure 5.15(a) and (b). As seen from the figure, superior convergence has been achieved by ZNN model (5.17), as compared to GNN model (5.19) (which, at the present form, seems unable to solve exactly this time-varying problem). Moreover, design parameter γ has remarkable effectiveness on the ZNN convergence rate. By calling "ZNNConvergennce" with $\gamma = 10$ and 100, we can generate two more figures and observe that the larger the design parameter γ is, the faster the recurrent neural network converges. This point could also be shown by monitoring the computational error $\|X(t) - X^*(t)\|$, i.e., Figure 5.16.

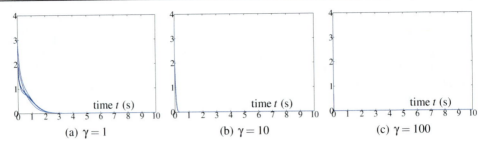

Figure 5.16. Convergence of computational error $\|X(t) - X^*(t)\|$ synthesized by ZNN (5.17) with different values of design parameter γ and using power-sigmoid functions. Reproduced from K. Chen, S. Yue and Y. Zhang, MATLAB simulation and comparison of Zhang neural network and gradient neural network for online solution of linear time-varying matrix equation $AXB - C = 0$, Figure 2, D.-S. Huang et al. (Eds.): ICIC 2008, LNAI 5227, pp. 68-75, 2008. ©Springer-Verlag Berlin Heidelberg 2008. With kind permission of Springer Science+Business Media.

In addition, it is worth mentioning that using power-sigmoid activation functions has a smaller steady-state residual error than using linear activation functions. Compared to the case of using linear activation functions, superior performance could also be achieved by using power-sigmoid functions under the same design parameters and conditions.

5.4.4. Simulink-Modeling Approach

While Subsection 5.4.1. presents the problem formulation, neural solvers and theoretical results of the ZNN model, in this subsection we investigate the modeling techniques for both ZNN (5.17) and GNN (5.19) to solve time-varying linear matrix equation (5.13).

For illustration, the following specific time-varying matrices $A(t)$, $B(t)$ and $C(t)$ are employed as a typical example:

$$A(t) = \begin{bmatrix} 2 & -\cos 3t & \sin 3t - 2 \\ \sin 3t + \cos 3t & 2 & 2 - \sin 3t \\ \sin 3t & \cos 3t & 2 - \sin 3t \end{bmatrix}, \quad B(t) = \begin{bmatrix} \sin 3t & -\cos 3t \\ \cos 3t & \sin 3t \end{bmatrix},$$

$C(t) = [2 + \sin 3t \cos 3t - 2\cos 3t, -\cos 3t + \sin^2 3t - 2\sin 3t; \sin 3t + 3\cos 3t - \sin 3t \cos 3t, 2 + 2\sin 3t - \sin^2 3t; \sin 3t + 2\cos 3t - \sin 3t \cos 3t, \cos 3t + 2\sin 3t - \sin^2 3t]$. Besides, the following theoretical solution $X^*(t)$ to matrix equation (5.13) with the above coefficients is shown here so as to verify the correctness of the neural-network solutions:

$$X^*(t) = \begin{bmatrix} \sin 3t & \cos 3t \\ -\cos 3t & \sin 3t \\ 0 & 1 \end{bmatrix}.$$

ZNN (5.17) and GNN (5.19) are then modeled based on MATLAB Simulink platform [21], with the overall ZNN and GNN models depicted in Figures 5.17 and 5.18, respectively. The detailed considerations are given below in constructing the two models.

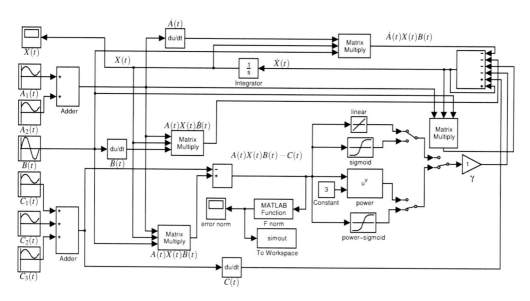

Figure 5.17. Overall model of ZNN model (5.19) for online equation $A(t)X(t)B(t) = C(t)$ solving. *Reproduced from N. Tan, K. Chen et al., Modeling, verification and comparison of Zhang neural net and gradient neural net for online solution of time-varying linear matrix equation, Figure 1, Proceedings of the 4th IEEE Conference on Industrial Electronics and Applications, pp. 3698-3703. ©[2009] IEEE. Reprinted, with permission.*

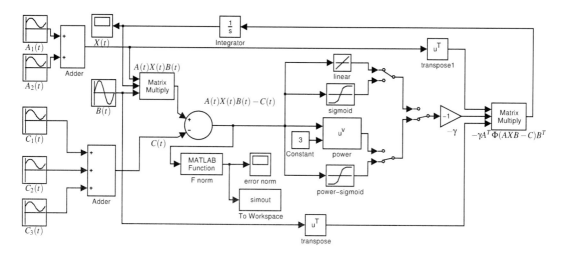

Figure 5.18. Overall model of GNN (5.17) exploited for online equation $A(t)X(t)B(t) = C(t)$ solving. *Reproduced from N. Tan, K. Chen et al., Modeling, verification and comparison of Zhang neural net and gradient neural net for online solution of time-varying linear matrix equation, Figure 2, Proceedings of the 4th IEEE Conference on Industrial Electronics and Applications, pp. 3698-3703. ©[2009] IEEE. Reprinted, with permission.*

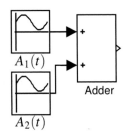

Figure 5.19. Matrix stream $A(t)$ constructed by adding $A_1(t)$ and $A_2(t)$. *Reproduced from N. Tan, K. Chen et al., Modeling, verification and comparison of Zhang neural net and gradient neural net for online solution of time-varying linear matrix equation, Figure 3, Proceedings of the 4th IEEE Conference on Industrial Electronics and Applications, pp. 3698-3703. ©[2009] IEEE. Reprinted, with permission.*

5.4.4.1. Generating Time-Varying Coefficients

For generating the three time-varying coefficient-matrices $A(t)$, $B(t)$ and $C(t)$, generally speaking, the *MATLAB Function* block (with a *Clock* block as its input) can be exploited. However, as coefficient-matrices $A(t)$, $B(t)$ and $C(t)$ consist of sine and cosine functions only, we could here use the *Sine Wave* block to generate them.

1) For matrix $A(t)$ generation, we can add simply $A_1(t)$ and $A_2(t)$ as depicted in Figure 5.19. The parameters are set as

- "Amplitude":
  † $A_1(t)$: $[0\ 1\ 1;\ 1\ 0\ 0;\ 1\ 1\ 0]$
  † $A_2(t)$: $[0\ 0\ 0;\ 1\ 0\ 1;\ 0\ 0\ 1]$

- "Bias":
  † $A_1(t)$: $[2\ 0\ 0;\ 0\ 2\ 2;\ 0\ 0\ 2]$
  † $A_2(t)$: $[0\ 0\ -2;\ 0\ 0\ 0;\ 0\ 0\ 0]$

- "Frequency": 3

- "Phase":
  † $A_1(t)$: $[0\ -pi/2\ 0;\ 0\ 0\ 0;\ 0\ pi/2\ 0]$
  † $A_2(t)$: $[0\ 0\ 0;\ pi/2\ 0\ pi;\ 0\ 0\ pi/2]$.

2) To generate time-varying coefficient matrix $B(t)$, the parameters could be set as (with no need for signals adding):

- "Amplitude": 1

- "Bias": 0

- "Frequency": 3

- "Phase": $[0\ -pi/2;\ pi/2\ 0]$.

3) Similar to time-varying matrix $A(t)$, the generation of time-varying matrix $C(t)$ could be achieved by adding $C_1(t)$, $C_2(t)$ and $C_3(t)$ with the following parameters:

- "Amplitude":
  - † $C_1(t)$: [0 1; 3 0; 1 1]
  - † $C_2(t)$: [0.5 − 0.5; 1 2; 2 2]
  - † $C_3(t)$: [2 2; −0.5 0.5; −0.5 0.5]

- "Bias":
  - † $C_1(t)$: [2 0; 0 2; 0 0]
  - † $C_2(t)$: [0 0.5; 0 0; 0 0]
  - † $C_3(t)$: [0 0; 0 − 0.5; 0 − 0.5]

- "Frequency":
  - † $C_1(t)$: [0 3; 3 0; 3 3]
  - † $C_2(t)$: [6 6; 3 3; 3 3]
  - † $C_3(t)$: [3 3; 6 6; 6 6]

- "Phase":
  - † $C_1(t)$: [0 − pi/2; −pi/2 0; 0 pi/2]
  - † $C_2(t)$: [0 pi/2; 0 0; pi/2 0]
  - † $C_3(t)$: [−pi/2 pi; 0 pi/2; 0 pi/2].

Besides, please note that the default option "Interpret vector parameters as 1-D" of all these blocks should be deselected.

5.4.4.2. Other Blocks Construction and Setting

In this subsubsection, more attention is paid to other blocks appearing in Figures 5.17 and 5.18 with appropriate parameters.

Four types of activation-function arrays are investigated in this section, where $p = 3$ and $\xi = 4$ are the default parameter-values used in the AF subsystem. 1) For the situation of using linear activation functions, the *purelin* block of the "Neural Network Blockset" could be used. 2) For the situation of using power functions, we could use the *Math Function* block by choosing "pow" in its function list with an input parameter set as 3. 3) We could similarly construct the subsystems of bipolar-sigmoid and power-sigmoid function arrays, which were mentioned in Subsection 2.5.1. and [28] with more details. To test the convergence performance using different activation functions, by double-clicking the *Manual Switch* blocks in Figures 5.17 and 5.18, we can readily choose the function array and test it.

To generate different initial states $X(0)$, we could set the "Initial condition" parameter of the *Integrator* block to be "4*(rand(3,2)-0.5*ones(3,2)". The default option "Elementwise" of *Product* blocks has to be changed to "Matrix" so as to perform the standard matrix multiplication. Besides, the option "Save format" of *simout* blocks should be "Array".

Before running the ZNN and GNN models, we had better pay attention to the following three important points by opening and using the "Configuration Parameters" dialog box: 1) max step size being 0.2; 2) relative tolerance being 1e-7 and absolute tolerance being 1e-7; 3) the check box in front of "States" as of the "Data Import/Export" option being selected. For the purpose of displaying the modeling results more clearly, we could also set the "StopFcn" as follows (which is of "Callbacks" in the "Model Properties" dialog box as started from the "File" pull-down menu).

```
figure(1)
for k=1:6
  j=[1 3 5 2 4 6];
  subplot(3,2,j(k)); axis([0 10 -2 2]); plot(tout, ...
    xout(:,k),'k'); hold on
end
subplot(3,2,1); axis([0 10 -2 2]); plot(tout, ...
  sin(3*tout),'k:'); text(5.5,-1.5,'time t (s)');
text(4.5,1.4,'X11');
subplot(3,2,3); axis([0 10 -2 2]); plot(tout, ...
  -cos(3*tout),'k:'); text(5.5,-1.5,'time t (s)');
text(4.5,1.4,'X21');
subplot(3,2,5); axis([0 10 -2 2]); plot(tout, ...
  zeros(length(tout),1),'k:');
text(5.5,-1.5,'time t (s)'); text(4.5,1.4,'X31');
subplot(3,2,2); axis([0 10 -2 2]);
plot(tout,cos(3*tout),'k:');
text(5.5,-1.5,'time t (s)'); text(4.5,1.4,'X12');
subplot(3,2,4); axis([0 10 -2 2]);
plot(tout,sin(3*tout),'k:');
text(5.5,-1.5,'time t (s)'); text(4.5,1.4,'X22');
subplot(3,2,6); axis([0 10 -2 2]);
plot(tout,ones(length(tout),1),'k:');
text(5.5,-1.5,'time t (s)'); text(4.5,1.4,'X32');
figure(2); axis([0 10 0 15]); plot(tout,simout,'k');
```

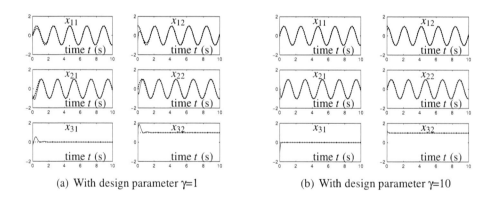

(a) With design parameter $\gamma=1$ (b) With design parameter $\gamma=10$

Figure 5.20. State trajectories of ZNN (5.17) using power-sigmoid functions with $\gamma = 1$ and 10, where dotted curves correspond to theoretical solution $X^*(t)$. *Reproduced from N. Tan, K. Chen et al., Modeling, verification and comparison of Zhang neural net and gradient neural net for online solution of time-varying linear matrix equation, Figure 4, Proceedings of the 4th IEEE Conference on Industrial Electronics and Applications, pp. 3698-3703. ©[2009] IEEE. Reprinted, with permission.*

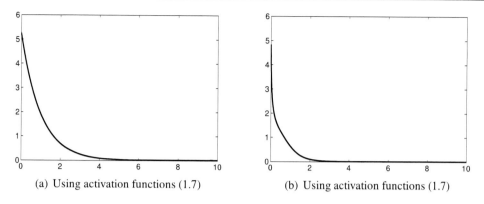

(a) Using activation functions (1.7)   (b) Using activation functions (1.7)

Figure 5.21. Residual error $\|A(t)X(t)B(t) - C(t)\|_F$ of ZNN (5.17) with $\gamma = 1$. *Reproduced from N. Tan, K. Chen et al., Modeling, verification and comparison of Zhang neural net and gradient neural net for online solution of time-varying linear matrix equation, Figure 5, Proceedings of the 4th IEEE Conference on Industrial Electronics and Applications, pp. 3698-3703. ©[2009] IEEE. Reprinted, with permission.*

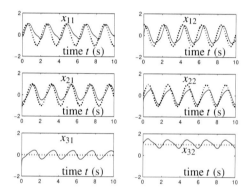

Figure 5.22. State trajectories of GNN (5.19) using activation functions (1.7) with $\gamma = 1$, where dotted curves correspond to $X^*(t)$. *Reproduced from N. Tan, K. Chen et al., Modeling, verification and comparison of Zhang neural net and gradient neural net for online solution of time-varying linear matrix equation, Figure 6, Proceedings of the 4th IEEE Conference on Industrial Electronics and Applications, pp. 3698-3703. ©[2009] IEEE. Reprinted, with permission.*

5.4.4.3. Verification Results

As illustrated in Figures 5.20 and 5.21 about the convergence of ZNN (5.17), the solution generated by the ZNN model could always converge to theoretical solution $X^*(t)$ exactly. In addition, Figure 5.20 shows that the convergence becomes much faster if design parameter γ increases. As seen from Figure 5.21, the residual error (i.e., $\|E(t)\|_F$) synthesized by using power-sigmoid functions vanishes nearly twice faster than that by linear functions under the same condition, e.g., $\gamma = 1$. Moreover, other modeling results substantiate as well that superior convergence can be achieved by using power-sigmoid functions under the same γ value, compared to three other types of activation functions. Meanwhile, in Figures 5.22 and 5.23, with the same value of γ, GNN state-trajectories perform less favorably and the residual error is considerably large (about 1.5). In addition, the simulation results further

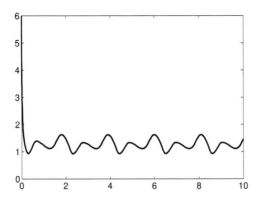

Figure 5.23. Convergence of residual error $\|A(t)X(t)B(t) - C(t)\|_F$ of GNN (5.19) using activation functions (1.7) with $\gamma = 1$. *Reproduced from N. Tan, K. Chen et al., Modeling, verification and comparison of Zhang neural net and gradient neural net for online solution of time-varying linear matrix equation, Figure 7, Proceedings of the 4th IEEE Conference on Industrial Electronics and Applications, pp. 3698-3703. ©[2009] IEEE. Reprinted, with permission.*

show that when the variation frequency of coefficient matrices increases, the GNN model (which is intrinsically designed for stationary problem solving) performs much unfavorably in handling the time-varying problem. On the other hand, ZNN model (5.17) could still converge well to the time-varying theoretical solution $X^*(t)$ exactly.

5.5. Conclusion

In this chapter, by following the design method recently proposed by Zhang *et al.*, a recurrent neural network has been developed and analyzed for the online solution of time-varying linear matrix equation. The neural network could globally exponentially converge to the exact solution of the three time-varying linear matrix equations (i.e., Sylvester equation, Lyapunov equation, $AXB = C$). By considering different types of activation functions and various implementation errors, such ZNN models have been investigated and simulated thoroughly. In addition, several important MATLAB/Simulation techniques have been introduced such as Kronecker product of matrices, MATLAB routine "ode45", and obtaining of matrix derivatives. For final circuit-implementation, the blocks modeling of ZNN and GNN is also investigated in this chapter and viewed as an important step. As compared with conventional GNN models, the new design approach and its resultant ZNN models could have much superior efficacy on time-varying problem solving. MATLAB-coding based simulation results and graphical modeling results have both further substantiated the feasibility of ZNN models having global exponential convergence to theoretical solutions.

References

[1] Zhang, Y; Wang, J. Recurrent neural networks for nonlinear output regulation. *Automatica*, 2001, vol. 37, no. 8, pp. 1161-1173.

[2] Zhang, Y; Wang, J. Global exponential stability of recurrent neural network for synthesizing linear feeback control systems via pole assignment. *IEEE Transactions on Neural Networks,* 2002, vol. 13, no. 3, pp. 633-644.

[3] Steriti, RJ; Fiddy, MA. Regularized image reconstruction using SVD and a neural network method for matrix inversion. *IEEE Transactions on Signal Processing,* 1993, vol. 41, no. 10, pp. 3074-3077.

[4] Carneiro, NCF; Caloba, LP. A new algorithm for analog matrix inversion. *Proceedings of the 38th Midwest Symposium on Circuits and Systems,* 1995, vol. 1, pp. 401-404.

[5] El-Amawy, A. A systolic architecture for fast dense matrix inversion. *IEEE Transactions on Computers,* 1989, vol. 38, no. 3, pp. 449-455.

[6] Jang, J; Lee, S; Shin, S. An optimization network for matrix inversion. *Neural Information Processing Systems,* New York: American Institute of Physics, 1988, pp. 397-401.

[7] Sturges, RH, Jr. Analog matrix inversion (robot kinematics). *IEEE Journal of Robotics and Automation,* 1988, vol. 4, no. 2, pp. 157-162.

[8] Wang, YQ; Gooi, HB. New ordering methods for space matrix inversion via diagonaliztion. *IEEE Transactions on Power Systems*, 1997, vol. 12, no. 3, pp. 1298-1305.

[9] Yeung, KS; Kumbi, F. Symbolic matrix inversion with application to electronic circuits. *IEEE Transactions on Circuits and Systems,* 1988, vol. 35, no. 2, pp. 235-238.

[10] Zhang, Y. Revisit the analog computer and gradient-based neural system for matrix inversion. *Proceedings of IEEE International Symposium on Intelligent Control*, 2005, pp. 1411-1416.

[11] Zhang, Y; Ge, SS. A general recurrent neural network model for time-varying matrix inversion. *Proceedings of the 42nd IEEE Conference on Decision and Control,* 2003, vol. 6, pp. 6169-6174.

References

[12] Zhang, Y; Ge, SS. Design and analysis of a general recurrent neural network model for time-varying matrix inversion. *IEEE Transactions on Neural Networks,* 2005, vol. 16, no. 6, pp. 1477-1490.

[13] Zhang, Y; Jiang, D; Wang, J. A recurrent neural network for solving Sylvester equation with time-varying coefficients. *IEEE Transactions on Neural Networks,* 2002, vol. 13, no. 5, pp. 1053-1063.

[14] Tank, D; Hopfield, JJ. Simple neural optimization networks: an A/D converter, signal decision circuit, and a linear programming circuit. *IEEE Transactions on Circuits and Systems,* 1986, vol. 33, no. 5, pp. 533-541.

[15] Wang, J. A recurrent neural network for real-time matrix inversion. Applied Mathematics and Computation, 1993, vol. 55, no. 1, pp. 89-100.

[16] Song, J; Yam, Y. Complex recurrent neural network for computing the inverse and pseudo-inverse of the complex matrix. *Applied Mathematics and Computation,* 1998, vol. 93, no. 2-3, pp. 195-205.

[17] Manherz, RK; Jordan, BW; Hakimi, SL. Analog methods for computation of the generalized inverse. *IEEE Transactions on Automatic Control,* 1968, vol. 13, no. 5, pp. 582-585.

[18] Mead, C. *Analog VLSI and Neural Systems.* Reading, MA: Addison-Wesley, 1989.

[19] Zhang, Y. A set of nonlinear equations and inequalities arising in robotics and its online solution via a primal neural network. *Neurocomputing,* 2006, vol. 70, no. 1-3, pp. 513-524.

[20] Zhang, Y. Towards piecewise-linear primal neural networks for optimization and redundant robotics. *Proceedings of IEEE International Conference on Networking, Sensing and Control,* 2006, pp. 374-379.

[21] The MathWorks Inc., Using Simulink, version 6.6, Natick, MA, 2007; available at http://www.mathworks.com/acce ss/helpdesk/help/toolbox/simulink.

[22] Zhang, Y. *Analysis and Design of Recurrent Neural Networks and Their Applications to Control and Robotic Systems.* Ph.D. Thesis, Chinese University of Hong Kong, 2002.

[23] Fernando, KV; Nicholson, H. Solution of Lyapunov equation for the state matrix. *Electronics Letters,* 1981, Vol. 17, no. 5, pp. 204-205.

[24] Sreeram, V; Agathoklis, P. Solution of Lyapunov equation with system matrix in companion form. *IEE Proceedings–D,* 1991, vol. 138, no. 6, pp. 529-534.

[25] Horn, RA; Johnson, CR. *Topics in Matrix Analysis.* Cambridge: Cambridge University Press, 1991, pp. 239-297.

[26] Ding, F; Chen, T. Gradient based iterative algorithms for solving a class of matrix equations. *IEEE Transactions on Automatic Control,* 2005, vol. 50, no. 8, pp. 1216-1221.

[27] Bazaraa, MS; Sherali, HD; Shetty, CM. *Nonlinear Programming-Theory and Algorithms.* New York: Wiley, 1993.

[28] Zhang, Y; Guo, X; Ma, W. Modeling and simulation of Zhang neural network for on-line linear time-varying equations solving based on MATLAB Simulink. *Proceedings of the 7th International Conference on Machine Learning and Cybernetics,* 2008, pp. 805-810.

Chapter 6

ZNN for Time-Varying Matrix Square Roots Finding

Abstract

A special kind of recurrent neural networks (RNN) has recently been proposed by Zhang *et al.* for online time-varying problems solving. Different from conventional gradient neural networks (GNN), such RNN (or termed specifically as Zhang neural networks, ZNN) are designed based on matrix-valued error functions, instead of scalar-valued energy functions. In addition, ZNN models are usually depicted in implicit dynamics rather than explicit dynamics. In this chapter, we generalize, develop, compare and simulate the ZNN and GNN models for online solution of time-varying matrix square roots. Important simulation and modeling techniques are thus investigated to facilitate the models' verification. Computer-verification results based on power-sigmoid activation functions further substantiate the superior ZNN convergence and efficacy on time-varying problems solving, as compared to the GNN model.

6.1. Introduction

The problem of solving for matrix square roots widely arises in many scientific and engineering areas; e.g., control theory [1], optimization [2], and signal processing [3]. Due to its fundamental roles, much more effort has been directed towards the solving algorithms of matrix square roots [1–4]. In general, the solution of matrix square roots could be achieved via matrix equations solving. There are two general types of solution to those matrix-equation problems. One is the numerical algorithms performed on digital computers [2,3], of which the minimal arithmetic operations are usually proportional to the cube of the matrix dimension n [5]. Due to the serial-processing nature of numerical algorithms performed on digital computers, they may not be efficient enough for large-scale and/or online problems solving. Being the second general type of solution, many parallel-processing computational schemes, including various dynamic RNN solvers, have been developed, analyzed, and implemented on specific architectures [6–13]. The RNN (or termed neural-dynamic) approach is thus now regarded as a powerful alternative to online computation because of its parallel distributed nature and, more importantly, the convenience of hardware realization [6].

Different from conventional gradient-based neural networks (or termed gradient neural networks, GNN) designed intrinsically for static problems solving [6], a special kind of RNN has recently been proposed by Zhang *et al.* [7–13] for real-time solution of time-varying problems. The resultant Zhang neural networks (ZNN) are designed based on matrix-valued error functions, instead of scalar-valued energy functions usually associated with GNN. In this chapter, a ZNN model is generalized, developed and investigated to solve online for the time-varying matrix square roots; in mathematics, to solve the time-varying quadratic matrix equation $X^2(t) = A(t)$ over time t. The ZNN efficacy on such time-varying matrix-roots solving is demonstrated well.

6.2. Problem Formulation and Neural Solvers

In this section, the problem formulation of time-varying matrix square roots is introduced firstly, and then the RNN design procedures are shown comparatively.

6.2.1. Problem Formulation

Let us consider the following time-varying matrix square root (TVMSR) problem (which is also a time-varying nonlinear-matrix-equation problem):

$$X^2(t) - A(t) = 0, \quad t \in [0, +\infty), \tag{6.1}$$

where $A(t) \in \mathbb{R}^{n \times n}$ denotes a smoothly time-varying positive-definite matrix, which, together with its time derivative $\dot{A}(t)$, are assumed to be known numerically or could be measured accurately. In addition, $X(t)$ is the time-varying unknown matrix to be solved for, and our objective in this work is to find $X(t) \in \mathbb{R}^{n \times n}$ so that (6.1) holds true for any $t \geq 0$.

Before solving (6.1), we provide the following preliminaries (see [1, 3, 10, 14, 15] as well) for the basis of further discussion.

Definition 6.2.1. Given a smoothly time-varying matrix $A(t) \in \mathbb{R}^{n \times n}$, if matrix $X(t) \in \mathbb{R}^{n \times n}$ satisfies the time-varying nonlinear matrix equation $X^2(t) = A(t)$, then $X(t) \in \mathbb{R}^{n \times n}$ is a time-varying square root of matrix $A(t) \in \mathbb{R}^{n \times n}$ [or say, $X(t)$ is a time-varying solution to nonlinear matrix equation (6.1)].

Square-Root Existence Condition 6.2.1. If smoothly time-varying matrix $A(t) \in \mathbb{R}^{n \times n}$ is positive-definite (in general sense [15]) at any time instant $t \in [0, +\infty)$, then there exists a time-varying matrix square root $X(t) \in \mathbb{R}^{n \times n}$ for $A(t)$.

6.2.2. Zhang Neural Network

To solve for time-varying matrix square root $A^{1/2}(t)$, by Zhang *et al.*'s method [7–13], the following matrix-valued error function could be defined firstly:

$$E(t) = X^2(t) - A(t) \in \mathbb{R}^{n \times n}.$$

Secondly, the error-function's time-derivative $\dot{E}(t) \in \mathbb{R}^{n \times n}$ could be made such that every entry $e_{ij}(t) \in \mathbb{R}$ of $E(t) \in \mathbb{R}^{n \times n}$ converges to zero, $i, j = 1, 2, \cdots, n$; in mathematics, to choose $\dot{e}_{ij}(t)$ such that $\lim_{t \to \infty} e_{ij}(t) = 0, \forall i, j \in \{1, 2, \cdots, n\}$. A general form of $\dot{E}(t)$ can be

$$\frac{dE(t)}{dt} = -\Gamma \Phi\big(E(t)\big), \tag{6.2}$$

where design parameter $\Gamma \in \mathbb{R}^{n \times n}$ is a positive-definite matrix used to scale the convergence rate of the neural network. For simplicity, we can use γ in place of Γ with $\gamma > 0 \in \mathbb{R}$. In addition, the activation-function array $\Phi(\cdot) : \mathbb{R}^{n \times n} \to \mathbb{R}^{n \times n}$ is a matrix-valued entry-to-entry mapping, in which each scalar-valued processing-unit $\phi(\cdot)$ is a monotonically-increasing odd activation function (for more details, see Section 1.5.).

Thirdly, expanding the ZNN design formula (6.2) leads to the following implicit dynamic equation of the ZNN model which solves online for the matrix square roots to nonlinear time-varying equation (6.1):

$$X(t)\dot{X}(t) + \dot{X}(t)X(t) = -\gamma \Phi\big(X^2(t) - A(t)\big) + \dot{A}(t). \tag{6.3}$$

In ZNN (6.3), starting from an initial condition $X(0) \in \mathbb{R}^{n \times n}$, $X(t)$ is the activation state matrix corresponding to theoretical time-varying matrix square root $X^*(t) := A^{1/2}(t)$ of $A(t)$. In addition, about ZNN (6.3), we have the following propositions on its convergence (as generalized from [3, 7–13]).

Proposition 6.2.1. *Consider a smoothly time-varying matrix $A(t) \in \mathbb{R}^{n \times n}$ in nonlinear equation (6.1), which satisfies Square-Root Existence Condition 6.2.1. If a monotonically-increasing odd activation-function array $\Phi(\cdot)$ is used, then*

- *neural-state matrix $X(t) \in \mathbb{R}^{n \times n}$ of ZNN (6.3), starting from randomly-generated positive-definite diagonal initial-state-matrix $X(0) \in \mathbb{R}^{n \times n}$, could converge to theoretical positive-definite time-varying matrix square root $X^*(t)$ of $A(t)$; and,*

- *neural-state matrix $X(t) \in \mathbb{R}^{n \times n}$ of ZNN (6.3), starting from randomly-generated negative-definite diagonal initial-state-matrix $X(0) \in \mathbb{R}^{n \times n}$, could converge to theoretical negative-definite time-varying matrix square root $X^*(t)$ of $A(t)$.* □

Proposition 6.2.2. *In addition to Proposition 6.2.1, if a linear activation-function array $\Phi(\cdot)$ is used, then*

- *neural-state matrix $X(t) \in \mathbb{R}^{n \times n}$ of ZNN (6.3), starting from randomly-generated positive-definite diagonal initial-state-matrix $X(0) \in \mathbb{R}^{n \times n}$, could exponentially converge to theoretical positive-definite time-varying matrix square root $X^*(t)$ of $A(t)$ with rate γ;*

- *neural-state matrix $X(t) \in \mathbb{R}^{n \times n}$ of ZNN (6.3), starting from randomly-generated negative-definite diagonal initial-state-matrix $X(0) \in \mathbb{R}^{n \times n}$, could exponentially converge to theoretical negative-definite time-varying matrix square root $X^*(t)$ of $A(t)$ with rate γ.*

As compared to the above linear-array situation, superior convergence can be achieved for ZNN (6.3) by using an array $\Phi(\cdot)$ made of power-sigmoid activation functions (1.7). □

6.2.3. Gradient Neural Network

For comparison, it is also worth pointing out here that almost all numerical algorithms and neural-dynamic computational schemes were designed intrinsically for static problems solving, e.g., with constant matrix $A \in \mathbb{R}^{n \times n}$ [rather than the time-varying matrix $A(t) \in \mathbb{R}^{n \times n}$ of the authors' interest]. These algorithms and schemes are generally related to the gradient-descent method in optimization described briefly as the procedure below.

Firstly, a scalar-valued norm-based lower-bounded energy function, e.g., $\|X^2(t) - A\|_F^2/2$ with Frobenius norm $\|A\|_F := \sqrt{\operatorname{trace}(A^T A)}$, is adopted such that its minimum point is the solution to nonlinear matrix equation $X^2 = A$.

Secondly, an algorithm could be designed by evolving along a descent direction of the above energy function until a minimum is reached. For example, the well-known typical descent direction is the negative gradient, i.e.,

$$-\frac{\partial(\|X^2(t) - A\|_F^2/2)}{\partial X} = -X^T(t)\left(X^2(t) - A\right) - \left(X^2(t) - A\right)X^T(t). \tag{6.4}$$

Thus, by using the gradient-design method, we could have the following conventional linear-activation GNN model,

$$\dot{X}(t) = -\gamma\left(X^T(t)(X^2(t) - A) + (X^2(t) - A)X^T(t)\right),$$

and the general form of nonlinear-activation GNN model,

$$\dot{X}(t) = -\gamma X^T(t)\Phi\left(X^2(t) - A\right) - \gamma\Phi\left(X^2(t) - A\right)X^T(t). \tag{6.5}$$

6.2.4. Models and Methods Comparison

Moreover, before ending this section, we would like to present the following important remarks about the comparison between ZNN (6.3) and GNN (6.5).

- ZNN model (6.3) is designed based on the elimination of every entry of the matrix-valued indefinite error function $E(t) = X^2(t) - A(t)$, with the value of $e_{ij}(t)$ being positive, negative, or unbounded. In contrast, GNN model (6.5) is designed based on the elimination of the scalar-valued energy function $\|X^2(t) - A\|_F^2/2$ which could only be positive or at least bounded below.

- ZNN model (6.3) is depicted in an implicit dynamics, i.e., $\dot{X}(t)X(t) + X(t)\dot{X}(t) = \cdots$, which might coincide well with systems in nature and in practice (e.g., in analogue electronic circuits and mechanical systems [7] owing to Kirchhoff's and Newton's laws, respectively). In contrast, GNN model (6.5) is depicted in an explicit dynamics, i.e., $\dot{X}(t) = \cdots$, which is usually associated with classic Hopfield-type or gradient-based artificial neural networks.

- ZNN model (6.3) methodically and systematically exploits the time-derivative information of problem-matrix $\dot{A}(t)$ during its real-time solving process. In contrast, GNN model (6.5) has not exploited such important information, and thus may not be effective on solving such time-varying problems.

- As analyzed in the above propositions, the neural state $X(t)$ computed by ZNN model (6.3) could exponentially converge to theoretical time-varying matrix square root $A^{1/2}(t)$. In contrast, GNN model (6.5) could only generate an approximate result to theoretical matrix square root $A^{1/2}(t)$ with much larger steady-state errors.

- Belonging to a predictive approach, ZNN model (6.3) and its design method make good use of the time-derivative information $\dot{A}(t)$, and thus they could be more effective on the system convergence to a "moving" theoretical solution. In contrast, belonging to the conventional tracking approach, GNN model (6.5) and its method act by adapting to the change of matrix $A(t)$ in a posterior passive manner, and thus they theoretically can not catch the exact solution which is on the move (i.e., time-varying).

- As shown in [16] and [17], the connection from Newton iteration to ZNN models could be established. That is, Newton iteration for solving static problems appears to be a special case of the discrete-time ZNN models (by considering only the use of linear activation functions and fixing the step-size to be 1).

- The derivation of ZNN models [such as (6.3)] might only need the less difficult knowledge of bachelors' mathematical course [e.g., see the scalar case of ZNN design formula (6.2)]. In contrast, the derivation of GNN models [such as (6.5)] requires more complicated mathematical knowledge of postgraduates' or even PhD's level [e.g., check the derivation of (6.4)].

6.3. MALATB-Coding Simulation Techniques

While Section 6.2. presents the problem formulation and its RNN solvers with theoretical and comparative results, the following simulation techniques [14] are investigated in this section to show the RNN-solution characteristics.

To simulate the implicit-dynamic systems such as ZNN model (6.3) associated with a given mass matrix [e.g., to solve for the matrix square roots of $A(t) \in \mathbb{R}^{2 \times 2}$ in this chapter], two important simulation techniques have to be emphasized as follows.

- MATLAB routine "ode45" with a mass-matrix property is introduced to solve the initial-value ordinary-differential-equation (ODE) problem [14].

- Matrix derivative [e.g., $\dot{A}(t)$] could be obtained by using routine "diff" and symbolic math toolbox.

6.3.1. ODE with Mass Matrix

In the simulation of ZNN (6.3), ODE routine "ode45" (or "ode23t") is preferred because it can solve the initial-value ODE problem with a mass matrix, e.g., $M(t,x)\dot{x} = z(t,x)$, where, in this work and here, x is the vectorization form of X in ZNN (6.3) [7–10], $z(t,x)$ is the vectorization form of $-\gamma\Phi(X^2(t) - A(t)) + \dot{A}(t)$, and the mass matrix $M(t,x) := I \otimes X(t) + X^T(t) \otimes I$, with symbol $\otimes$ denoting the Kronecker product of two matrices [7–10]. The following program code [14] using routine "kron" could be used to generate such a mass matrix, $M(t,x)$.

```
function output=MatrixM(t,x,gamma)
xMatrix=reshape(x,2,2); % back to matrix
mass=kron(eye(2),xMatrix)+kron(xMatrix',eye(2));
output=mass;
```

To solve the ODE problem with a mass matrix, routine "odeset" should also be noted. Its "Mass" property is assigned to be the function "MatrixM", which returns the value of mass matrix $M(t,x)$. Note that, because $M(t,x)$ strongly depends on neural state $X(t)$, the value of the "MStateDep" property of routine "odeset" should be set "strong" [14]. For example, the following program code can be used in the simulation of the initial-value ODE system resulting from ZNN (6.3) with a state-dependent mass matrix $M(t,x)$ and initial state $x(0)$.

```
tspan=[0 8]; x0=diag(2*rand(2,1));
x0=reshape(x0,4,1); % vectorization
options=odeset('Mass',@MatrixM,'MStateDep','strong');
[t,x]=ode45(@ZNNrighthandside,tspan,x0,options,gamma);
```

where the function "ZNNrighthandside", to be shown in Subsection 6.3.3., returns the evaluation of the vectorization form of the right-hand side of ZNN (6.3).

Besides, it is worth mentioning here that the simulation of the GNN model does not rely on this mass-matrix technique, as it has a simple identity matrix as its mass matrix on the left hand side of the dynamic equation (i.e., a so-called explicit form). Hence, the program code to simulate the initial-value ODE system as of GNN (6.5) is

```
tspan=[0 8]; x0=diag(2*rand(2,1));
x0=reshape(x0,4,1); options=odeset();
[t,x]=ode45(@GNNrighthanside,tspan,x0,options,gamma);
```

6.3.2. Obtaining Matrix Derivatives

While matrix $A(t)$ can be generated by the user-defined function "MatrixA", its time derivative $\dot{A}(t)$ required by ZNN (6.3) should be obtained. Without loss of generality, we can use MATLAB routine "diff" to generate $\dot{A}(t)$.

In order to get the time derivative of $A(t)$, a symbolic object "u" is constructed firstly, and then the code "D=diff(MatrixA(u))" [14] is used to generate the analytical form of $\dot{A}(t)$. Finally, evaluating such an analytical form of $\dot{A}(t)$ with a numerical t will generate the required $\dot{A}(t)$. The code for generating $\dot{A}(t)$ is as follows.

```
syms u; D=diff(MatrixA(u));
u=t; diffA=eval(D)
```

Note that, as mentioned before, without using the symbolic "u", the command "diff" "(MatrixA(t))" returns the row difference of "MatrixA(t)", which is not the desired time derivative of $A(t)$.

6.3.3. RNN Right-Hand Side

As mentioned in the previous subsection, the following code "ZNNrighthandside" defines a function which evaluates the vectorization form of the right-hand side of ZNN (6.3). In other words, it returns the value of $z(t,x)$ in the implicit dynamic-equation $M(t,x)\dot{x}(t) = z(t,x)$, where $z(t,x)$ was defined before with different types of activation functions (note that power-sigmoid activation functions are preferred in this section).

```
function y=ZNNrighthandside(t,x,gamma)
if nargin==2, gamma=1; end
syms u; dotA=diff(MatrixA(u));
u=t; dA=eval(dotA); dA=reshape(dA,4,1);
x=reshape(x,2,2); % back to matrix
xx=x*x;
xx=reshape(xx,4,1); % vectorization
mAv=reshape(MatrixA(t),4,1);
y=-gamma*AFMpowersigmoid(xx-mAv)+dA;
```

For comparison, GNN (6.5) is exploited as well to solve the time-varying nonlinear matrix equation (6.1). Similarly, we could define a function "GNNrighthandside" which evaluates the vectorization form of the right-hand side of GNN (6.5). In other words, it returns the value of the right-hand side $g(t,x)$ in the explicit dynamic equation $\dot{x}(t) = g(t,x)$, where, in this GNN situation, $g(t,x)$ is the vectorization form of $-\gamma X^T(t)\Phi(X^2(t) - A(t)) - \gamma\Phi(X^2(t) - A(t))X^T(t)$ with different types of activation functions introduced (note that linear activation functions are used here).

```
function y=GNNrighthandside(t,x,gamma)
if nargin==2, gamma=1; end
x=reshape(x,2,2); % return to matrix
A=MatrixA(t);
yy=-gamma*x'*AFMlinear(x*x-A)-gamma*AFMlinear(x*x-A)*x';
y=reshape(yy,4,1);
```

6.3.4. An Illustrative Verification Example

In this subsection, for illustrative purposes, a computer-verification example [10] is presented to show the performance and efficacy of ZNN model (6.3) on solving the time-varying nonlinear matrix equation (6.1). For comparison, the conventional GNN model (6.5) is simulated as well for the same illustrative example. Let us consider equation (6.1) with the following time-varying matrix $A(t)$:

$$A(t) = \begin{bmatrix} 4 + \sin 2t \cos 2t & 3\cos 2t \\ 3\sin 2t & \sin 2t \cos 2t + 1 \end{bmatrix}.$$

Simple manipulations may verify that a theoretical time-varying matrix square root $A^{1/2}(t)$ of $A(t)$ could be

$$X^*(t) := A^{1/2}(t) = \begin{bmatrix} 2 & \cos 2t \\ \sin 2t & 1 \end{bmatrix},$$

which is used to check the correctness of the neural-network solutions $X(t)$ of ZNN (6.3) and GNN (6.5).

Using the aforementioned simulation techniques, we could obtain the ZNN state $X(t)$ [i.e., corresponding to the matrix square root $A^{1/2}(t)$ of $A(t)$], which is illustrated in Figure 6.1(a) generated by the following program code.

```
function ZNNconvergence(gamma)
tspan=[0 8];
options=odeset('Mass',@MatrixM,'MStateDep','strong');
for iter=1:8
    x0=diag(2*rand(2,1));
    x0=reshape(x0,4,1);
[t,x]=ode45(@ZNNrighthandside,tspan,x0,options,gamma);
subplot(2,2,1);plot(t,x(:,1)); hold on
subplot(2,2,3);plot(t,x(:,2)); hold on
subplot(2,2,2);plot(t,x(:,3)); hold on
subplot(2,2,4);plot(t,x(:,4)); hold on
end
```

As illustrated in Figure 6.1(a), starting from randomly-generated positive-definite diagonal initial-state $X(0) \in [0,2]^{2 \times 2}$, the neural state $X(t)$ of ZNN model (6.3) could converge to theoretical (positive-definite) time-varying matrix square root $A^{1/2}(t)$ rapidly and accurately, where design parameter $\gamma = 1$. For comparison, the solution of GNN (6.5) using linear activation functions is shown in Figure 6.1(b) generated by the following code.

```
function GNNconvergence(gamma)
tspan=[0 8]; options=odeset();
for iter=1:8
    x0=diag(2*rand(2,1));
    x0=reshape(x0,4,1);
[t,x]=ode45(@GNNrighthandside,tspan,x0,options,gamma);
subplot(2,2,1);plot(t,x(:,1)); hold on
subplot(2,2,3);plot(t,x(:,2)); hold on
subplot(2,2,2);plot(t,x(:,3)); hold on
subplot(2,2,4);plot(t,x(:,4)); hold on
end
```

It is evident that GNN state $X(t)$ (denoted by solid curves) could only approximately fit with theoretical solution $A^{1/2}(t)$ (denoted by dash-dotted curves). In other words, there always exist appreciable steady-state errors between the GNN solution and the theoretical time-varying solution $A^{1/2}(t)$ to (6.1). It is worth mentioning here that, if initial $X(0)$ starts from a negative-definite diagonal matrix, neural state $X(t)$ will converge towards the negative-definite solution of (6.1).

To monitor the RNN convergence, we can also exploit and show the computational error, $\|X(t) - X^*(t)\|_F$. The program code for doing so is presented in the shadow boxes below. By running the program code "ZNNnormError", we could have Figure 6.2 which shows that, starting from randomly-generated positive-definite diagonal initial-state-matrix $X(0) \in [0,2]^{2 \times 2}$, the neural state $X(t)$ of ZNN (6.3) could converge to theoretical solution

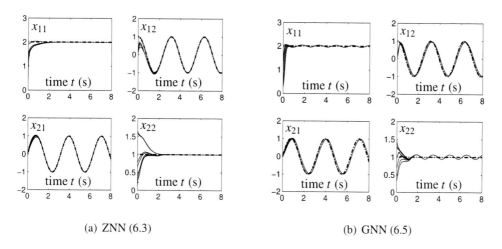

Figure 6.1. Online solution of time-varying matrix square root $A^{1/2}(t)$ by RNN [i.e., ZNN (6.3) and GNN (6.5)] with $\gamma = 1$, where the theoretical solution $A^{1/2}(t)$ is denoted in dash-dotted curves, while the neural-network solutions are denoted in solid curves.

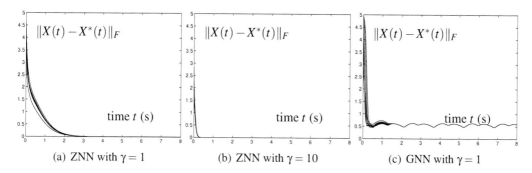

Figure 6.2. Convergence of computational error $\|X(t) - X^*(t)\|_F$ of ZNN (6.3) and GNN (6.5) applied to time-varying matrix square roots finding.

$A^{1/2}(t)$, as the computational error $\|X(t) - X^*(t)\|_F$ converges to zero correspondingly. Furthermore, such a convergence can be expedited by increasing γ. For example, if γ increases to 10^3, the convergence time is within 3.5 milliseconds; and, if γ increases to 10^6, the convergence time is 3.5 microseconds. For comparison, the computational error of GNN (6.5) using linear activation functions is shown in Figure 6.2(c). Evidently, the steady-state GNN-solution error is relatively large and can not vanish to zero (both theoretically and practically). This is because the time-derivative information of $A(t)$ has not been utilized in this gradient-based approach.

```
function ZNNnormError(gamma)
if nargin<1
    gamma=1;
end
total=8;
for iter=1:total
    x0=diag(2*rand(2,1));
    x0=reshape(x0,4,1);
    normError(x0,gamma);
end
```

```
function normError(x0,gamma)
tspan=[0 8];
options=odeset('Mass',@MatrixM,'MStateDep','strong');
[t,x]=ode45(@ZNNrighthandside,tspan,x0,options,gamma);
At=[2*ones(length(t),1) sin(2*t) cos(2*t) ones(length(t),1)]';
err=x'-At;
total=length(t);
for i=1:total,
    nerr(i)=norm(err(:,i));
end
figure(2); plot(t,nerr); hold on
```

In summary, Figures 6.1 and 6.2 show that the neural state $X(t)$ of ZNN (6.3) exponentially converge to theoretical time-varying matrix square root $A^{1/2}(t)$ of $A(t)$, whereas GNN (6.5) could only approximately approach it.

6.4. Simulink-Modeling Techniques

While the previous sections present theoretical results and MATLAB-coding techniques, we investigate in this section the system-modeling techniques for ZNN and GNN models.

6.4.1. Basic Blocks

As we know, a comprehensive block library exists in MATLAB Simulink (see [11–13, 18–21] as well), which includes various kinds of basic blocks (e.g., sinks, sources, linear/nonlinear components, and connectors). By interconnecting these basic function blocks, the models of ZNN (6.3) and GNN (6.5) can be constructed readily as depicted in Figures 6.3 and 6.4. In the following part, the basic blocks used in these models are described briefly.

- The *Constant* block, specified by its parameter "Constant value", is used to generate a constant scalar or matrix. For example, a 3-dimensional identity matrix can be generated by setting the parameter "Constant value" to be "[1 0 0; 0 1 0; 0 0 1]".

- The *Product* block, specified as in the standard matrix-product mode, is used to multiply the matrices and vectors involved in neural-network systems.

- The *Gain* block can be used to scale the neural-network convergence rate, e.g., acting as the scaling parameter γ.

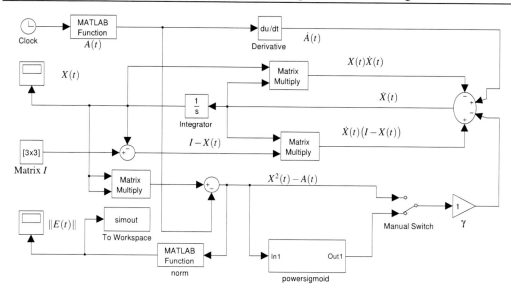

Figure 6.3. Overall ZNN Simulink model of (6.3) applied to the TVMSR problem solving. *Reproduced from Y. Zhang and W. Yang, Time-varying matrix square roots solving via Zhang neural network and gradient neural network: modeling, verification and comparison, Figure 1, W. Yu et al. (Eds.): ISNN 2009, Part I, LNCS 5551, pp. 11-20, 2009. ©Springer-Verlag Berlin Heidelberg 2009. With kind permission of Springer Science+Business Media.*

- The *Subsystem* block could be used to construct the sigmoid or power-sigmoid activation-function array, which make the whole system more readable.

- The *MATLAB Fcn* block is used to generate matrix $A(t)$ with the *Clock* block's output being its input or can be used to compute the matrix norm.

- The *Math Function* block can perform various common mathematical operations, for example, to generate the transpose of a matrix in our context.

- The *To Workspace* block, with its option "Save format" set to be "Array", is used to save the modeling results and data to the workspace.

- The *Integrator* block is used to make continuous-time integration on the input signals over a period of time. In this work, by setting its "Initial condition" as "$diag(2 * rand(3,1))$", we firstly generate a diagonal positive-definite initial state matrix $X(0)$ with its diagonal elements randomly distributed in [0,2], and then integrate $\dot{X}$ with this *Integrator* block.

In addition, based on the activation-function arrays presented in Subsection 2.5.2., we could establish the models of ZNN (6.3) and GNN (6.5) depicted in Figures 6.3 and 6.4, and verify the superior ZNN convergence in time-varying problems solving, especially when using a power-sigmoid activation function array.

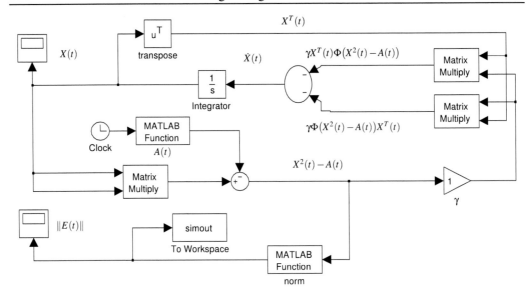

Figure 6.4. Overall GNN Simulink model of (6.5) applied to the TVMSR problem solving. *Reproduced from Y. Zhang and W. Yang, Time-varying matrix square roots solving via Zhang neural network and gradient neural network: modeling, verification and comparison, Figure 2, W. Yu et al. (Eds.): ISNN 2009, Part I, LNCS 5551, pp. 11-20, 2009. ©Springer-Verlag Berlin Heidelberg 2009. With kind permission of Springer Science+Business Media.*

6.4.2. Parameters and Options Setting

As for the RNN models established and shown in Figures 6.3 and 6.4, we may have to modify and config some of the default modeling parameters by opening the dialog box entitled "Configuration Parameters" and setting the modeling/simulation options as follows:

- Solver: "ode45" (or "ode23t");

- Max step size: "0.2";

- Min step size: "auto";

- Absolute tolerance: "auto";

- Relative tolerance: "1e-6" (i.e., 10^{-6});

- Algebraic loop: "none".

Other parameters and options could be set to be the default values or depending on specific modeling tasks. In addition, the check box in front of "States" as of the option "Data Import/Export" should be selected, which is for the purpose of better displaying the RNN-modeling results and is associated with the code "StopFcn" in Figure 6.5 (as of "Callbacks" in the dialog box entitled "Model Properties" which is started from the "File" pull-down menu).

Before ending this subsection, it is worth pointing out further that, by using convenient click-and-drag mouse operations (instead of the above-presented compilation of simulation

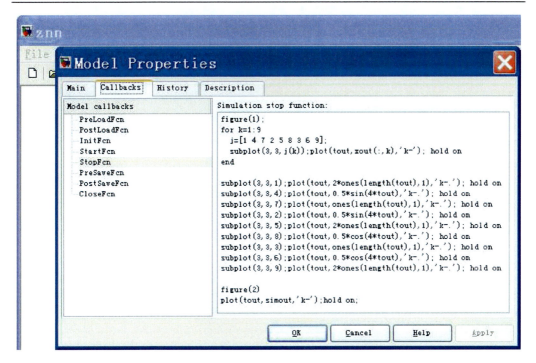

Figure 6.5. Necessary "StopFcn" code of "Callbacks" in dialog box "Model Properties".
Reproduced from Y. Zhang and W. Yang, Time-varying matrix square roots solving via Zhang neural network and gradient neural network: modeling, verification and comparison, Figure 3, W. Yu et al. (Eds.): ISNN 2009, Part I, LNCS 5551, pp. 11-20, 2009. ©Springer-Verlag Berlin Heidelberg 2009. With kind permission of Springer Science+Business Media.

code), the RNN models can be built up systematically and easily by means of MATLAB Simulink function blocks. Programming efforts could thus be reduced drastically [11–13]. Moreover, the system-modeling might be an important and necessary step for the final hardware realization of recurrent neural networks, in addition to mathematical analysis and simulative verification. Furthermore, by using the Simulink HDL Coder (with HDL denoting Hardware Description Language), the recurrent neural networks developed in Simulink environment could further be extended to the HDL code and then to the FPGA and ASIC realization [18–21].

6.4.3. Modeling and Verification Results

To verify the performance, efficacy and superiority of ZNN (6.3) in comparison with GNN (6.5), we consider the following time-varying matrix $A(t)$ with its time-varying theoretical square root $X^*(t)$ given below for comparison purposes:

$$A(t) = \begin{bmatrix} 5+0.25s^2 & 2s+0.5c & 4+0.25s \times c \\ 2s+0.5c & 4.25 & 2c+0.5s \\ 4+0.25s \times c & 2c+0.5s & 5+0.25c^2 \end{bmatrix}, \quad X^*(t) = \begin{bmatrix} 2 & 0.5s & 1 \\ 0.5s & 2 & 0.5c \\ 1 & 0.5c & 2 \end{bmatrix},$$

where s and c denote $\sin(4t)$ and $\cos(4t)$, respectively. The ZNN and GNN models depicted in Figures 6.3 and 6.4 are now applied to solving the TVMSR problem (6.1).

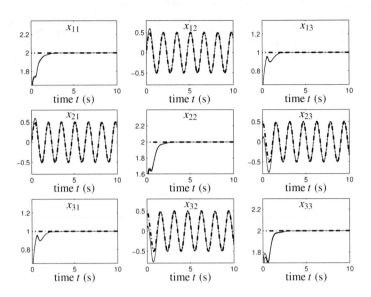

(a) Neural state matrix $X(t)$ of ZNN (6.3) using a power-sigmoid processing array

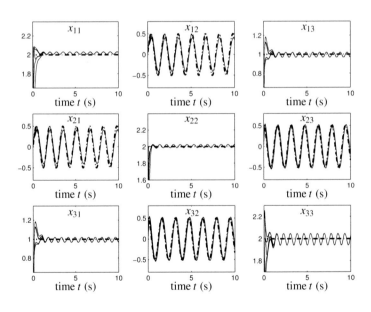

(b) Neural state matrix $X(t)$ of GNN (6.5) using a linear processing array

Figure 6.6. Online solution of time-varying matrix square root $A^{1/2}(t)$ by ZNN (6.3) and GNN (6.5) with $\gamma = 1$, where the theoretical solution is denoted in dash-dotted curves. *Reproduced from Y. Zhang and W. Yang, Time-varying matrix square roots solving via Zhang neural network and gradient neural network: modeling, verification and comparison, Figure 4, W. Yu et al. (Eds.): ISNN 2009, Part I, LNCS 5551, pp. 11-20, 2009. ©Springer-Verlag Berlin Heidelberg 2009. With kind permission of Springer Science+Business Media.*

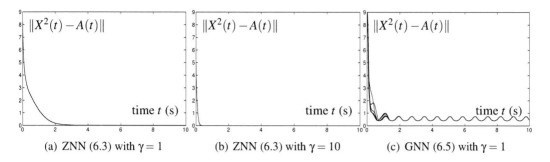

Figure 6.7. Residual-error profile of RNN models during time-varying square roots finding. *Reproduced from Y. Zhang and W. Yang, Time-varying matrix square roots solving via Zhang neural network and gradient neural network: modeling, verification and comparison, Figure 5, W. Yu et al. (Eds.): ISNN 2009, Part I, LNCS 5551, pp. 11-20, 2009. ©Springer-Verlag Berlin Heidelberg 2009. With kind permission of Springer Science+Business Media.*

As illustrated in Figure 6.6(a), starting from randomly-generated positive-definite diagonal initial-state $X(0) \in [0,2]^{3\times 3}$, the neural state matrix $X(t)$ of ZNN (6.3) with design parameter $\gamma = 1$ could converge rapidly to the theoretically time-varying matrix square root $X^*(t)$ in an error-free manner. In contrast, in Figure 6.6(b), GNN (6.5) does not track the theoretical solution $X^*(t)$ well, instead with quite large solution errors.

To monitor and show the solution process of ZNN model (6.3) and GNN model (6.5), residual error $\|X^2(t) - A(t)\|$ could also be exploited (which, for many engineering applications, might be the only and preferable choice). The left two sub-graphs of Figure 6.7 show that, starting from randomly-generated positive-definite diagonal initial-state $X(0) \in [0,2]^{3\times 3}$, the residual error of ZNN (6.3) converges to zero exactly, and that superior performance can be achieved by increasing the value of γ from 1 to 10. In comparison, as shown in Figure 6.7(c), the residual error of GNN model (6.5) is relatively much larger (never vanishing to zero) and oscillating.

6.5. Conclusion

By following Zhang *et al.*'s RNN-design method, a special kind of neural networks (i.e, ZNN) has been developed, compared, simulated and modeled for the online solution of time-varying matrix square roots problem. To do so, several important simulation and modeling techniques have been introduced for the verification of the neural networks, such as, "ode45" (or "ode23t") with a mass matrix, the obtaining of matrix derivatives, and overall Simulink model. Computer-verification results have substantiated further the effectiveness and efficiency of ZNN models for the online time-varying matrix square roots finding.

References

[1] Higham, NJ. Stable iterations for the matrix square root. *Numerical Algorithms*, 1997, vol. 15, no. 2, pp. 227-242.

[2] Long, J; Hu, X; Zhang, L. Newton's method with exact line search for the square root of a matrix. *Proceedings of International Symposium on Nonlinear Dynamics*, 2007, pp. 1-5.

[3] Hasan, MA; Hasan, AA; Rahman, S. Fixed point iterations for computing square roots and the matrix sign function of complex matrices. *Proceedings of IEEE Conference on Decision and Control*, 2000, pp. 4253-4258.

[4] Johnson, CR; Okubo, K; Reams, R. Uniqueness of matrix square roots and an application. *Linear Algebra and its Application*, 2001, vol. 323, no. 1, pp. 51-60.

[5] Zhang, Y; Leithead, WE; Leith, DJ. Time-series Gaussian process regression based on Toeplitz computation of $O(N^2)$ operations and $O(N)$-level storage. *Proceedings of the 44th IEEE Conference on Decision and Control*, 2005, pp. 3711-3716.

[6] Zhang, Y. Revisit the analog computer and gradient-based neural system for matrix inversion. *Proceedings of IEEE International Symposium on Intelligent Control*, 2005, pp. 1411-1416.

[7] Zhang, Y; Jiang, D; Wang, J. A recurrent neural network for solving Sylvester equation with time-varying coefficients. *IEEE Transactions on Neural Networks*, 2002, vol. 13, no. 5, pp. 1053-1063.

[8] Zhang, Y; Ge, SS. Design and analysis of a general recurrent neural network model for time-varying matrix inversion. *IEEE Transactions on Neural Networks*, 2005, vol. 16, no. 6, pp. 1477-1490.

[9] Zhang, Y; Peng, H. Zhang neural network for linear time-varying equation solving and its robotic application. *Proceedings of the 6th International Conference on Machine Learning and Cybernetics*, 2007, pp. 3543-3548.

[10] Zhang, Y; Yang, Y. Simulation and comparison of Zhang neural network and gradient neural network solving for time-varying matrix square roots. *Proceedings of Intelligent Information Technology Application*, 2008, vol. 2, pp. 966-970.

[11] Zhang, Y; Tan, N; Cai, B; Chen, Z. MATLAB Simulink modeling of Zhang neural network solving for time-varying pseudoinverse in comparison with gradient neural network. *Proceedings of Intelligent Information Technology Application*, 2008, vol. 1, pp. 39-43.

[12] Zhang, Y; Guo, X; Ma, W. Modeling and simulation of Zhang neural network for online linear time-varying equations solving based on MATLAB Simulink. *Proceedings of the 7th International Conference on Machine Learning and Cybernetics*, 2008, pp. 805-810.

[13] Ma, W; Zhang, Y; Wang, J. MATLAB Simulink modeling and simulation of Zhang neural networks for online time-varying Sylvester equation solving. *Proceedings of International Joint Conference on Neural Networks*, 2008, pp. 286-290.

[14] The MathWorks Inc. *MATLAB 7.0*. MA: Natick, 2004.

[15] Zhang, Y. On the LVI-based primal-dual neural network for solving online linear and quadratic programming problems. *Proceedings of American Control Conference*, 2005, pp. 1351-1356.

[16] Zhang, Y; Ma, W; Yi, C. The link between Newton iteration for matrix inversion and Zhang neural network (ZNN). *Proceedings of IEEE International Conference on Industrial Technology*, 2008, pp. 1-6.

[17] Zhang, Y; Cai, B; Liang, M; Ma, W. On the variable step-size of discrete-time Zhang neural network and Newton iteration for constant matrix inversion. *Proceedings of Intelligent Information Technology Application*, 2008, vol. 1, pp. 34-38.

[18] The MathWorks Inc. *Simulink 7 Getting Started Guide*. MA: Natick, 2008.

[19] Shanblatt, MA; Foulds, B. A Simulink-to-FPGA implementation tool for enhanced design flow. *Proceedings of IEEE International Conference on Microelectronic Systems Education*, 2005, pp. 89-90.

[20] Grout, IA. Modeling, simulation and synthesis: from Simulink to VHDL generated hardware. *Proceedings of the 5th World Multi-Conference on Systemics, Cybernetics and Informatics*, 2001, vol. 15, pp. 443-448.

[21] Grout, IA; Keane, K. A MATLAB to VHDL conversion toolbox for digital control. *Proceedings of IFAC Symposium on Computer Aided Control Systems Design*, 2000, pp. 164-169.

Part III

Return to Scalar or Constant Problems

Chapter 7

Zhang Dynamics for the Scalar-Valued Nonlinear Problems

Abstract

Different from gradient-based dynamics (GD), a special class of neural dynamics has been found, developed, generalized and investigated by Zhang *et al.*, e.g., for online solution of time-varying nonlinear equations (in the form of $f(x(t),t) = 0$) and/or static nonlinear equations (in the form of $f(x) = 0$). The resultant Zhang dynamics (ZD) is designed based on the elimination of an indefinite error-function (instead of the elimination of a square-based positive or at least lower-bounded energy-function usually associated with GD and/or Hopfield-type neural networks). For comparative purposes, the gradient-based dynamics is also developed and exploited for online solving such two forms of nonlinear equations. Computer-simulation results substantiate further the theoretical analysis and efficacy of the ZD models for online solution of nonlinear time-varying and/or static equations.

7.1. Introduction

The solution of time-varying nonlinear equations in the form of $f(x(t),t) = 0$ and/or static (or termed, constant) nonlinear equation in the form of $f(x) = 0$ is widely encountered in science and engineering fields. Many computational methods, models, and/or numerical algorithms [1–9] have thus been presented and investigated for solving such nonlinear equations. However, it may not be efficient enough for most numerical algorithms because of their serial-processing nature performed on digital computers [10]. Recently, due to the in-depth research in neural networks, the dynamic-system approach using recurrent neural models has become one of the important parallel-processing methods for solving online optimization and algebraic problems [11–23]. However, it is worth mentioning that most reported computational-schemes are related to the gradient-descent method or other methods intrinsically designed for constant problems (or say, time-invariant problems, static problems, or stationary problems) solving [11, 18–21, 24], which may then be applied directly to the time-varying environments.

Different from the above gradient-based dynamic (GD) approach, a new class of neural-dynamics (ND) has recently been proposed by Zhang *et al.* [12, 13, 25–27] (formally since March 2001) for time-varying problems solving (e.g., time-varying Sylvester equation solving, matrix inversion and optimization). By following and generalizing the Zhang *et al.*'s design method, a neural-dynamics has recently been introduced to handle time-varying and/or static nonlinear equations. The proposed Zhang dynamics (or termed, ZD for presentation convenience) is designed based on the elimination of an indefinite error-function (instead of the elimination of a square-based positive or at least lower-bounded energy-function usually associated with conventional gradient-based, Hopfield-type and/or Lyapunov design-&-analysis methods [1,7,8,11–13,19–21,25–28]). For comparative purposes, the conventional gradient-based dynamics (GD) is investigated as well in this chapter for time-varying and/or static nonlinear equations solving. Theoretical analysis and simulative results both show the efficacy and accuracy of the resultant Zhang dynamics (ZD) on those two kinds of equations solving.

7.2. Time-Varying Scalar-Valued Nonlinear Equations

Our objective in this section is to find $x(t) \in \mathbb{R}$ in real time t such that the following smoothly time-varying nonlinear equation holds true:

$$f\big(x(t),t\big) = 0. \tag{7.1}$$

The existence of the theoretical time-varying solution $x^*(t)$ at any time instant $t \in [0, +\infty)$ is assumed for discussion purposes. The design procedures of GD and ZD are investigated comparatively for solving equation (7.1) in this section. In addition, some basic types of activation functions (such as linear, sigmoid, power, and power-sigmoid activation-functions) are analyzed and discussed for the convergence of the ZD-model.

7.2.1. Gradient-Based Dynamics

Gradient-descent method is a conventional approach frequently used, e.g., to solve the constant situation of equation (7.1). To solve the time-varying nonlinear equation (7.1), the gradient-descent method requires us to define a norm or square-based error function (or termed, energy function) such as $\mathcal{E}(x(t),t) = f^2(x(t),t)$. Then, a typical continuous-time adaptation rule based on the negative-gradient information leads to the following differential equation (which we might term as gradient dynamics and/or implement as a gradient neural dynamics [11, 14–16, 29]):

$$\dot{x}(t) := \frac{dx}{dt} = -\frac{\gamma}{2} \frac{\partial \mathcal{E}}{\partial x} = -\gamma f\big(x(t),t\big) \frac{\partial f}{\partial x},$$

where γ is a positive design-parameter used to scale the convergence rate of the presented neural dynamics, and $x(t)$, starting from an initial condition $x(0) = x_0 \in \mathbb{R}$, is the activation state corresponding to the theoretical solution $x^*(t)$. As an extension to the above design approach and under the inspiration of [13], we could obtain the general nonlinear neural-dynamic model by using a general nonlinear activation function $\phi(\cdot)$ as follows:

$$\dot{x}(t) = -\gamma \phi \Big(f\big(x(t),t\big) \Big) \frac{\partial f}{\partial x}. \tag{7.2}$$

Similar to conventional neural-dynamic approaches, the design parameter γ in (7.2) and in the following text, being an inductance parameter or the reciprocal of a capacitance parameter in hardware implementation, could be set as large as the hardware permits (e.g., in analog circuits or VLSI [11]) or selected appropriately (e.g., between 10^3 and 10^8) for simulative and/or experimental purposes.

7.2.2. Zhang Dynamics

Following the new design method of Zhang *et al.* [12, 13, 25, 30–33], we could construct the following indefinite error-function so as to set up a ZD model to solve online the time-varying nonlinear equation (7.1):

$$e(t) := f(x(t), t).$$

Then, the error-function's time-derivative $\dot{e}(t)$ should be chosen and forced mathematically such that the error function $e(t)$ converges exponentially to zero. Specifically, we can describe our general choice of $\dot{e}(t)$ in the following form (which we term as ZD design formula):

$$\frac{de(t)}{dt} = -\gamma\phi(e(t)),$$

or equivalently, we have

$$\frac{df}{dt} = -\gamma\phi(f(x(t), t)), \tag{7.3}$$

where γ is a positive design-parameter used to scale the convergence rate, and $\phi(\cdot): \mathbb{R} \to \mathbb{R}$ denotes a general nonlinear activation-function. By expanding the above ZD design formula (7.3), we could thus have

$$\frac{\partial f}{\partial x}\frac{dx}{dt} + \frac{\partial f}{\partial t} = -\gamma\phi(f(x(t), t)),$$

leading to the following differential equation (which we might term and/or implement as Zhang dynamics [12, 13, 25, 30–33]):

$$\frac{\partial f}{\partial x}\dot{x}(t) = -\gamma\phi(f(x(t), t)) - \frac{\partial f}{\partial t} \quad \text{and/or} \quad \dot{x}(t) = -\left(\gamma\phi(f(x(t), t)) + \frac{\partial f}{\partial t}\right)/\frac{\partial f}{\partial x} \tag{7.4}$$

where $x(t)$, starting from an initial condition $x(0) = x_0 \in \mathbb{R}$, is the activation state corresponding to the theoretical time-varying solution $x^*(t)$ of equation (7.1).

In view of dynamic equations (7.2) and (7.4), different choices for $\phi(\cdot)$ may lead to different performance of the neural dynamics. Generally speaking, any monotonically increasing odd activation function $\phi(\cdot)$ could be used for the construction of the neural dynamics. Four basic types of activation functions are investigated in this chapter and can be seen in Section 1.5. for more details.

7.2.3. Convergence Analysis

While the previous subsections present the general frameworks about GD and ZD for solving nonlinear equation $f(x(t), t) = 0$, detailed design consideration and theoretical results are given in this subsection.

For ZD model (7.4), we could have the following propositions on its exponential convergence properties.

Proposition 7.2.1. *Consider a solvable nonlinear time-varying equation $f(x(t),t) = 0$, with $f(\cdot)$ continuously differentiable. If a monotonically-increasing odd function $\phi(\cdot)$ is used, the neural state $x(t)$ of ZD (7.4), starting from randomly generated initial state $x(0) = x_0$, could converge to a theoretical time-varying solution $x^*(t)$ of $f(x(t),t) = 0$.*

Proof. We can define a Lyapunov function candidate $V(x(t),t) = f^2(x(t),t)/2 \geq 0$ for the neural system (7.4) with its time derivative

$$\dot{V}(x(t),t) := \frac{dV(x(t),t)}{dt} = f(x(t),t)\left(\frac{\partial f}{\partial x}\frac{dx}{dt} + \frac{\partial f}{\partial t}\right) = -\gamma f(x(t),t)\phi\big(f(x(t),t)\big) \quad (7.5)$$

Because a monotonically-increasing odd activation-function is used, we could have $\phi(-f(x(t),t)) = -\phi(f(x(t),t))$ and

$$\phi\big(f(x(t),t)\big) \begin{cases} > 0, & \text{if } f(x(t),t) > 0, \\ = 0, & \text{if } f(x(t),t) = 0, \\ < 0, & \text{if } f(x(t),t) < 0. \end{cases}$$

Then we could obtain

$$f(x(t),t)\phi\big(f(x(t),t)\big) \begin{cases} > 0, & \text{if } f(x(t),t) \neq 0, \\ = 0, & \text{if } f(x(t),t) = 0, \end{cases}$$

which guarantees the final negative-definiteness of $\dot{V}(x(t),t)$; i.e. $\dot{V}(x(t),t) < 0$ for $f(x(t),t) \neq 0$ [equivalently, $x(t) \neq x^*(t)$], whereas $\dot{V}(x(t),t) = 0$ for $f(x(t),t) = 0$ [equivalently, $x(t) = x^*(t)$]. By Lyapunov theory [34], residual error $e(t) = f(x(t),t)$ could converge to zero; equivalently, the neural state $x(t)$ of ZD (7.4) could converge to a theoretical solution $x^*(t)$ with $f(x^*(t),t) = 0$ starting from some randomly-generated sufficiently-close initial states [note that $\partial f/\partial x$ and $\partial f/\partial t$ appear in the derivation of equation (7.5)]. The proof is complete. $\square$

Proposition 7.2.2. *Let $x^*(t)$ denote a theoretical time-varying solution of nonlinear equation $f(x(t),t) = 0$, where $f(\cdot)$ is a continuously-differentiable function (specifically, with at least first-order derivatives) at an interval containing $x^*(t)$. In addition to Proposition 7.2.1, the neural state $x(t)$ of ZD model (7.4) could converge to $x^*(t)$, provided that the initial state $x(0) = x_0$ is close enough to $x^*(t)$. In addition, ZD model (7.4) possesses the following properties.*

1) *If the linear activation-function is used, the exponential convergence with rate γ [in terms of residual error $e(t) = f(x(t),t) \to 0$] could be achieved for ZD model (7.4).*

2) *If the bipolar-sigmoid activation-function is used, the superior convergence can be achieved for error range $e(t) = f(x(t),t) \in [-\delta, \delta]$, $\exists \delta > 0$, as compared to the situation of using linear activation function described in Property 1.*

Zhang Dynamics for the Scalar-Valued Nonlinear Problems

3) *If the power activation function is used, then superior convergence can be achieved for error ranges $(-\infty, -1)$ and $(1, +\infty)$, as compared to the situation of using the linear activation function described in Property 1.*

4) *If the power-sigmoid activation function is used, superior convergence can be achieved for the whole error range $e(t) = f(x(t)) \in (-\infty, +\infty)$, as compared to the situation of using the linear activation function described in Property 1.*

Proof. We now come to prove the additional convergence properties of ZD (7.4) to $x^*(t)$ by considering the mentioned several types of activation functions $\phi(\cdot)$.

1) For the case of using linear activation function, it follows from (7.3) that $df(x(t),t)/dt = -\gamma f(x(t),t)$, which yields $f(x(t),t) = \exp(-\gamma t)f(x(0),0)$. This proves the exponential convergence rate γ of ZD (7.4) in the sense of residual error $e(t) = f(x(t),t) \to 0$. Moreover, we could show that the state $x(t)$ of ZD model (7.4) converges to the theoretical time-varying solution $x^*(t)$ of nonlinear equation $f(x(t),t) = 0$ by the following procedure. From Taylor's theorem [1], given some α between $x(t)$ and $x^*(t)$, we have

$$
\begin{aligned}
f(x) =& f\big(x(t) - x^*(t) + x^*(t), t\big) \\
=& f\big(x^*(t),t\big) + \big(x(t) - x^*(t)\big)\frac{\partial f\big(x(t),t\big)}{\partial x}\Big|_{x=x^*} \\
& + \frac{\big(x(t) - x^*(t)\big)^2}{2!}\frac{\partial f^{(2)}\big(x(t),t\big)}{\partial x^2}\Big|_{x=x^*} \\
& + \cdots + \frac{\big(x(t) - x^*(t)\big)^n}{n!}\frac{\partial f^{(n)}\big(x(t),t\big)}{\partial x^n}\Big|_{x=x^*} \\
& + \frac{\big(x(t) - x^*(t)\big)^{n+1}}{(n+1)!}\frac{\partial f^{(n+1)}\big(x(t),t\big)}{\partial x^{n+1}}\Big|_{x=\alpha}.
\end{aligned}
$$

Thus, in view of $f(x(t),t) = \exp(-\gamma t)f(x(0),0)$ and $f(x^*(t),t) = 0$, by omitting the higher-order terms, we could have

$$
\big(x(t) - x^*(t)\big)\frac{\partial f\big(x(t),t\big)}{\partial x}\Big|_{x=x^*} \approx \exp(-\gamma t)f\big(x(0),0\big),
$$

which, if $\partial f(x(t),t)/\partial x|_{x(t)=x^*(t)} \neq 0$, yields

$$
\|x(t) - x^*(t)\| \approx \exp(-\gamma t)\Big\|\frac{f\big(x(0),0\big)}{\partial f\big(x(t),t\big)/\partial x}\Big|_{x=x^*}\Big\|,
$$

where symbol $\|\cdot\|$ denotes the Euclidean norm for a vector (which, in our present situation, denotes the absolute value of a scalar argument in this chapter). More generally, even if $\partial f^{(k)}(x(t),t)/\partial x^k|_{x(t)=x^*(t)} = 0, \forall k = 1, 2, \cdots, (n-1)$ and $\partial f^{(n)}(x(t),t)/\partial x^n|_{x(t)=x^*(t)} \neq 0$, we could still have

$$
\|x(t) - x^*(t)\| \approx \exp(-\gamma t/n)\Big\|\Big(\frac{f\big(x(0),0\big)n!}{\partial f^{(n)}\big(x(t),t\big)/\partial x^n}\Big|_{x=x^*}\Big)^{\frac{1}{n}}\Big\|.
$$

This implies that the state $x(t)$ of ZD (7.4) could converge to the theoretical solution $x^*(t)$ in an exponential manner, provided that $x(0) = x_0$ is close enough to $x^*(t)$.

2) For the case of using bipolar-sigmoid function $\phi(u) = \left(1 - \exp(-\xi u)\right)/\left(1 + \exp(-\xi u)\right)$, we know that there exists an error range $e(t) = f(x(t),t) \in [-\delta,\delta]$ with $\delta > 0$ such that $\|\phi(f(x(t),t))\| > \|f(x(t),t)\|$ [11, 13]. So, by reviewing the proof of Proposition 7.2.1 [especially, equation (7.5)], we know that the superior convergence can be achieved by ZD model (7.4) with bipolar-sigmoid activation function for such an error range, as compared to Property 1.

3) For the pth power activation-function $\phi(u) = u^p$, the solution to differential equation (7.3) becomes

$$f(x(t),t) = f(x(0),0)\left((p-1)f^{p-1}(x(0),0)\gamma t + 1\right)^{-\frac{1}{p-1}}.$$

Besides, for $p = 3$, residual error $f(x(t),t) = f(x(0),0)/\sqrt{2f^2(x(0),0)\gamma t + 1}$. Evidently, as $t \to \infty$, $f(x(t),t) \to 0$. By reviewing the proof of Proposition 7.2.1, especially, the Lyapunov function candidate $V(x) = f^2(x(t),t)/2$ and its time derivative $\dot{V}(x,t) = -\gamma f(x,t)\phi(f(x,t))$ in (7.5), in the error range $\|f(x,t)\| \gg 1$, we could have $f^{p+1}(x,t) \gg f^2(x,t) \gg \|f(x,t)\|$. In other words, the deceleration magnitude of power activation function could be much greater than that of linear function. This implies that when using power function, a much faster convergence could be achieved by ZD (7.4) for such error ranges in comparison with Property 1 of using linear activation function [11, 13].

4) It follows from the above analysis that, to achieve superior convergence, a high-performance neural dynamics could be developed by switching power activation function to sigmoid or linear activation function at the switching points $e(t) = f(x,t) = \pm 1$. Thus, if the power-sigmoid activation function is used with suitable design parameters ξ and p, superior convergence can be achieved by ZD model (7.4) for the whole error range $(-\infty, +\infty)$, as compared to Property 1 of using the linear activation function. $\qquad\square$

Remark 7.2.1. It is worth comparing here the two design methods of GD model (7.2) and ZD model (7.4), both of which are exploited for the online solution of the nonlinear time-varying equation (7.1). The differences lie in the following facts.

1) The design of GD model (7.2) is based on the elimination of the square-based energy-function $\mathcal{E}(t) = f^2(x(t),t)$. In contrast, the design of ZD model (7.4) is based on the elimination of an indefinite error-function $e(x(t),t) = f(x(t),t)$, which could be positive, negative, bounded or even unbounded.

2) The design of GD model (7.2) is based on a method intrinsically for nonlinear equations but with constant coefficients. It is thus only able to approximately approach the theoretical solution of a time-varying problem. In contrast, the design of ZD (7.4) is based on a new method intrinsically for time-varying problems solving. It is thus able to exponentially converge to an exact/theoretical solution of the time-varying nonlinear equation.

3) GD model (7.2) has not exploited the time-derivative information of $f(\cdot)$ [i.e., $\partial f(x(t),t)/\partial t$], and thus is not effective enough on solving such a time-varying nonlinear-equation problem. In contrast, ZD model (7.4) methodically and systematically exploits the time-derivative information of function $f(\cdot)$ during its real-time solving process. This seems to be the reason why ZD model (7.4) could exponentially converge to the exact/theoretical solution of such a time-varying nonlinear-equation problem. $\qquad\square$

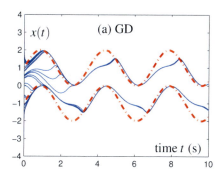 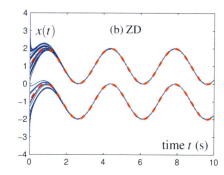

Figure 7.1. Solution-performance comparison of GD (7.2) and ZD (7.4) with design parameter $\gamma = 1$ for solving nonlinear time-varying equation (7.6), where dash lines in red denote the theoretical solutions.

7.2.4. Simulation Studies

While the previous subsections present the theoretical results about the GD model (7.2) and ZD model (7.4), this subsection substantiates them by showing the following computer-simulation results and observations (which are all based on the power-sigmoid activation function depicted in equation (1.7) with $\xi = 4$ and $p = 3$).

For illustration and comparison purposes, both neural dynamics [i.e., GD (7.2) and ZD (7.4)] are exploited for solving online the time-varying nonlinear equation $f(x(t),t) = 0$ with the following illustrative example:

$$f(x,t) = x^2 - 2\sin(1.8t)x + \sin^2(1.8t) - 1 = 0. \tag{7.6}$$

The above is actually equal to $f(x,t) = (x - \sin(1.8t) - 1)(x - \sin(1.8t) + 1) = 0$, and its time-varying theoretical solutions could be written down as $x_1^*(t) = \sin(1.8t) - 1$ and $x_2^*(t) = \sin(1.8t) + 1$. The computer-simulation results of applying GD model (7.2) and ZD model (7.4) to solve (7.6) are illustrated in Figures 7.1 and 7.2. Figure 7.1 shows the solution-performance comparison of GD (7.2) and ZD (7.4) with $\gamma = 1$ during the solution procedure of nonlinear time-varying equation (7.6), where dash lines in red denote the theoretical time-varying solutions. As seen from Figure 7.1(a), starting from randomly-generated initial states within $[-4,4]$, the neural state $x(t)$ of GD (7.2) does not fit well with the theoretical time-varying solutions $x_1^*(t)$ and/or $x_2^*(t)$. Simply put, the steady-state error of GD model (7.2) is considerably large. In contrast, the neural state $x(t)$ of ZD model (7.4), also starting from randomly-generated initial state within $[-4,4]$, could converge to one of the theoretical solutions [i.e., $x_1^*(t)$ or $x_2^*(t)$], which could be shown clearly in Figure 7.1(b). This is because the time-derivative information of time-varying coefficients in (7.6) has been fully utilized in Zhang et al's neural-dynamic design method and model.

Furthermore, in order to further investigate the convergence performance, we could also monitor the residual error $\|x^2 - 2\sin(1.8t)x + \sin^2(1.8t) - 1\|$ during the equation-solving process by both neural models. It is seen from Figure 7.2(a) and other simulation data that, by using GD (7.2) to online solution of time-varying nonlinear equation (7.6), its residual error $\|x^2 - 2\sin(1.8t)x + \sin^2(1.8t) - 1\|$ is rather large, and its steady-state residual error $\lim_{t\to\infty} \|x^2 - 2\sin(1.8t) + \sin^2(1.8t) - 1\|$ is about 0.3146 and 0.0279 (as

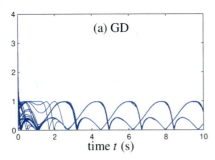 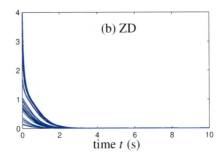

Figure 7.2. Residual error $\|x^2 - 2\sin(1.8t)x + \sin^2(1.8t) - 1\|$ of GD (7.2) and ZD (7.4) with design parameter $\gamma = 1$ for solving nonlinear time-varying equation (7.6).

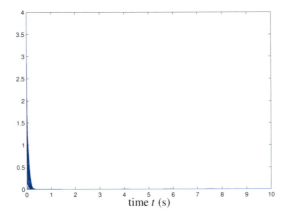

Figure 7.3. Residual error $\|x^2 - 2\sin(1.8t)x + \sin^2(1.8t) - 1\|$ of ZD (7.4) with design parameter $\gamma = 10$ for solving nonlinear time-varying equation (7.6).

computed at $t = 100$ seconds), which respectively correspond to the usage of design parameter $\gamma = 1$ and $\gamma = 10$. In contrast, as seen form Figures 7.2(b) and 7.3 as well as other simulation data that, by applying ZD (7.4) to solve time-varying nonlinear equation (7.6) under the same simulation conditions, its residual error $\|x^2 - 2\sin(1.8t)x + \sin^2(1.8t) - 1\|$ could converge to zero within 5 seconds. More specifically, the steady-state residual error $\lim_{t \to \infty} \|x^2 - 2\sin(1.8t) + \sin^2(1.8t) - 1\|$ is about 1.046×10^{-12} and 1.199×10^{-13}, which respectively correspond to the usage of design parameter $\gamma = 1$ and $\gamma = 10$. Note that such a maximum steady-state residual error should theoretically be zero but are numerically nonzero because of the finite-arithmetic simulation performed on finite-memory digital computers with floating-point relative accuracy being 2.220×10^{-16} [MathWorks Inc., Version 7.0.4.365 (R14)]. In addition, it is worth pointing out that, as shown in Figures 7.2(b) and 7.3, the convergence time for ZD (7.4) can be expedited from 5 seconds to 0.5 seconds, as design parameter γ is increased from 1 to 10. Moreover, if $\gamma = 10^4$, the convergence time is 0.5 milliseconds. This may show that ZD (7.4) has exponential-convergence property, which could be expedited by increasing the value of design parameter γ.

We could thus summarize that the ZD model is more effective and efficient for solving online nonlinear time-varying equations, as compared to the conventional GD model.

7.3. The Constant Nonlinear Equations

Let us consider the constant (or termed, static) nonlinear equation of a general form,

$$f(x) = 0, \qquad (7.7)$$

where $f(\cdot) : \mathbb{R} \to \mathbb{R}$ is assumed continuously-differentiable. Our starting point in this section is to exploit GD and ZD to find a root $x \in \mathbb{R}$ such that the above nonlinear equation (7.7) holds true. For ease of presentation, let x^* denote such a theoretical solution (also termed, root, zero) of nonlinear equation (7.7). In the ensuing subsections, gradient dynamics (GD) and Zhang dynamics (ZD) are both generalized, modeled and exploited comparatively to solve nonlinear equation (7.7).

7.3.1. Gradient Dynamics

The conventional neural-dynamic schemes generally belong to the gradient-descent method in optimization [11, 17–19, 24, 29, 35]. For readers' convenience, we show the GD design procedure below.

Firstly, a square-based nonnegative energy-function such as $\mathcal{E}(x) = f^2(x)/2$ can be constructed so that its minimum points are the roots of the nonlinear equation $f(x) = 0$.

Secondly, the solution algorithm is designed to evolve along a descent direction of this energy function until a minimal point is reached. The typical descent direction is the negative gradient of this energy function $\mathcal{E}(x)$.

Thirdly, by using the negative-gradient method to construct a dynamic model for solving nonlinear equation $f(x) = 0$, we could have the following dynamic equation as of the conventional linear gradient-dynamic solver:

$$\frac{dx(t)}{dt} = -\gamma \frac{\partial \mathcal{E}(x)}{\partial x} = -\gamma f(x) \frac{\partial f(x)}{\partial x} = -\gamma f(x) f'(x).$$

As an extension of the above GD design method and model (which is inspired by [13, 25, 26, 32, 36, 37] as well), we can generalize a nonlinear form of gradient-based dynamics as the following by exploiting nonlinear activation function $\phi(\cdot)$:

$$\dot{x}(t) = -\gamma \phi\big(f(x)\big) f'(x), \qquad (7.8)$$

where design parameter $\gamma > 0$ is used to adjust the convergence rate of the neural-dynamics, $x(t)$, starting from randomly-generated initial condition $x(0) = x_0 \in \mathbb{R}$, is the neural-activation state corresponding to theoretical solution x^*, and activation function $\phi(\cdot) : \mathbb{R} \to \mathbb{R}$ is defined to be the same as that mentioned in Section 7.2..

7.3.2. Zhang Dynamics

To monitor and control the solution process of nonlinear equation (7.7), following Zhang *et al.*'s design method (having been proposed since March 2001, e.g., in [12, 13, 25, 32, 35–37]), we could firstly define the following indefinite error-function (of which the word "indefinite" means that such an error-function can be negative or even lower-unbounded):

$$e(t) := f\big(x(t)\big),$$

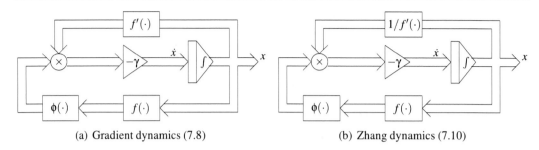

(a) Gradient dynamics (7.8) (b) Zhang dynamics (7.10)

Figure 7.4. Block diagrams of neural-dynamic solvers for online solution of $f(x) = 0$. *Reproduced from Y. Zhang, C. Yi et al., Comparison on gradient-based neural dynamics and Zhang neural dynamics for online solution of nonlinear equations, Figure 1, L. Kang et al. (Eds.): ISICA 2008, LNCS 5370, pp. 269-279, 2008. ©Springer-Verlag Berlin Heidelberg 2008. With kind permission of Springer Science+Business Media.*

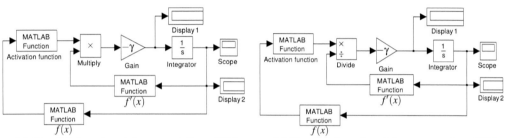

(a) Simulink-based model representation of GD (b) Simulink-based model representation of ZD

Figure 7.5. Simulink-based models representation about GD (7.8) and ZD (7.10). *Reproduced from Y. Zhang, P. Xu et al., Further studies on Zhang neural-dynamics and gradient dynamics for online nonlinear equations solving, Figure 1, Proceedings of 2009 IEEE International Conference on Automation and Logistics, pp. 566-571. ©[2009] IEEE. Reprinted, with permission.*

where, evidently, if the error-function $e(t)$ converges to zero, then $x(t)$ converges to theoretical solution (i.e., root) x^*.

Next, let the time derivative $\dot{e}(t)$ of error function $e(t)$ be chosen and forced mathematically such that $e(t)$ converges to zero; in math, we have to choose $\dot{e}(t)$ such that $\lim_{t \to \infty} e(t) = 0$. A suggested ZD (or termed ZNN, Zhang neural networks) design-formula for simply choosing a general form of the error-function's time-derivative $\dot{e}(t)$ can be the following (see [12, 13, 25, 32, 35–37] as well):

$$\frac{de(t)}{dt} = -\gamma \phi(e(t)), \text{ i.e., } \frac{df(x)}{dt} = -\gamma \phi(f(x)), \tag{7.9}$$

where design-parameter $\gamma > 0$ is used to control the convergence rate of the neural-dynamic solution, and activation function $\phi(\cdot) : \mathbb{R} \to \mathbb{R}$, which is defined the same as before, is assumed monotonically-increasing and odd. Expanding the ZD-design formula (7.9), we could have the following differential equation of Zhang dynamics:

$$\frac{\partial f(x)}{\partial x}\frac{dx}{dt} = -\gamma\phi\big(f(x)\big), \text{ i.e., } f'(x)\dot{x} = -\gamma\phi\big(f(x)\big),$$

$$\text{or equivalently [if } f'(x) \neq 0], \dot{x}(t) = -\gamma\frac{\phi\big(f(x)\big)}{f'(x)}, \tag{7.10}$$

where $x(t)$, starting from randomly-generated initial condition $x(0) \in \mathbb{R}$, is the state and output of ZD (7.10) corresponding to theoretical root x^* of nonlinear equation (7.7). Moreover, it follows from GD (7.8) and ZD (7.10) that, different GD and/or ZD performances can be achieved by using different design-parameter γ and activation function $\phi(\cdot)$.

In addition, the block-diagram representation of neural-dynamics GD (7.8) and ZD (7.10) could be shown in Figure 7.4. According to the two neural-dynamic equations and block-diagrams, the Simulink-based models representation of GD (7.8) and ZD (7.10) can also be seen from Figure 7.5. As seen from the two figures, the difference between GD (7.8) and ZD (7.10) for this specific problem $f(x) = 0$ solving is evidently interesting: GD has a multiplication term of $f'(x)$, whereas ZD has a division term of $f'(x)$. In addition to the difference, in view of the fact that both GD and ZD solvers could work effectively for this problem solving, people may ask why and how $f'(x)$ is needed as well as how GD and ZD solvers differ from each other. For more detailed, complete and systematic comparisons, please also see the preceding chapters and/or refer to [11–13, 18, 19, 24–26, 32, 35–37].

7.3.3. Theoretical Analysis

While the previous subsections present the general frameworks about gradient-based dynamics and Zhang dynamics for solving nonlinear equation $f(x) = 0$, detailed design consideration and theoretical results are given in this subsection. Following the Definitions 1.4.1 and 1.4.2 presented in Section 1.4., we could have the following propositions for Zhang dynamics (7.10).

Proposition 7.3.1. *Consider a solvable nonlinear equation $f(x) = 0$, where $f(\cdot)$ is a continuously differentiable function. If a monotonically-increasing odd activation function $\phi(\cdot)$ is employed, then the neural state $x(t)$ of Zhang dynamics (7.10) starting from randomly-generated initial state $x(0) = x_0 \in \mathbb{R}$ could converge to the theoretical solution x^* of nonlinear equation $f(x) = 0$ depicted in (7.7).*

Proof. We can define a Lyapunov function candidate $V(x) = f^2(x)/2 \geq 0$, and its time derivative along the system trajectory of Zhang dynamics (7.10) becomes

$$\frac{dV(x)}{dt} = f(x)f'(x)\frac{dx}{dt} = -\gamma f(x)\phi\big(f(x)\big). \tag{7.11}$$

Because a monotonically-increasing odd function is used as an activation function, we could have $\phi\big(-f(x)\big) = -\phi\big(f(x)\big)$, and

$$\phi\big(f(x)\big) \begin{cases} > 0, & \text{if } f(x) > 0, \\ = 0, & \text{if } f(x) = 0, \\ < 0, & \text{if } f(x) < 0. \end{cases}$$

Hence we have

$$f(x)\phi(f(x)) \begin{cases} > 0, & \text{if } f(x) \neq 0, \\ = 0, & \text{if } f(x) = 0, \end{cases}$$

which guarantees the final negative-definiteness of $\dot{V}(x)$; i.e., $\dot{V}(x) < 0$ for $f(x) \neq 0$ (equivalently, $x \neq x^*$) and $\dot{V}(x) = 0$ for $f(x) = 0$ (equivalently, $x = x^*$). By Lyapunov theory [34], residual error $e(t) = f(x(t))$ could converge to zero. That is, the neural state $x(t)$ of ZD (7.10) could converge to a theoretical solution x^* with $f(x^*) = 0$ starting from some randomly-generated sufficiently-close initial states [note that $f'(x)$ appears in the derivation of (7.11)]. The proof is now complete. $\qquad\square$

Proposition 7.3.2. *Let x^* denote a theoretical solution to problem $f(x) = 0$, where $f(\cdot)$ is a continuously-differentiable function (specifically, with at least first-order derivatives at some interval containing x^*). In addition to Proposition 7.3.1, the neural state $x(t)$ of Zhang dynamics (7.10) could converge to x^* if initial state x_0 is close enough to x^*. Moreover, Zhang dynamics (7.10) possesses the following properties.*

1) *If the linear activation function is used, then exponential convergence with rate γ [in terms of residual error $e(t) = f(x(t))$] can be achieved for (7.10).*

2) *If the bipolar-sigmoid activation-function is used, superior convergence can be achieved for error range $e(t) = f(x(t)) \in [-\delta, \delta]$, $\exists \delta > 0$, as compared to the situation of using the linear activation function described in Property 1.*

3) *If the power activation function is used, then superior convergence can be achieved for error ranges $(-\infty, -1)$ and $(1, +\infty)$, as compared to the situation of using the linear activation function described in Property 1.*

4) *If the power-sigmoid activation function is used, superior convergence can be achieved for the whole error range $e(t) = f(x(t)) \in (-\infty, +\infty)$, as compared to the situation of using the linear activation function in Property 1.*

Proof. We now come to prove the additional convergence properties of Zhang dynamics (7.10) by using the mentioned several types of activation functions $\phi(\cdot)$.

1) For the situation of using the linear activation function, it follows from equation (7.9) that $df(x)/dt = -\gamma f(x)$, which yields $f(x(t)) = \exp(-\gamma t)f(x_0)$. This proves the exponential convergence rate γ of Zhang dynamics (7.10) in the sense of its residual error $f(x(t)) \to 0$. Furthermore, we could also show the state convergence of Zhang dynamics to the theoretical solution x^* via the following procedure. According to Taylor's theorem [1], we could have

$$\begin{aligned} f(x) =& f(x - x^* + x^*) \\ =& f(x^*) + (x - x^*)f'(x)|_{x=x^*} + \frac{(x-x^*)^2}{2!}f''(x)|_{x=x^*} + \cdots \\ & + \frac{(x-x^*)^n}{n!}f^{(n)}(x)|_{x=x^*} + \frac{(x-x^*)^{n+1}}{(n+1)!}f^{(n+1)}(x)|_{x=\alpha}, \end{aligned}$$

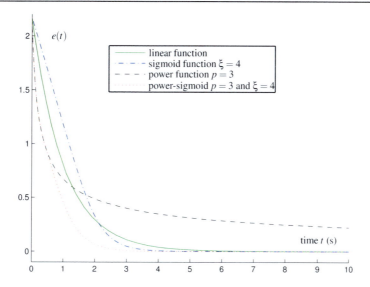

Figure 7.6. Convergence behavior of error $e(t)$ with different activation functions. *Reproduced from Y. Zhang, C. Yi et al., Comparison on gradient-based neural dynamics and Zhang neural dynamics for online solution of nonlinear equations, Figure 2, L. Kang et al. (Eds.): ISICA 2008, LNCS 5370, pp. 269-279, 2008. ©Springer-Verlag Berlin Heidelberg 2008. With kind permission of Springer Science+Business Media.*

for some α existing between $x(t)$ and x^*. Thus, in view of the $f(x(t)) = \exp(-\gamma t)f(x_0)$ and $f(x^*) = 0$, by omitting higher-order terms, we could have $(x - x^*)f'(x)|_{x=x^*} \approx \exp(-\gamma t)f(x_0)$, which, if $f'(x)|_{x=x^*} \neq 0$, yields

$$\|x - x^*\| \approx \exp(-\gamma t)f(x_0)/f'(x)|_{x=x^*}$$

and shows the exponential convergence of Zhang dynamics (7.10) in terms of the neural state $x(t) \to x^*$ as well. In addition, note that, even if $f^{(k)}(x)|_{x=x^*} = 0$, $\forall k = 1, 2, \cdots, (n-1)$, and $f^{(n)}(x)|_{x=x^*} \neq 0$, we still have

$$\|x - x^*\| \approx \exp(-\gamma t/n)\left(f(x_0)n!/f^n(x)|_{x=x^*}\right)^{1/n},$$

which shows again the exponential convergence of Zhang dynamics (7.10) in terms of neural state $x(t) \to x^*$, provided that initial state x_0 is close enough to x^*.

2) For the situation of using bipolar-sigmoid function $\phi(u) = (1 - \exp(-\xi u))/(1 + \exp(-\xi u))$, we know that there exists an error range $e(t) = f(x(t)) \in [-\delta, \delta]$ with $\delta > 0$ such that $\|\phi(f(x))\| > \|f(x)\|$ [11, 13]. So, by reviewing the proof of Proposition 7.3.1 [especially, equation (7.11)], we know further that superior convergence can be achieved by ZD (7.10) with bipolar-sigmoid activation function for such an error range, as compared to Property 1 of using the linear function.

3) For the pth power activation function $\phi(u) = u^p$, it follows from (7.9) that $df(x(t))/dt = -\gamma f^p(t)$, and its general form of trajectory is written down as

$$f(x(t)) = f(x_0)\left((p-1)f^{p-1}(x_0)\gamma t + 1\right)^{-\frac{1}{p-1}}.$$

Specifically, for $p = 3$, the residual error $f(x(t)) = f(x_0)/\sqrt{2f^2(x_0)\gamma t + 1}$. Evidently, as $t \to \infty$, $f(x) \to 0$. Look back to the proof of Proposition 7.3.1 [specifically, equation (7.11)]: as for Lyapunov function candidate $V(x) = f^2(x)/2$ and its time derivative $\dot{V}(x) = -\gamma f(x)\phi(f(x))$, we could have $f(x)\phi(f(x)) = f^{p+1}(x) \gg f^2(x)$ for error ranges $f(x) \ll -1$ and/or $f(x) \gg 1$. This implies that when using the power activation function, a much faster convergence can be achieved by ZD (7.10) for such error ranges in comparison with Property 1 [11, 13].

4) It follows from the above analysis (especially, Properties 2 and 3) that, in order to achieve superior convergence, a high-performance neural dynamics could be developed by switching the power activation function to the bipolar-sigmoid activation function at switching points $f(x) = \pm 1$. Thus, if the power-sigmoid activation function is used with suitable design parameters $\xi \geq 1$ and $p \geq 3$, superior convergence could be achieved for ZD (7.10) theoretically over the whole error range $(-\infty, +\infty)$, as compared to the linear-activation-function situation. $\square$

For graphical interpretation, the convergence behavior of residual error $e(t)$ is illustrated in Figure 7.6 by using different activation functions, where $\gamma = 1$. Note that, to draw all curves in the same one plot, small values of design parameters of nonlinear activation functions (such as $\xi = 4$ and $p = 3$) have to be used.

Remark 7.3.1. It is worth comparing here the two design methods of neural dynamics; namely, Zhang dynamics (7.10) and gradient-based dynamics (7.8). They are both exploited for online solution of nonlinear equation (7.7). However, gradient-based dynamics (7.8) is designed based on the elimination of square-based error-function $\mathcal{E}(x) = f^2(x)$ as well as the gradient-descent method. In contrast, Zhang dynamics (7.10) is designed based on the elimination of an indefinite error-function $e(t) = f(x)$ itself, which might be positive, negative, bounded or even (lower) unbounded. $\square$

7.3.4. Computer-Simulation Studies

While the previous subsections present GD (7.8), ZD (7.10) and related analysis results, in this subsection several illustrative examples are presented to investigate and show the convergent performance of the two kinds of neural dynamics, which are based on the use of the power-sigmoid activation function with design parameters $\xi = 4$ and $p = 3$.

Example 1. Consider the following nonlinear equation in the quadratic form:

$$f(x) = x^2/2 - 2x + 1.875 = 0. \tag{7.12}$$

For comparative purposes, the theoretical solutions to the above nonlinear equation could be written down as $x_1^* = 1.5$ and $x_2^* = 2.5$. These are used here to verify the theoretical results discussed in the previous subsections.

As seen from Figure 7.7, starting from randomly-generated initial states selected within $[-5, 5]$, the activation state $x(t)$ of the investigated neural-dynamics (7.8) and (7.10) could both converge to a theoretical solution, either x_1^* or x_2^* as denoted by dotted lines in red.

Example 2. Let us consider the following nonlinear equation:

$$f(x) = (x - 4)^{10}(x - 1) = 0. \tag{7.13}$$

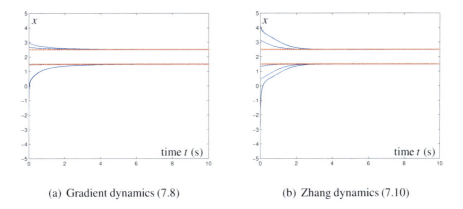

(a) Gradient dynamics (7.8)   (b) Zhang dynamics (7.10)

Figure 7.7. Online solution of nonlinear equation (7.12) by randomly-initialized GD (7.8) and ZD (7.10) with $\gamma = 1$, where dotted lines in red denote the theoretical solutions. *Reproduced from Y. Zhang, C. Yi et al., Comparison on gradient-based neural dynamics and Zhang neural dynamics for online solution of nonlinear equations, Figure 3, L. Kang et al. (Eds.): ISICA 2008, LNCS 5370, pp. 269-279, 2008. ©Springer-Verlag Berlin Heidelberg 2008. With kind permission of Springer Science+Business Media.*

To check the ND-solution correctness, the theoretical roots to the above nonlinear equation can be given simply as $x_1^* = 4$ (a multiple root of order 10) and $x_2^* = 1$ (a simple root).

It can be seen from the left subgraph of Figure 7.8 that the neural state $x(t)$ of ZD (7.10), starting from 200 initial states randomly generated within [-2,6], could converge to a theoretical root of nonlinear equation (7.13), either $x_1^* = 4$ or $x_2^* = 1$. In contrast, by applying gradient-based neural dynamics (7.8), some less-correct solutions (in other words, different from $x_1^* = 4$ and $x_2^* = 1$) have also been generated, which can be seen clearly in the right subgraph of Figure 7.8. In addition, the 3-dimensional convergent performance of ZD (7.10) and GD (7.8) solving nonlinear equation (7.13) is shown in Figure 7.9. It can be seen from the figure that, starting from initial states [e.g., $x(0) = 0.5$ here] close enough to the simple root $x^* = 1$, the neural states $x(t)$ of both ZD (7.10) and GD (7.8) converge to theoretical root $x^* = 1$. On the other hand, for the case of multiple root $x^* = 4$ of order 10, if the initial states [e.g., $x(0) = 2.5$ or $x(0) = 5.5$ here] are close enough to this multiple root, the neural state $x(t)$ of ZD (7.10) could converge well to the theoretical root $x_1^* = 4$; but, when GD (7.8) is applied under the same conditions, the GD neural-state $x(t)$ will mostly converge to a less-correct solution (or to say, an approximate solution with $0 < |x(40) - 4| < 1$).

Moreover, we have also investigated the convergent performance of the two dynamics solving nonlinear equation similar to (7.13) but with a multiple root of order less than 10 (specifically, 3 and 2); i.e., $f(x) = (x - 4)^3(x - 1) = 0$ and $f(x) = (x - 4)^2(x - 1) = 0$. The results are presented now and here. It follows from Figures 7.10 and 7.11 that, starting from 200 initial states randomly generated within [-2,6] and sufficiently close to simple root $x^* = 1$, the neural state $x(t)$ of both ZD (7.10) and GD (7.8) could converge well to such a theoretical root $x^* = 1$. On the other hand, for the case of the multiple root $x^* = 4$, the ZD neural-state $x(t)$ can (almost) always converge well to such a theoretical root $x^* = 4$; but,

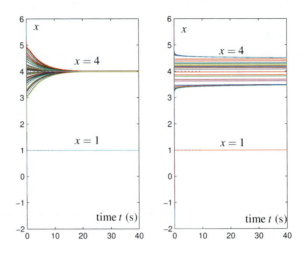

Figure 7.8. Online solution to $(x-4)^{10}(x-1) = 0$ by ZD (7.10) and GD (7.8) with design parameter $\gamma = 1$. Left: ZD (7.10); Right: GD (7.8). *Reproduced from Y. Zhang, P. Xu et al., Further studies on Zhang neural-dynamics and gradient dynamics for online nonlinear equations solving, Figure 2, Proceedings of 2009 IEEE International Conference on Automation and Logistics, pp. 566-571. ©[2009] IEEE. Reprinted, with permission.*

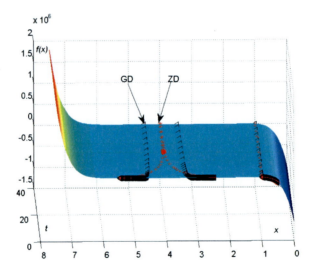

Figure 7.9. 3-dimensional convergent performance of ZD (7.10) and GD (7.8) with design parameter $\gamma = 1$ for online solution of $(x-4)^{10}(x-1) = 0$. *Reproduced from Y. Zhang, P. Xu et al., Further studies on Zhang neural-dynamics and gradient dynamics for online nonlinear equations solving, Figure 3, Proceedings of 2009 IEEE International Conference on Automation and Logistics, pp. 566-571. ©[2009] IEEE. Reprinted, with permission.*

Zhang Dynamics for the Scalar-Valued Nonlinear Problems

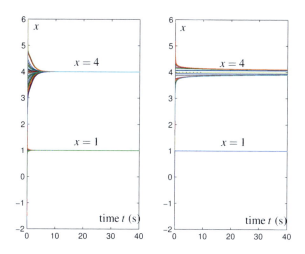

Figure 7.10. Online solution to $(x-4)^3(x-1) = 0$ by ZD (7.10) and GD (7.8) with design parameter $\gamma = 1$. Left: ZD (7.10); Right: GD (7.8). *Reproduced from Y. Zhang, P. Xu et al., Further studies on Zhang neural-dynamics and gradient dynamics for online nonlinear equations solving, Figure 4, Proceedings of 2009 IEEE International Conference on Automation and Logistics, pp. 566-571. ©[2009] IEEE. Reprinted, with permission.*

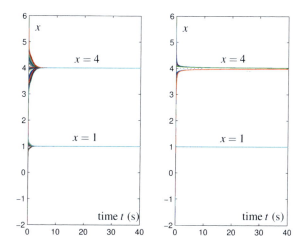

Figure 7.11. Online solution to $(x-4)^2(x-1) = 0$ by ZD (7.10) and GD (7.8) with design parameter $\gamma = 1$. Left: ZD (7.10); Right: GD (7.8). *Reproduced from Y. Zhang, P. Xu et al., Further studies on Zhang neural-dynamics and gradient dynamics for online nonlinear equations solving, Figure 5, Proceedings of 2009 IEEE International Conference on Automation and Logistics, pp. 566-571. ©[2009] IEEE. Reprinted, with permission.*

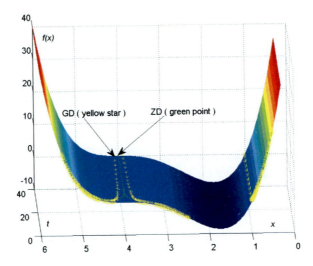

Figure 7.12. 3-dimensional convergent performance of ZD (7.10) and GD (7.8) with design parameter γ = 1 for online solution of $(x-4)^3(x-1) = 0$. *Reproduced from Y. Zhang, P. Xu et al., Further studies on Zhang neural-dynamics and gradient dynamics for online nonlinear equations solving, Figure 6, Proceedings of 2009 IEEE International Conference on Automation and Logistics, pp. 566-571. ©[2009] IEEE. Reprinted, with permission.*

similar to Figure 7.8, GD (7.8) may still generate less correct (or approximate) solutions. In addition, the corresponding 3-dimensional convergent performance of the two kinds of neural-dynamics is shown in Figures 7.12 and 7.13 (note that the three initial states involved are 0.5, 2.5 and 5.5).

It can be seen through Figures 7.8 to 7.13 that, for the case of the multiple root (i.e., $x^* = 4$), the neural state $x(t)$ of ZD (7.10) can (almost) always converge accurately to the theoretical root regardless of its order. The efficacy is possibly owing to the division term of $f'(x)$ exploited in ZD (7.10). In contrast, GD (7.8) may generate less correct results (or to say, approximate solutions). Moreover, comparing the right subgraphs of Figures 7.8, 7.10 and 7.11 (corresponding to Figures 7.9, 7.12 and 7.13, respectively), we can observe that, with the increase of the order of the multiple root $x^* = 4$, the convergent performance of GD (7.8) to it becomes (much) worse. This increasing inability, as we have investigated, is possibly owing to the multiplication term of $f'(x)$ exploited in GD (7.8). For more information, please expand these ND equations for such specific functions $f(x)$.

Example 3. Let us consider the following nonlinear equation:

$$f(x) = 0.01(x+7)(x-1)(x-8) + \sin x + 2.4 = 0, \qquad (7.14)$$

with its three theoretical roots denoted by "root 1", "root 2" and "root 3" in Figure 7.14. By using ZD (7.10) to solve the above nonlinear equation (7.14), we can see from Figure 7.14 that, starting from different initial states close enough, the neural state $x(t)$ could converge to their corresponding theoretical roots. For example, if initial state $x(0)$ is -9.8 or -5.0 (corresponding to "initial state 1" or "initial state 2" in the figure), the neural state $x(t)$ of ZD (7.10) converges to the "root 1" as depicted in Figure 7.14. Similarly, starting from

Zhang Dynamics for the Scalar-Valued Nonlinear Problems

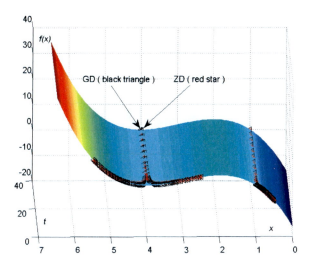

Figure 7.13. 3-dimensional convergent performance of ZD (7.10) and GD (7.8) with design parameter $\gamma = 1$ for online solution of $(x-4)^2(x-1) = 0$. *Reproduced from Y. Zhang, P. Xu et al., Further studies on Zhang neural-dynamics and gradient dynamics for online nonlinear equations solving, Figure 7, Proceedings of 2009 IEEE International Conference on Automation and Logistics, pp. 566-571. ©[2009] IEEE. Reprinted, with permission.*

initial state $x(0) = 1.0$ or 9.5 (corresponding to "initial state 5" and "initial state 6" in the figure), the neural state $x(t)$ would respectively converge to "root 2" and "root 3" as depicted in Figure 7.14.

Correspondingly, let us see another simulation result, i.e., Figure 7.15. Starting from 50 initial states randomly generated within [-10,10] and sufficiently close to the theoretical roots (i.e., "root 1", "root 2" and "root 3"), the neural states $x(t)$ of both ZD (7.10) and GD (7.8) converge well to them. But, for some initial states [e.g., $x(0) = -4$ or 0.8 corresponding to "initial state 3" or "initial state 4", respectively, in Figure 7.14] which are close to a local minimum point, GD (7.8) will generate a wrong solution, whereas, for this situation, the ZD-state $x(t)$ will move towards the local minimum point and then stop with the following warning information.

```
Warning: Failure at t=1.479876e-005. Unable
to meet integration tolerances without redu
-cing the step size below the smallest valu
-e allowed (2.710505e-020) at time t.
> In ode15s at 741   In znnvsgnnft at 20
```

This warning may remind users to re-choose a suitable initial state with which to restart the ZD-solution process for correct results. In contrast, as mentioned before, for this situation, GD (7.8) will yield a wrong solution [i.e., the local minimum point with $f'(x) = 0$], which may mislead the users into obtaining a "solution" of the nonlinear equation.

Example 4. Let us consider the following nonlinear equation:

$$f(x) = \cos x + 3 = 0, \qquad (7.15)$$

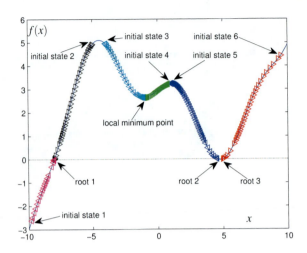

Figure 7.14. Convergent performance of the neural state of ZD (7.10) for online solution of nonlinear equation (7.14) starting from different initial states. *Reproduced from Y. Zhang, P. Xu et al., Further studies on Zhang neural-dynamics and gradient dynamics for online nonlinear equations solving, Figure 8, Proceedings of 2009 IEEE International Conference on Automation and Logistics, pp. 566-571. ©[2009] IEEE. Reprinted, with permission.*

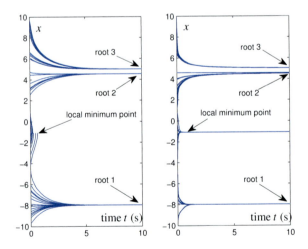

Figure 7.15. Online solution of $0.01(x+7)(x-1)(x-8)+\sin x+2.4=0$ by ZD (7.10) and GD (7.8) with parameter $\gamma = 1$. Left: ZD (7.10); Right: GD (7.8). *Reproduced from Y. Zhang, P. Xu et al., Further studies on Zhang neural-dynamics and gradient dynamics for online nonlinear equations solving, Figure 9, Proceedings of 2009 IEEE International Conference on Automation and Logistics, pp. 566-571. ©[2009] IEEE. Reprinted, with permission.*

Zhang Dynamics for the Scalar-Valued Nonlinear Problems

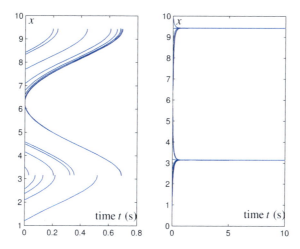

Figure 7.16. Online solution of ZD (7.10) (left) and GD (7.8) (right) with $\gamma = 1$ for nonlinear equation $\cos x + 3 = 0$ which evidently has no theoretical roots. *Reproduced from Y. Zhang, P. Xu et al., Further studies on Zhang neural-dynamics and gradient dynamics for online nonlinear equations solving, Figure 10, Proceedings of 2009 IEEE International Conference on Automation and Logistics, pp. 566-571. ©[2009] IEEE. Reprinted, with permission.*

for another interesting observation. Evidently, the above nonlinear equation (7.15) has no solution (i.e., no root, no zero). As can be seen from Figure 7.16, by applying ZD (7.10), the neural state $x(t)$, starting from 20 initial states within [0,10], moves towards some values and then stops with a warning as the above. In contrast, GD (7.8) generates some misleading results again. The specific values involved, as we investigate, are related to the local minimum points satisfying $f'(x) = -\sin x = 0$; i.e., $x = k\pi$ with $k = 0, \pm 1, \pm 2, \cdots$. In addition, it is worth mentioning an anonymous reviewer's understanding on this point; i.e., according to his/her experience, the warning given by the simulation tool may not mean "(it) remind(s) users to re-choose a suitable initial state"; on the contrary, it may mean "(we) are using the wrong tools for the job". For $f(x) = \cos x + 3 = 0$ depicted in (7.15), there is evidently no solution (or to say, no root), and thus no correct tools could complete the root-finding job. So, we believe, with many thanks to the reviewer, that a warning given by the simulation tool in our context means "it reminds users to re-choose a suitable initial state or tells users no solution around (the current initial state)". Besides, the authors suggest the readers checking the simulation manuals for more and exact information when conducting such simulations, in addition to having a right understanding of the warning messages.

In summary, from the above illustrative examples and [33], we have the following observations and facts about the convergent performance of ZD (7.10) and GD (7.8) exploited for the online solution of nonlinear equation $f(x) = 0$.

1) If the randomly-generated initial states are close enough to a theoretical root, for the case of a simple root, ZD (7.10) and GD (7.8) could both effectively converge to the theoretical simple root; and, for the case of a multiple root, with the increase of the order of the multiple root, the neural state $x(t)$ of ZD (7.10) could still converge

well to such a theoretical root [in contrast, GD (7.8) more probably yields wrong (or approximate) solutions].

2) For a nonlinear equation containing local minimum point(s) (e.g., see Example 3), if an initial state is close enough to the local minimum point, then the neural state $x(t)$ of ZD (7.10) would move towards the local-minimum point and then stop with a warning. This reminds users to re-choose another initial state and to start the ZD solver again if necessary. However, in this local-minimum-point situation, GD (7.8) generates wrong results, misleading users into thinking that a correct root is now obtained.

3) For the case of a nonlinear equation having no solution, the neural state of ZD (7.10) moves towards some value(s) [possibly the local minimum point(s) satisfying $f'(x) = 0$] and then stops with the warning information. In contrast, in this no-solution situation, GD (7.8) would generate wrong/misleading solution(s) again, which appears to be similar to the above observation.

7.4. Conclusion

By following Zhang *et al.*'s method, in this chapter, a special kind of neural dynamics has been generalized, developed, modeled and compared for online solution of time-varying or constant nonlinear equations. Differing from conventional gradient-based neural-dynamics, the resultant Zhang neural-dynamics has been elegantly driven by an indefinite error-function [instead of usually-exploited lower-bounded energy-function(s)]. For comparative purposes, the gradient-based neural-dynamics has also been applied to the online solution of those two kinds of nonlinear equations. In some senses, both theoretical analysis and computer simulation results demonstrate the efficacy and superiority of ZD models on time-varying and/or static nonlinear equations solving as well.

References

[1] Mathews, JH; Fink, KD. *Numerical Methods Using MATLAB, Fourth Edition.* New Jersey: Pretice Hall, 2004.

[2] Yang, L; Zhu, Z; Wang, Y. Exact solutions of nonlinear equations. *Physics Letters A,* 1999, vol. 260, no. 1-2, pp. 55-59.

[3] Babolian, E; Biazar, J. Solution of nonlinear equations by modified Adomian decomposition method. *Applied Mathematics and Computation,* 2002, vol. 132, no. 1, pp. 167-172.

[4] Babolian, E; Biazar, J; Vahidi, AR. Solution of a system of nonlinear equations by Adomian decomposition method. *Applied Mathematics and Computation,* 2004, vol. 150, no. 3, pp. 847-854.

[5] Chun, C. Construction of Newton-like iteration methods for solving nonlinear equations. *Numerische Mathematik,* 2006, vol. 104, pp. 297-315.

[6] Ujevic, N. A method for solving nonlinear equations. *Applied Mathematics and Computation,* 2006, vol. 174, no. 2, pp. 1416-1426.

[7] Lin, C. *Numerical Computation Methods.* Beijing: Science Press, 2005, pp. 18-53.

[8] Feng, J; Che, G; Nie, Y. *Principles of Numerical Analysis.* Beijing: Science Press, 2001, pp. 182-222.

[9] Sharma, JR. A composite third order Newton-Steffensen method for solving nonlinear equations. *Applied Mathematics and Computation,* 2005, vol. 169, no. 1, pp. 242-246.

[10] Zhang, Y; Leithead, WE; Leith, DJ. Time-series Gaussian process regression based on Toeplitz computation of $O(N^2)$ operations and $O(N)$-level storage. *Proceedings of IEEE Conference on Decision and Control,* 2005, pp. 3711-3716.

[11] Zhang, Y. Revisit the analog computer and gradient-based neural system for matrix inversion. *Proceedings of IEEE international symposium on intelligent control,* 2005, pp. 1411-1416.

[12] Zhang, Y; Jiang, D; Wang, J. A recurrent neural network for solving Sylvester equation with time-varying coefficients. *IEEE Transactions on Neural Networks,* 2002, vol. 13, no. 5, pp. 1053-1063.

[13] Zhang, Y; Ge, SS. Design and analysis of a general recurrent neural network model for time-varying matrix inversion. *IEEE Transactions on Neural Networks,* 2005, vol. 16, no. 6, pp. 1477-1490.

[14] Zhang, Y. A set of nonlinear equations and inequalities arising in robotics and its online solution via a primal neural network. *Neurocomputing,* 2006, vol. 70, no. 1-3, pp. 513-524.

[15] Steriti, RJ; Fiddy, MA. Regularized image reconstruction using SVD and a neural network method for matrix inversion. *IEEE Transactions on Signal Processing,* 1993, vol. 41, no. 10, pp. 3074-3077.

[16] Tank, D; Hopfield, JJ. Simple neural optimization networks: an A/D converter, signal decision circuit, and a linear programming circuit. *IEEE Transactions on Circuits and Systems,* 1986, vol. 33, no. 5, pp. 533-541.

[17] Jang, J; Lee, S; Shin, S. *An Optimization Network for Matrix Inversion, in: Neural Information Processing Systems (ed. Anderson, DZ),* New York: American Institute of Physics, 1988, pp. 397-401.

[18] Wang, J. A recurrent neural network for real-time matrix inversion. *Applied Mathematics and Computation,* 1993, vol. 55, no. 1, pp. 89-100.

[19] Zhang, Y. Towards piecewise-linear primal neural networks for optimization and redundant robotics. *Proceedings of IEEE International Conference on Networking, Sensing and Control,* 2006, pp. 374-379.

[20] Zhang, Y; Wang, J. Global exponential stability of recurrent neural networks for synthesizing linear feedback control systems via pole assignment. *IEEE Transactions on Neural Networks,* 2002, vol. 13, no. 3, pp. 633-644.

[21] Zhang, Y; Wang, J. A dual neural network for convex quadratic programming subject to linear equality and inequality constraints. *Physics Letters A,* 2002, vol. 298, no. 4, pp. 271-278.

[22] Cichocki, A; Unbehauen, R. Neural network for solving systems of linear equations and related problems. *IEEE Transactions on Circuits and Systems,* 1992, vol. 39, no. 2, pp. 124-138.

[23] Cichocki, A; Unbehauen, R. *Neural Networks for Optimization and Signal Processing.* Chichester: Wiley, 1993.

[24] Zhang, Y; Chen, K; Ma, W; Li, X. MATLAB simulation of gradient-based neural network for online matrix inversion. *LNAI Proceedings of International Conference on Intelligent Computing,* 2007, vol. 4682, pp. 98-109.

[25] Zhang, Y; Chen, Z; Chen, K; Cai, B. Zhang neural network without using time-derivative information for constant and time-varying matrix inversion. *Proceeding of international joint conference on neural networks,* 2008, pp. 142-146.

References

[26] Zhang, Y; Tan, N; Cai, B; Chen, Z. MATLAB Simulink modeling of Zhang neural network solving for time-varying pseudoinverse in comparison with gradient neural network. *Proceedings of the 2nd International Symposium on Intelligent Information Technology Application,* 2008, vol. 1, pp. 39-43.

[27] Zhang, Y; Li, Z. Zhang neural network for online solution of time-varying convex quadratic program subject to time-varying linear-equality constraints. *Physics Letters A,* 2009, vol. 373, no. 18-19, pp. 1639-1643.

[28] Mead, C. *Analog VLSI and Neural Systems.* Reading, MA: Addison-Wesley, 1989.

[29] Manherz, RK; Jordan, BW; Hakimi, SL. Analog methods for computation of the generalized inverse. *IEEE Transactions on Automatic Control,* 1968, vol. 13, no.5, pp. 582-585.

[30] Zhang, Y; Chen, K. Comparison on Zhang neural network and gradient neural network for time-varying linear matrix equation $AXB = C$ solving. *Proceeding of IEEE international conference on industrial technology,* 2008, pp. 1-6.

[31] Zhang, Y; Guo, X; Ma, W. Modeling and simulation of Zhang neural network for online linear time-varying equations solving based on MATLAB Simulink. *Proceedings of the seventh international conference on machine learning and cybernetics,* 2008, pp. 12-15.

[32] Zhang, Y; Peng, H. Zhang neural network for linear time-varying equation solving and its robotic application. *Proceedings of the sixth international conference on machine learning and cybernetics,* 2007, pp. 19-22.

[33] Zhang, Y; Ma, W; Yi, C. The link between Newton iteration for matrix inversion and Zhang neural network. *Proceeding of IEEE international conference on industrial technology,* 2008, pp. 1-6.

[34] Zhang, Y. *Dual Neural Networks: Design, Analysis, and Application to Redundant robotics, in: Progress in neurocomputing research (ed: Kang, GB).* New York: Nova Science Publishers, 2007, pp. 41-81.

[35] Zhang, Y; Yang, Y. Simulation and comparison of Zhang neural network and gradient neural network solving for time-varying matrix square roots. *Proceedings of the 2nd International Symposium on Intelligent Information Technology Application,* 2008, vol. 2, pp. 966-970.

[36] Zhang, Y; Fan, Z; Li, Z. Zhang neural network for online solution of time-varying Sylvester equation. *LNCS Proceedings of International Symposium on Intelligence Computation and Applications,* 2007, vol. 4683, pp. 276-285.

[37] Chen, K; Yue, S; Zhang, Y. MATLAB simulation and comparison of Zhang neural network and gradient neural network for online time-varying matrix equation $AXB - C = 0$. *Proceedings of International Conference on Intelligent Computing,* 2008, pp. 68-75.

Chapter 8

ZNN and GNN Models for Linear Constant Problems Solving

Abstract

In this chapter, two recurrent neural networks, Zhang neural network (ZNN) and gradient neural network (GNN), are generalized and developed to solve online a set of simultaneous linear constant problems. Firstly, we investigate the convergence, network architecture and electronic implementation of ZNN model for linear equations solving. Secondly, rather than previously-presented asymptotical convergence, global exponential convergence is proved for GNN models used to solve linear equations. Moreover, superior convergence could be achieved by using power-sigmoid activation-functions, as compared with the situation of using linear activation-functions. Finally, this chapter investigates the simulation of GNN for online solution of the matrix-inverse problem. Several important techniques are employed to simulate such a neural-network system. In addition to investigating the singular case, this chapter also presents an application example on inverse-kinematic control of redundant manipulators via online pseudoinverse solution. Computer-simulation results substantiate further the analysis and efficacy of ZNN and GNN models for online linear constant problems solving.

8.1. Introduction

The problem solving of matrix and/or vector equations is widely encountered in various science and engineering fields, as it is usually an essential part in many solutions and applications; e.g., as preliminary steps for optimization [1, 2], signal-processing [3], electromagnetic systems [4], and robot inverse kinematics [5–7]. Our recent research has been directed towards the online solution of algebraic equations, which especially includes matrix inversion and linear equations solving. Numerical algorithms performed on digital computers are evidently a powerful tool for solving such problems [8–10], but their minimal arithmetic operations are usually proportional to the cube of the matrix dimension. Another problem-solving approach is the neural dynamic method, which has recently been investigated widely owing to its parallel distributed processing nature and convenience of analogue-circuit implementation [11–15].

It is worth mentioning that almost all the aforementioned neural dynamic computation schemes were designed intrinsically for constant matrices rather than time-varying matrices. They are in general related to the gradient descent method in optimization [1], where a scalar-valued cost function like $\|Ax - b\|^2$ or $\|AX - I\|^2$ is first constructed such that its minimum point is the theoretical solution of the problem, and then an algorithm is designed to evolve along a descent direction of this cost function until a minimum is reached. The typical descent direction is the negative gradient.

For online solution of linear equations, Wang presented a gradient neural network and its electronic circuit model [11]. Raida improved the electronic-circuit implementation for such a Wang neural network [15]. Asymptotical convergence was presented as well in [11] for such a gradient neural network. It is worth pointing out that a pure asymptotical convergence just implies that the GNN states only theoretically approach the solution (as time $t \to +\infty$), which may not be acceptable in practice [13]. As a result, the analysis on global exponential convergence is desirable for GNN models. In this chapter, we generalize Wang's neural network by using different activation-functions and provide detailed analysis on global exponential convergence of the GNN models.

Different from conventional gradient neural networks [11–15], a new recurrent neural network (RNN) model is presented based on a vector-valued error-monitoring function instead of the usually norm-based scalar-valued error functions [12, 13] (this special kind of RNN has been proposed by Zhang *et al.* for solving online time-varying matrix problems since March 2001, the formal name is Zhang neural network, and ZNN for short). In addition, different from generally explicit-dynamic formulations, such a ZNN model is depicted also via an implicit dynamic equation, and its electronic implementation appears to be as simple as that of conventional RNNs.

8.2. ZNN for Linear Time-Invariant Equations Solving

In this section, ZNN is presented for solving online linear time-invariant (LTI) equations, which has been developed based ingeniously on a vector-valued error-function rather than a scalar-valued norm-based function. Theoretical analysis and simulation results both substantiate the efficacy of such a ZNN model for online LTI equations solving.

8.2.1. ZNN Model for Linear Time-Varying (LTV) Equations Solving

Consider the linear time-varying (LTV) equations

$$A(t)x(t) = b(t), \tag{8.1}$$

where $A(t) \in \mathbb{R}^{n \times n}$ denotes a smoothly time-varying nonsingular coefficient matrix, $b(t) \in \mathbb{R}^n$ is a time-varying coefficient vector, and $x(t)$ is the unknown vector to be solved online. Assume that time-derivatives $\dot{A}(t)$ and $\dot{b}(t)$ are known (or at least measurable via analogue differentiation circuits or finite-difference approximation). According to Zhang *et al.*'s design method [12, 13], based on the vector-valued error-function $e(t) = A(t)x(t) - b(t) \in \mathbb{R}^n$ [instead of the scalar-valued norm-based error-function $\mathcal{E}_1(t) = \|Ax - b\|_2^2/2$ usually associated with gradient-based methods for time-invariant problems solving], the ZNN model

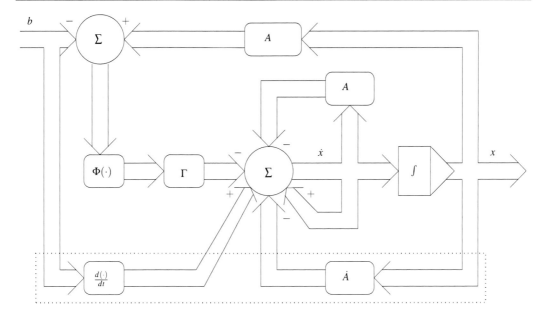

Figure 8.1. Block diagram of network architecture of ZNN models (8.2) and (8.5). *Reproduced from C. Yi and Y. Zhang, Analogue recurrent neural network for linear algebraic equation solving, Figure 1, Electronics Letters, pp. 1078-1079. ©[2008] IEEE. Reprinted, with permission.*

solving the LTV equations (8.1) can be designed as

$$A(t)\dot{x}(t) = -\dot{A}(t)x(t) - \Gamma\Phi\big(A(t)x(t) - b(t)\big) + \dot{b}(t), \tag{8.2}$$

where design parameter (i.e., learning rate) $\Gamma \in \mathbb{R}^{n \times n}$ is a positive-definite matrix used to adjust the convergence rate. Generally, Γ is set as γI, where $\gamma > 0$ should be selected as large as hardware permits and/or for simulative purposes, and I denotes the appropriately-dimensioned identity-matrix. Besides, $\Phi(\cdot) : \mathbb{R}^n \to \mathbb{R}^n$ is a vector-form activation-function array of the ZNN model. In general, any monotonically-increasing odd activation-function $\phi(\cdot)$, being the ith element of $\Phi(\cdot)$ (with $i = 1, 2, 3, \cdots, n$), could be used for the construction of the neural network (e.g., linear activation-function and power-sigmoid activation-function). In this subsection, the linear activation-function (1.4) is employed exemplarily; i.e., (8.2) can be simplified as $A(t)\dot{x}(t) = -\dot{A}(t)x(t) - \gamma(A(t)x(t) - b(t)) + \dot{b}(t)$. Figure 8.1 shows the block diagram for the architecture of ZNN (8.2), where argument t is dropped owing to presentation convenience. As the time-derivative information of coefficients in (8.1) is utilized elegantly, this neural-dynamic approach could solve online the time-varying linear equations (8.1) effectively and accurately [12, 13].

8.2.2. Motivation for Simplification

Evidently, the above ZNN model (8.2) is depicted in implicit dynamics, i.e., $A(t)\dot{x}(t) = \cdots$, which coincides well with systems in nature and in practice. For example, the implicit dynamics (or implicit systems) frequently arise in analogue electronic circuits and systems owing to Kirchhoff's laws. Furthermore, implicit systems appear to have higher abilities in representing dynamic systems, compared to explicit systems [e.g., Wang neural network

(8.4) presented comparatively in the following subsection]. This is in the sense that implicit dynamics can preserve physical parameters in the coefficient matrix on the left-hand side of the system as well, e.g., $A(t)$ in (8.2). Thus, motivated by the efficacy and superiority of the implicit ZNN model (8.2), in the remainder of this section, such a ZNN model is simplified and applied to the linear time-invariant equations solving, together with the network architecture and electronic implementation discussed.

8.2.3. Linear Time-Invariant (LTI) Equations Solving

Corresponding to the aforementioned LTV equations (8.1), let us consider the following usually-investigated LTI algebraic equations [8, 11]:

$$Ax = b, \tag{8.3}$$

where coefficient matrix $A := [a_{ij}] \in \mathbb{R}^{n \times n}$ and vector $b = [b_1, b_2, \cdots, b_n]^T \in \mathbb{R}^n$ are time-invariant hereafter (simply put, constant), and $x \in \mathbb{R}^n$ is the unknown vector to be solved. For solving linear system (8.3), many methods have been developed. For example, the recursive numerical-algorithm presented in [8] can find the solution of a total n equations after the nth step recursion by using Kaczmarz's method, whereas in [11] the gradient-based neural-dynamic model was proposed by Wang for solving (8.3):

$$\dot{x} = -\gamma A^T A x + \gamma A^T b. \tag{8.4}$$

It is worth pointing out comparatively that, by using Zhang *et al.*'s method [12,13], the ZNN model (8.2) designed for solving the time-varying linear equation (8.1) could be exploited to solve (8.3) with the following new simplified dynamics:

$$A\dot{x} = -\gamma \Phi(Ax - b). \tag{8.5}$$

As mentioned above, conventional GNN models are described by explicit dynamics, e.g., Wang's network (8.4). In contrast, our ZNN model (8.5) is depicted by an implicit dynamical equation, and its network architecture is illustrated in Figure 8.1 but without the dotted-rectangle part. It has no such time-derivative term because the derivatives of constant coefficients (i.e., A and b) of (8.3) equal to zero. In addition, symbols A and b in Figure 8.1 correspond to the constant coefficients A and b in linear equations (8.3) and the ZNN model (8.5). Moreover, it follows from the ZNN model (8.5) that its ith neuron's dynamic-equation can be written as the following explicit dynamics by using derivative and self feedback:

$$\dot{x}_i = -\gamma \phi \left(\sum_{j=1}^{n} a_{ij} x_j - b_i \right) - \left(\sum_{j=1, j \neq i}^{n} a_{ij} \dot{x}_j \right) + (1 - a_{ii}) \dot{x}_i, \ i = 1, 2, \cdots, n \tag{8.6}$$

Figure 8.2 shows the architecture of the electronic realization of (8.6), which is connected in both feedforward and feedback manner (that is partially why model (8.5) is termed recurrent network). In addition, for the ZNN model (8.5) solving the LTI equation (8.3), we can have the following proposition on its global exponential convergence.

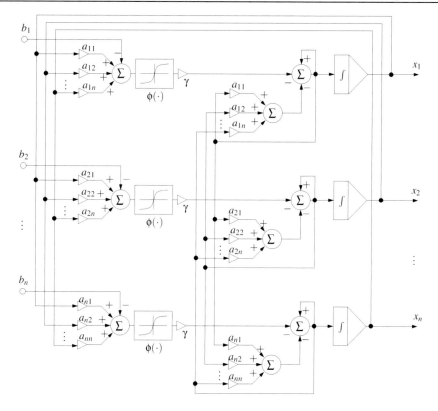

Figure 8.2. Circuit schematics of ZNN model (8.5) as derived from (8.6). *Reproduced from C. Yi and Y. Zhang, Analogue recurrent neural network for linear algebraic equation solving, Figure 2, Electronics Letters, pp. 1078-1079. ©[2008] IEEE. Reprinted, with permission.*

Proposition 8.2.1. *Consider a constant nonsingular coefficient-matrix $A \in \mathbb{R}^{n \times n}$ and coefficient vector $b \in \mathbb{R}^n$. If a monotonically-increasing odd activation-function array $\Phi(\cdot)$ is used, then the neural state $x(t)$ of ZNN model (8.5), starting from any initial state $x(0) \in \mathbb{R}^n$, will converge (in an exponential manner of residual error $Ax - b$) to the unique solution of linear equations (8.3) [12, 13].* □

8.2.4. Illustrative Example

For comparative and illustrative purposes, we can consider the same example (i.e., Hilbert matrix) as in [8]:

$$A = \begin{bmatrix} 1 & \frac{1}{2} & \cdots & \frac{1}{n} \\ \frac{1}{2} & \frac{1}{3} & \cdots & \frac{1}{n+1} \\ \cdots & \cdots & \cdots & \cdots \\ \frac{1}{n} & \frac{1}{n+1} & \cdots & \frac{1}{2n-1} \end{bmatrix}, \text{ or simply, } A = [a_{ij}] \text{ with } a_{ij} = \frac{1}{i+j-1},$$

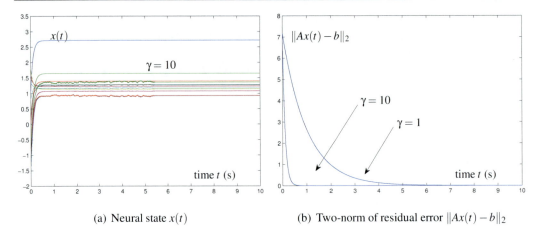

(a) Neural state $x(t)$ (b) Two-norm of residual error $\|Ax(t)-b\|_2$

Figure 8.3. Trajectories of $x(t)$ and $\|Ax(t)-b\|_2$ synthesized by ZNN model (8.5). *Reproduced from C. Yi and Y. Zhang, Analogue recurrent neural network for linear algebraic equation solving, Figure 3, Electronics Letters, pp. 1078-1079.* ©*[2008] IEEE. Reprinted, with permission.*

which could be highly ill-conditioned [8], in addition to $b = [b_1, b_2, b_3, \cdots, b_n]^T$ with $b_i = \sum_{j=1}^{n} a_{ij} e^{1/j}$, and here $e \simeq 2.718281828$. As we may know [8], if $n > 10$, many methods (e.g., Gauss-Jordan elimination method or LU factorization method) could not obtain effectively and accurately the numerical solution of LTI equation (8.3) with the above coefficients A and b. In comparison, by the Proposition 8.2.1, the neural-state $x(t)$ of ZNN (8.5) could globally converge to the unique solution of (8.3), which is shown in Figure 8.3(a) (with $\gamma = 10$ and $n = 12$). It is worth mentioning that if γ is increased to 10^4, the convergence time is within 6 ms. Figure 8.3(b) illustrates the convergence properties of the two-norm of the corresponding residual-error $\|Ax(t)-b\|_2$, which always converges to zero. In addition, the maximal two-norm value of steady-state residual-error, $\lim_{t \to +\infty} \|Ax(t)-b\|_2$, is about 1.1322×10^{-15} by using the resultant ZNN model (8.5) (with design parameter $\gamma = 10$ and measured as of $t = 10$ s). The above substantiates the efficacy of ZNN model (8.5) as an alternative new method for solving online the LTI equation (8.3).

8.3. GNN for Linear Equations Solving

In this section, we generalize Wang neural network (8.4) by using linear and/or power-sigmoid activation-function arrays $\Phi(\cdot)$ into the general GNN models, and then provide detailed analysis on global exponential convergence of such models. The aforementioned global exponential convergence implies that a recurrent neural network could converge arbitrarily fast to the solution [16]. In contrast, the asymptotical convergence only theoretically guarantees that a recurrent neural network could converge to the solution as time t tends to infinity [13]. For a better understanding on the difference and significance of exponential convergence, we show Figure 8.4 which gives a clear comparison with asymptotical convergence. So, in the remainder of this section, we focus on the exponential convergence and stability of the GNN models.

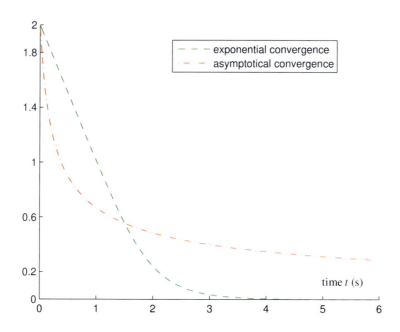

Figure 8.4. Comparison between asymptotical convergence and exponential convergence. *Reproduced from Y. Zhang and K. Chen, Global exponential convergence and stability of Wang neural network for solving online linear equations, Figure 1, Electronics Letters, pp. 145-146. ©[2008] IEEE. Reprinted, with permission.*

8.3.1. Problem Formulation and GNN Solver

In order to solve linear equation (8.3) (starting from this subsection, we only discuss LTI equations, so linear equations actually means LTI equations) in parallel and in real time t, Wang neural network (8.4) [11] could be rewritten as the following vector-form differential equation:

$$\dot{x}(t) = -\gamma A^T \left(Ax(t) - b \right), \tag{8.7}$$

where initial state $x_0 := x(0) \in \mathbb{R}^n$ and A^T denotes the transpose of matrix A. The neural-network architecture of GNN model (8.7) is shown in Figure 1.2. By using a monotonically-increasing odd activation-function array $\Phi(\cdot)$, GNN model (8.7) can be generalized as

$$\dot{x}(t) = -\gamma A^T \Phi \left(Ax(t) - b \right), \tag{8.8}$$

where $\Phi(\cdot) : \mathbb{R}^n \to \mathbb{R}^n$ denotes an activation-function vector mapping (or to say, an activation-function vector array) of recurrent neural networks. Note that array $\Phi(\cdot)$ is made of n monotonically-increasing odd activation-functions $\phi(\cdot)$; e.g., the linear activation-function (1.4) and the power-sigmoid activation-function (1.7) depicted in Figure 1.3.

It is worth mentioning that, the power-sigmoid activation-function, constructed as a combination of power and sigmoid functions, could make the neural network achieve superior convergence and robustness properties [13]. Besides, the block diagram of GNN model (8.8) is depicted in Figure 8.5 (in addition to Figure 1.2) for a better understanding on it.

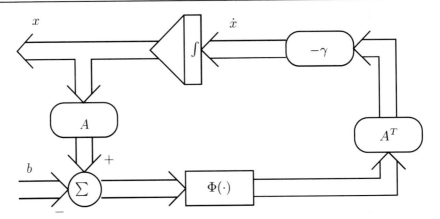

Figure 8.5. Block diagram of GNN model (8.8) solving $Ax = b$.

8.3.2. Theoretical Analysis

While Subsection 8.3.1. presents the GNN models [including the generalized one (8.8)], in this subsection we analyze their global exponential convergence and superior convergence properties of using different activation-function arrays.

Theorem 8.3.1. *Consider nonsingular constant matrix $A \in \mathbb{R}^{n \times n}$ in linear equations (8.3). If a linear or power-sigmoid activation-function array $\Phi(\cdot)$ is used, then state vector $x(t)$ of GNN model (8.8) starting from any initial state $x_0 \in \mathbb{R}^n$ will exponentially converge to the theoretical solution $x^* = A^{-1}b$ of linear equations $Ax(t) = b$. In addition, the exponential convergence rate is at least $\alpha\gamma$ with α denoting the minimum eigenvalue of A^TA. Moreover, the GNN model (8.8) using power-sigmoid activation-functions has better convergence than the GNN model using linear activation-functions [equivalently, GNN (8.7)].*

Proof. To analyze GNN models (8.7) and (8.8), let us define the solution error $\tilde{x}(t) = x(t) - x^*$, and let $\|x\|_2 = \sqrt{x^T x}$ denote the two norm of vector x. By substituting $x(t) = \tilde{x}(t) + x^*$ and $x^* = A^{-1}b$ into GNN (8.8), we have

$$\dot{\tilde{x}}(t) = -\gamma A^T \Phi(A\tilde{x}(t)). \qquad (8.9)$$

A Lyapunov function candidate can thus be defined as $v(x) = \|\tilde{x}(t)\|_2^2/2 = (\tilde{x}^T \tilde{x})/2 \geq 0$. Evidently, $v(x)$ is positive definite in the sense that $v(x) > 0$ for any $\tilde{x}(t) \neq 0$ and $v(x) = 0$ for $\tilde{x}(t) = 0$ only [of which the latter corresponds to $x(t) = x^*$]. In addition, $v(x) \to \infty$ as $\|\tilde{x}(t)\|_2 \to \infty$. Moreover, the time derivative of $v(x)$ along the system trajectory (8.9) is derived as the following:

$$\frac{dv}{dt} = \tilde{x}^T \dot{\tilde{x}} = \tilde{x}^T \left(-\gamma A^T \Phi(A\tilde{x})\right) = -\gamma (A\tilde{x})^T \Phi(A\tilde{x}). \qquad (8.10)$$

In view of processing-array $\Phi(\cdot)$ constituted by activation-functions $\phi(\cdot)$ which are defined in Section 1.5. and shown in Figure 1.3, we know that

$$\begin{aligned}(A\tilde{x})^T \Phi(A\tilde{x}) &= \sum_{i=1}^n [A\tilde{x}]_i \phi([A\tilde{x}]_i) \geq \sum_{i=1}^n [A\tilde{x}]_i [A\tilde{x}]_i \\ &= \sum_{i=1}^n [A\tilde{x}]_i^2 = (A\tilde{x})^T (A\tilde{x}) = \|A\tilde{x}(t)\|_2^2.\end{aligned} \qquad (8.11)$$

Then, combining (8.10) and (8.11), we have

$$\frac{dv}{dt} \leq -\gamma \|A\tilde{x}\|_2^2 \leq 0. \tag{8.12}$$

It follows that $\dot{v}(t)$ is negative definite in the sense that $\dot{v}(t) < 0$ for any $\tilde{x}(t) \neq 0$ and $\dot{v}(t) = 0$ for $\tilde{x}(t) = 0$ only (corresponding to $x(t) = x^*$), due to matrix A being nonsingular and design-parameter $\gamma > 0$.

By Lyapunov stability theory [16, 17], we have that $\tilde{x}(t) \to 0$ as time t approaches $+\infty$; equivalently, neural state $x(t)$ of GNN (8.8) is globally convergent to the theoretical inverse x^*. Furthermore, by assuming $\alpha > 0$ to be the minimum eigenvalue of $A^T A$, it follows from $\tilde{x}^T A^T A \tilde{x} \geq \alpha \tilde{x}^T \tilde{x}$ and (8.12) that

$$\dot{v} \leq -\gamma \|A\tilde{x}\|_2^2 = -\gamma \tilde{x}^T A^T A \tilde{x} \leq -\alpha\gamma \|\tilde{x}\|_2^2 \leq -2\alpha\gamma v.$$

Thus, $v(t) \leq \exp(-2\alpha\gamma t)v(0)$, which, together with $v(0) = \|\tilde{x}(0)\|_2^2/2$, yields the following:

$$\|\tilde{x}(t)\|_2 \leq \|\tilde{x}(0)\|_2 \exp(-\alpha\gamma t), \ \ t \in [0, +\infty).$$

The proof on global exponential convergence of GNN model (8.8) [including GNN (8.7)] is thus complete (with convergence rate being $\alpha\gamma$ at least).

We now come to prove the additional superior convergence of GNN model (8.8) by using power-sigmoid activation-functions, as compared to the situation of using linear activation-functions. On one hand, in the linear situation, since $\Phi(e) = e$ and $\phi(e_i) = e_i$ [with e_i the ith element of residual-error vector $e := Ax(t) - b$], GNN model (8.8) reduces to Wang neural network (8.4) or (8.7) [11]. Equation (8.9) thus becomes $\dot{\tilde{x}}(t) = -\gamma A^T(A\tilde{x}(t))$, and the time derivative of Lyapunov function $v(x)$ becomes

$$\frac{dv}{dt} = -\gamma(A\tilde{x})^T(A\tilde{x}) = -\gamma\|A\tilde{x}\|_2^2. \tag{8.13}$$

On the other hand, in the power-sigmoid situation, let us review equations (8.9) through (8.12). In (8.11), simply put, $[A\tilde{x}]_i \phi([A\tilde{x}]_i) \geq [A\tilde{x}]_i^2, \forall i \in \{1, 2, \cdots, n\}$. However, for most of the situations except $[A\tilde{x}]_i = \pm 1 = \phi([A\tilde{x}]_i)$,

$$[A\tilde{x}]_i \phi([A\tilde{x}]_i) > [A\tilde{x}]_i^2 \text{ (for } [A\tilde{x}]_i \in (-1, 1) \text{ or around } \pm 1);$$
$$[A\tilde{x}]_i \phi([A\tilde{x}]_i) \gg [A\tilde{x}]_i^2 \text{ (for } |[A\tilde{x}]_i| \gg 1).$$

In view of the above analysis and (8.12), we know by comparing with the linear situation (8.13) that the convergence speed in the power-sigmoid situation, $\dot{v}$, is at least equal to but usually faster (or much faster) than that in the linear situation (8.13). This means that the GNN model (8.8) using power-sigmoid activation-functions has a superior convergence to the GNN model (8.8) using linear activation-functions [equivalently, GNN (8.7)]. $\quad\square$

Remark 8.3.1. Nonlinearity always exists, which is one of the main motivations for us to investigate power-sigmoid or other kinds of activation-functions. Even if the linear activation-function is preferred to use, the nonlinear phenomenon may appear in its hardware implementation; e.g., in the form of saturation and/or inconsistency of the linear slope and in digital realization due to truncation and round-off errors [18]. $\quad\square$

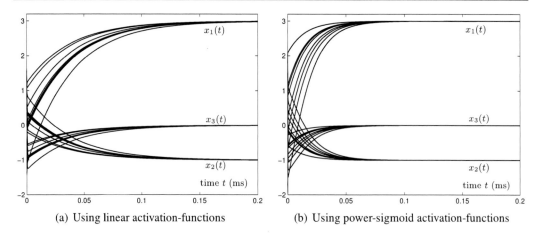

Figure 8.6. Solving $Ax = b$ by GNN (8.8) using different activation-functions with $\gamma = 10^6$ and starting from random initial states $x(0)$.

Remark 8.3.2. One more advantage of using the power-sigmoid activation-function over the linear activation-function lies in the extra parameters ξ and p. When there is an upper bound on γ due to hardware implementation, the new parameters ξ and p will be other effective factors expediting the neural-network convergence. □

Remark 8.3.3. By following the reviewers' inspiring and constructive comments, an approach proposed in [19], being another parallel-processing computational method for solving online linear equations, is investigated and compared as follows. i) The approach proposed in [19], which employs a signum activation-function array, could be viewed as a special case of GNN (8.8). However, as the signum function is discontinuous, it may introduce extra issues of solution-existence, uniqueness, chattering and stability analysis [20]. ii) In contrast, by using linear or power-sigmoid activation-function arrays, our GNN model (8.8) could be proved globally stable and has no such extra solution-issues. Moreover, global exponential convergence of our GNN model (8.8) is also proved, being evidently more efficient and desirable. □

Moreover, we could have the following theoretical results on global stability of the Wang neural network (8.7) even for singular coefficient matrix A.

Theorem 8.3.2. *Given singular matrix $A \in \mathbb{R}^{n \times n}$, the Wang neural network (8.7) is still globally stable.*

Proof. We can construct the Lyapunov function candidate

$$v(t) = \|Ax - b\|_2^2 \geq 0. \tag{8.14}$$

Then, along the system trajectory (8.7), the time-derivative of $v(t)$ could be obtained as the following

$$\frac{dv}{dt} = \frac{\partial v^T}{\partial x} \frac{dx}{dt} = (Ax - b)^T A(-\gamma A^T (Ax - b)) = -\gamma (Ax - b)^T AA^T (Ax - b) \leq 0. \tag{8.15}$$

Thus, by Lyapunov stability theory, the Wang neural network (8.7) is globally stable. □

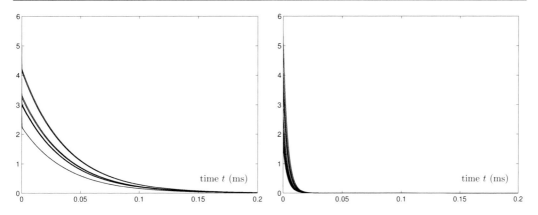

Figure 8.7. Solution error $\|x(t)-x^*\|_2$ of GNN (8.8) using linear activation-functions: in left graph $\gamma = 10^6$ and in right graph $\gamma = 10^7$.

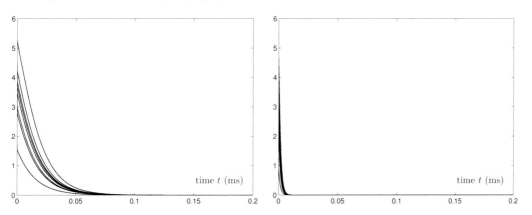

Figure 8.8. Solution error $\|x(t)-x^*\|_2$ of GNN (8.8) using power-sigmoid functions: in left graph $\gamma = 10^6$ and in right graph $\gamma = 10^7$.

8.3.3. Illustrative Examples

For illustrative purposes, let us consider the nonsingular coefficient matrix A and vector b in linear equation (8.3) as follows:

$$A = \begin{bmatrix} 1 & 2 & 0 \\ 0 & -1 & -1 \\ 1 & 1 & -2 \end{bmatrix}, \quad b = \begin{bmatrix} 1 \\ 1 \\ 2 \end{bmatrix},$$

for which we could verify that $x^* = [3, -1, 0]^T$ in order to compare the correctness of the neural-network solutions.

As seen from Figures 8.6 through 8.8, starting with any initial states randomly selected from $[-2, 2]^3$, state vector $x(t)$ of GNN model (8.8) always converges to the theoretical solution $x^* = A^{-1}b$, where design parameters $\xi = 5$ and $p = 3$ are used for simulative purposes. In addition, when using power-sigmoid activation-functions, GNN model (8.8) could converge roughly twice faster than that of using linear activation-functions. Furthermore, as shown in Figures 8.7 and 8.8, the convergence of GNN models could be expedited by

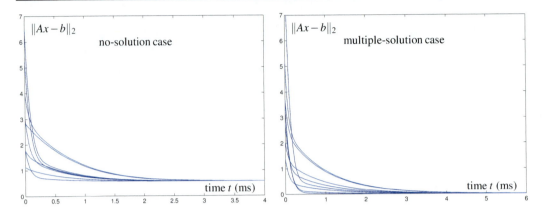

Figure 8.9. Energy function convergence of Wang neural network implying global stability for both no-solution case and multiple-solution case. *Reproduced from Y. Zhang and K. Chen, Global exponential convergence and stability of Wang neural network for solving online linear equations, Figure 3, Electronics Letters, pp. 145-146.* ©*[2008] IEEE. Reprinted, with permission.*

increasing γ. For example, as shown in Figure 8.8, the convergence time is approximately 0.1 millisecond when $\gamma = 10^6$, whereas the convergence time decreases to approximately 0.01 millisecond when γ increases to 10^7. These computer-simulation results substantiate the theoretical analysis and effectiveness of generalized GNN model (8.8).

In addition, to show the aforementioned global stability of the Wang neural network (8.7), let us consider the following singular coefficient matrix A (with $\alpha = 0$) and two possible situations of coefficient vector b for linear equation $Ax = b$:

$$A = \begin{bmatrix} 1 & -1 & 0 \\ -1 & 2 & 1 \\ 0 & 1 & 1 \end{bmatrix}, \quad b^{(1)} = \begin{bmatrix} 1 \\ 1 \\ 1 \end{bmatrix}, \quad b^{(2)} = \begin{bmatrix} 0 \\ 1 \\ 1 \end{bmatrix},$$

where, if $b := b^{(1)}$, then there is no theoretical solution to $Ax = b$ because of $rank([A\ b^{(1)}]) = 3 \neq 2 = rank(A)$; and, if $b := b^{(2)}$, then there are an infinity number of theoretical solutions to $Ax = b$ because of $rank([A\ b^{(2)}]) = 2 = rank(A)$. From Figure 8.9, we can see that, starting from random initial states $x(0)$, the Wang neural network (8.7) is always stable for both the no-solution case and the multiple-solution case of $Ax = b$.

8.4. GNN for Constant Matrix Inversion

To solve for a matrix inverse, the GNN system design is based on the equation, $AX - I = 0$, with $A \in \mathbb{R}^{n \times n}$. We can define a scalar-valued energy function such as $\mathcal{E}_2(t) = \|AX(t) - I\|_F^2/2$. Then, we use the negative of the gradient $\partial \mathcal{E}_2/\partial X = A^T(AX(t) - I)$ as the descent direction. As a result, the classic linear GNN model is shown as follows:

$$\dot{X}(t) = -\gamma \frac{\partial \mathcal{E}_2}{\partial X} = -\gamma A^T (AX(t) - I), \quad X(0) = X_0, \tag{8.16}$$

where design parameter $\gamma > 0$, being an inductance parameter or the reciprocal of a capacitive parameter, is set as large as the hardware permits, or selected appropriately for experiments.

Manherz *et al* [21] proposed using two approaches to implement (8.16); namely, the transformer network and the electronic analog computer. By following the use of analog dividers introduced by Sturges [5], a circuit implementation was generalized for matrix inversion related to system (8.16) [22]. Other analog implementations/simulations given, like in [3, 23–29], usually also assumed to be under the ideal conditions. However, nonlinearity always exists; for example, the tolerance of electronic components, the finite gain and bandwidth of operational amplifiers, as discussed in [22]. Even if the linear activation-function is used in the linear model (8.16), the nonlinear phenomenon may appear in its hardware implementation, e.g., in the form of saturation and/or inconsistency of the linear slope. Then, we may ask ourselves: what kind of results could we have in the presence of various implementation errors and different activation-function arrays?

Our recent research on matrix inversion has actually changed one of its focuses onto inverting online the time-varying matrix; e.g., [12, 13, 30]. When handling time-varying matrices, different kinds of activation-functions and various implementation errors are investigated in [13, 30] to pursue the superior convergence and robustness properties of a general ZNN model, as compared to its linear model [12]. The research shows that under the same design specification, a better performance can be achieved when using a power-sigmoid activation-function array. As inspired by revising [13], in this section, different kinds of activation-functions (linear, sigmoid, power, and power-sigmoid functions) are examined for qualitatively analyzing the superior convergence and robustness of the traditional gradient-neural-network system.

As proposed in [31], the following general GNN model is an extension to the above gradient design approach with a nonlinear activation-function array Φ:

$$\dot{X}(t) = -\gamma A^T \Phi \left(AX(t) - I \right) \tag{8.17}$$

where $X(t)$, starting from an initial condition $X(0) = X_0 \in \mathbb{R}^{n \times n}$, is the activation state matrix corresponding to the theoretical inverse A^{-1} of matrix A.

8.4.1. Main Theoretical Results

In view of GNN equation (8.17), different choices of Φ may lead to different performance. In general, any strictly-monotonically-increasing odd activation-function $\phi(\cdot)$, being an element of matrix mapping Φ, may be used for the construction of the neural network. In order to demonstrate the main ideas, four types of activation-functions presented in Section 1.5. are investigated in our simulation. Other types of activation-functions can be generated by these four basic types.

Following the analysis results of [13, 31], the convergence results of using different activation-functions are qualitatively presented as follows.

Proposition 8.4.1. *[12, 13, 30–32] For a nonsingular matrix $A \in \mathbb{R}^{n \times n}$, any strictly monotonically-increasing odd activation-function array $\Phi(\cdot)$ can be used for constructing the GNN model* (8.17).

1) *If the linear activation-function is used, then the global exponential convergence is achieved for GNN model* (8.17) *with convergence rate proportional to the product of γ and the minimum eigenvalue of $A^T A$.*

2) *If the bipolar sigmoid activation-function is used, then the superior convergence can be achieved for error range $[-\delta, \delta]$, $\exists \delta \in (0, 1)$, as compared to the linear activation-function case. This is because the error signal $e_{ij} = [AX - I]_{ij}$ in (8.17) is amplified by the bipolar sigmoid function for error range $[-\delta, \delta]$.*

3) *If the power activation-function is used, then the superior convergence can be achieved for error ranges $(-\infty, -1]$ and $[1, +\infty)$, as compared to the linear-activation-function case. This is because the error signal $e_{ij} = [AX - I]_{ij}$ in (8.17) is amplified by the power activation-function for error ranges $(-\infty, -1]$ and $[1, +\infty)$.*

4) *If the power-sigmoid activation-function is used, then superior convergence can be achieved for the whole error range $(-\infty, +\infty)$, as compared to the linear-activation-function case. This is in view of Properties 2 and 3.* $\square$

In the realization of neural networks, there are always some errors involved [5, 12, 18, 22, 30, 33]. For example, for the linear activation-function, its imprecise implementation may look more like a sigmoid or piecewise-linear function because of the finite gain and frequency dependency of operational amplifiers and multipliers. On the other hand, for the power-sigmoid activation-function shown in Figure 1.3(d), its imprecise implementation may deteriorate into a linear function due to the incapacity of electronic components, fortunately giving a lower bound of the performance of such an imprecise model. This is one of the motivations that we study and prefer the power-sigmoid activation-function. Numerically speaking, if the resultant imprecise implementation of an activation function is still a monotonically-increasing odd function, the asymptotic convergence could still be achieved for the imprecise model.

In addition to the imprecise implementation of activation functions, there are two more implementation errors for GNN models (8.16) and (8.17). One is the small perturbation of A, an implementation error resulting from truncating/roundoff errors in digital realization or high-order residual errors of circuit components in analog realization of A in system (8.17). Thus, the imprecise system may become

$$\dot{X} = -\gamma(A + \Delta_A)^T \Phi\big((A + \Delta_A)X(t) - I\big), \tag{8.18}$$

where the additive term Δ_A is such that $\|\Delta_A\| \leq \varepsilon_1$, $\exists \varepsilon_1 \geq 0$. Another implementation error of system (8.17) is the model-implementation error Δ_R due to the imprecise implementation of system dynamics, such that the system (8.17) becomes

$$\dot{X} = -\gamma A^T \Phi\big(AX(t) - I\big) + \Delta_R, \tag{8.19}$$

where $\|\Delta_R\| \leq \varepsilon_2$, $\exists \varepsilon_2 \geq 0$. Following the analysis and results of [3, 5, 12, 16, 21, 22, 25, 26, 28–30, 34, 35], we have the following observations.

Proposition 8.4.2. *[12, 13, 30–32] Consider the perturbed GNN model (8.18) where the additive term Δ_A exists such that $\|\Delta_A\| \leq \varepsilon_1$, $\exists \varepsilon_1 \geq 0$, then the steady-state residual error $\lim_{t \to \infty} \|X(t) - A^{-1}\|$ is uniformly upper bounded by some positive scalar, provided that the resultant matrix $A + \Delta_A$ is still nonsingular.* $\square$

Proposition 8.4.3. *[12, 13, 30–32] Consider the imprecise implementation* (8.19). *The steady-state residual error* $\lim_{t \to \infty} \|X(t) - A^{-1}\|$ *is uniformly upper bounded by some positive scalar, provided that the design parameter* γ *is large enough (the so-called design-parameter requirement). Moreover, the steady state residual error* $\lim_{t \to \infty} \|X(t) - A^{-1}\|$ *can be made to zero as* γ *tends to positive infinity.* $\square$

As additional results to the above lemmas, we have the following general observations.

1. For large entry error (e.g., $|e_{ij}| > 1$ with $e_{ij} := [AX - I]_{ij}$), the power activation-function could amplify the error signal ($|e_{ij}^p| > \cdots > |e_{ij}^3| > |e_{ij}| > 1$), thus able to automatically remove the design-parameter requirement.

2. For small entry error (e.g., $|e_{ij}| < 1$), the use of sigmoid activation-functions has better convergence and robustness than the use of linear activation-functions, because of the larger slope of the sigmoid function near the origin.

Thus, using the power-sigmoid activation-function in (1.7) is theoretically a better choice than other activation-functions for superior convergence and robustness.

8.4.2. Kronecker Product and Vectorization

While Subsection 8.4.1. presents the main theoretical results of the GNN models, this subsection investigate the MATLAB simulation techniques about the Kronecker product and vectorization in order to show the characteristics of such a neural network.

The dynamic equations of GNN models (8.17) and (8.19) are all described in matrix form which could not be simulated directly. To simulate such neural systems, the Kronecker product of matrices and vectorization technique are introduced in order to transform the matrix-form differential equations to vector-form differential equations.

Based on the definition in Appendix A.2. of Kronecker product and vectorization technique, for simulation proposes, the matrix differential equation (8.17) can be transformed to a vector differential equation. We thus obtain the following theorem.

Theorem 8.4.1. *The matrix-form differential equation* (8.17) *can be reformulated as the following vector-form differential equation:*

$$\text{vec}(\dot{X}) = -\gamma(I \otimes A^T)\Phi\big((I \otimes A)\,\text{vec}(X) - \text{vec}(I)\big), \tag{8.20}$$

where activation-function mapping $\Phi(\cdot)$ *in* (8.20) *is defined the same as in* (8.17) *except that its dimensions are changed hereafter as* $\Phi(\cdot) : \mathbb{R}^{n^2 \times 1} \to \mathbb{R}^{n^2 \times 1}$.

Proof. For readers' convenience, we repeat the matrix-form differential equation (8.17) here as $\dot{X} = -\gamma A^T \Phi(AX(t) - I)$.

By vectorizing equation (8.17) based on the Kronecker product and the above $\text{vec}(\cdot)$ operator, the left hand side of (8.17) is $\text{vec}(\dot{X})$, and the right hand side of equation (8.17) is

$$
\begin{aligned}
\text{vec}\left(-\gamma A^T \Phi\big(AX(t) - I\big)\right) & \\
= -\gamma \text{vec}\left(A^T \Phi\big(AX(t) - I\big)\right) & \\
= -\gamma(I \otimes A^T)\,\text{vec}\left(\Phi(AX(t) - I)\right). &
\end{aligned}
\tag{8.21}
$$

Note that, as shown in Subsection 2.5.2., the definition and coding of the activation-function array $\Phi(\cdot)$ are very flexible and could be a vectorized mapping from $\mathbb{R}^{n^2 \times 1}$ to $\mathbb{R}^{n^2 \times 1}$. We thus have

$$
\begin{aligned}
&\text{vec}\left(\Phi(AX(t) - I)\right) \\
&= \Phi\left(\text{vec}\left(AX(t) - I\right)\right) \\
&= \Phi\left(\text{vec}(AX) + \text{vec}(-I)\right) \\
&= \Phi\left((I \otimes A)\,\text{vec}(X) - \text{vec}(I)\right).
\end{aligned} \tag{8.22}
$$

Combining equations (8.21) and (8.22) yields the vectorization of the right hand side of matrix-form differential equation (8.17):

$$
\text{vec}\left(-\gamma A^T \Phi\left(AX(t) - I\right)\right) = -\gamma(I \otimes A^T)\Phi\left((I \otimes A)\,\text{vec}(X) - \text{vec}(I)\right).
$$

Clearly, the vectorization of both sides of matrix-form differential equation (8.17) should be equal, which generates the vector-form differential equation (8.20). The proof is thus complete. $\qquad\square$

Based on Remark 4.4.1 about the routines "kron" and "vec", the following code is used to define a function returns the evaluation of the right-hand side of matrix-form GNN model (8.17). In other words, it also returns the evaluation of the right-hand side of vector-form GNN model (8.20). Note that $I \otimes A^T = (I \otimes A)^T$.

```
function output=GnnRightHandSide(t,x,gamma)
if nargin==2, gamma=1; end
A=MatrixA; n=size(A,1); IA=kron(eye(n),A);
% The following generates the vectorization of
% identity matrix I
vecI=reshape(eye(n),n^2,1);
% The following calculates the right hand side
% of equations (2) and (5)
output=-gamma*IA'*Powersigmoid(IA*x-vecI);
```

Note that we can change "Powersigmoid" in the above MATLAB code to "Sigmoid" (or "Linear") for using different activation-function arrays.

8.4.3. Illustrative Examples

For illustration, let us consider the following constant nonsingular matrix:

$$
A = \begin{bmatrix} 1 & 0 & 1 \\ 1 & 1 & 0 \\ 1 & 1 & 1 \end{bmatrix}, \quad A^T = \begin{bmatrix} 1 & 1 & 1 \\ 0 & 1 & 1 \\ 1 & 0 & 1 \end{bmatrix}, \quad A^{-1} = \begin{bmatrix} 1 & 1 & -1 \\ -1 & 0 & 1 \\ 0 & -1 & 1 \end{bmatrix}.
$$

For example, matrix A can be given in the following MATLAB code.

```
function A=MatrixA(t)
A=[1 0 1;1 1 0;1 1 1];
```

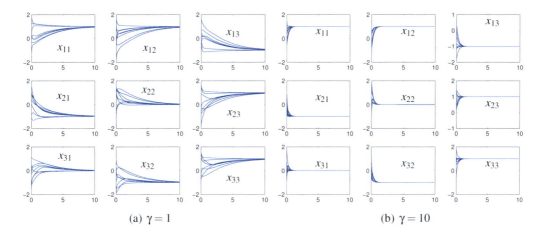

Figure 8.10. Online matrix inversion by GNN model (8.17). *Reproduced from Y. Zhang, K. Chen et al., MATLAB simulation of gradient-based neural network for online matrix inversion, Figure 1, D.-S. Huang et al. (Eds.): ICIC 2007, LNAI 4682, pp. 98-109, 2007. ©Springer-Verlag Berlin Heidelberg 2007. With kind permission of Springer Science+Business Media.*

The gradient neural network (8.17) is thus in the following specific form

$$\begin{bmatrix} \dot{x}_{11} & \dot{x}_{12} & \dot{x}_{13} \\ \dot{x}_{21} & \dot{x}_{22} & \dot{x}_{23} \\ \dot{x}_{31} & \dot{x}_{32} & \dot{x}_{33} \end{bmatrix} = -\gamma \begin{bmatrix} 1 & 1 & 1 \\ 0 & 1 & 1 \\ 1 & 0 & 1 \end{bmatrix} \Phi \left(\begin{bmatrix} 1 & 0 & 1 \\ 1 & 1 & 0 \\ 1 & 1 & 1 \end{bmatrix} \begin{bmatrix} x_{11} & x_{12} & x_{13} \\ x_{21} & x_{22} & x_{23} \\ x_{31} & x_{32} & x_{33} \end{bmatrix} - \begin{bmatrix} 1 & 0 & 0 \\ 0 & 1 & 0 \\ 0 & 0 & 1 \end{bmatrix} \right).$$

8.4.3.1. Simulation of Convergence

To simulate gradient neural network (8.17) starting from eight random initial states, we firstly define a function "GnnConvergence" as follows.

```
function GnnConvergence(gamma)
tspan=[0 10]; n=size(MatrixA,1);
for i=1:8
  x0=4*(rand(n^2,1)-0.5*ones(n^2,1));
  [t,x]=ode45(@GnnRightHandSide,tspan,x0,[],gamma);
  for j=1:n^2
    k=mod(n*(j-1)+1,n^2)+floor((j-1)/n);
    subplot(n,n,k); plot(t,x(:,j)); hold on
  end
end
```

To show the convergence of the gradient neural model (8.17) using power-sigmoid activation-function with $\xi = 4$ and $p = 3$ and using the design parameter $\gamma := 1$, the MATLAB command is "GnnConvergence(1)", which generates Figure 8.10(a). Similarly, the MATLAB command "GnnConvergence(10)" can generate Figure 8.10(b).

To monitor the network convergence, we can also use and show the norm of the computational error $\|X(t) - A^{-1}\|_F$. The MATLAB codes are given below, i.e., the user-defined functions "NormError" and "GnnNormError".

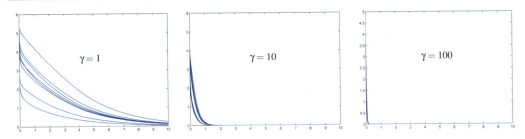

Figure 8.11. Convergence of $\|X(t) - A^{-1}\|_F$ using power-sigmoid activation-function. *Reproduced from Y. Zhang, K. Chen et al., MATLAB simulation of gradient-based neural network for online matrix inversion, Figure 2, D.-S. Huang et al. (Eds.): ICIC 2007, LNAI 4682, pp. 98-109, 2007.* ©*Springer-Verlag Berlin Heidelberg 2007. With kind permission of Springer Science+Business Media.*

```
function NormError(x0,gamma)
tspan=[0 10]; options=odeset();
[t,x]=ode45(@GnnRightHandSide,tspan,x0,options,gamma);
Ainv=inv(MatrixA);
B=reshape(Ainv,size(Ainv,1)^2,1);
total=length(t); x=x';
for i=1:total, nerr(i)=norm(x(:,i)-B); end
plot(t,nerr); hold on
```

```
function GnnNormError(gamma)
if nargin<1, gamma=1; end
total=8; n=size(MatrixA,1);
for i=1:total
    x0=4*(rand(n^2,1)-0.5*ones(n^2,1));
    NormError(x0,gamma);
end
text(2.4,2.2,['gamma=' int2str(gamma)]);
```

By calling "GnnNormError" three times with different γ values, we can generate Figure 8.11. It shows that starting from any initial state randomly selected in $[-2,2]$, the state matrices of ZNN (8.17) all converge to the theoretical inverse A^{-1}, where the computational errors $\|X(t) - A^{-1}(t)\|_F$ all converge to zero. Such a convergence can be expedited by increasing γ. For example, if γ is increased to 10^3, the convergence time is within 30 milliseconds; and, if γ is increased to 10^6, the convergence time is within 30 microseconds.

8.4.3.2. Simulation of Robustness

Similar to the transformation of the matrix-form differential equation (8.17) to a vector-form differential equation (8.20), the perturbed GNN model (8.19) can be vectorized as follows:

$$\text{vec}(\dot{X}) = -\gamma(I \otimes A^T)\Phi\big((I \otimes A)\text{vec}(X) - \text{vec}(I)\big) + \text{vec}(\Delta_R). \quad (8.23)$$

To show the robustness characteristics of gradient neural networks, the following

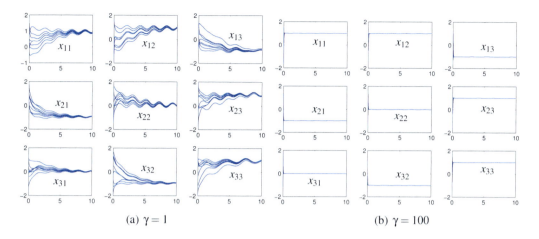

Figure 8.12. Online matrix inversion by GNN (8.19) with large implementation errors. *Reproduced from Y. Zhang, K. Chen et al., MATLAB simulation of gradient-based neural network for online matrix inversion, Figure 3, D.-S. Huang et al. (Eds.): ICIC 2007, LNAI 4682, pp. 98-109, 2007. ©Springer-Verlag Berlin Heidelberg 2007. With kind permission of Springer Science+Business Media.*

model-implementation error is added in a sinusoidal form (with $\varepsilon_2 = 0.5$):

$$\Delta_R = \varepsilon_2 \begin{bmatrix} \cos(3t) & -\sin(3t) & 0 \\ 0 & \sin(3t) & \cos(3t) \\ 0 & 0 & \sin(2t) \end{bmatrix}.$$

The following code is used to define the function "GnnRightHandSideImprecise" for ODE solvers, which returns the evaluation of the right-hand side of the perturbed GNN (8.19), in other words, the right-hand side of the vector-form differential equation (8.23).

```
function output=GnnRightHandSideImprecise(t,x,gamma)
if nargin==2, gamma=1; end
e2=0.5;
deltaR=e2*[cos(3*t) -sin(3*t) 0; 0 sin(3*t) cos(3*t); ...
    0 0 sin(2*t)];
vecR=reshape(deltaR,9,1);
vecI=reshape(eye(3),9,1);
IA=kron(eye(3),MatrixA);
output=-gamma*IA'*Powersigmoid(IA*x-vecI)+vecR;
```

To use the sigmoid (or linear) activation-function, we only need to change "Powersigmoid" to "Sigmoid" (or "Linear") in the above MATLAB code. Based on the above function "GnnRightHandSideImprecise" and the function below (i.e.,"GnnRobust"), MATLAB commands "GnnRobust(1)" and "GnnRobust(100)" can generate Figure 8.12.

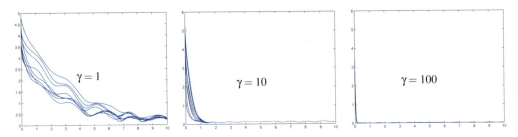

Figure 8.13. Convergence of computational error $\|X(t)-A^{-1}\|_F$ by perturbed GNN (8.19). *Reproduced from Y. Zhang, K. Chen et al., MATLAB simulation of gradient-based neural network for online matrix inversion, Figure 4, D.-S. Huang et al. (Eds.): ICIC 2007, LNAI 4682, pp. 98-109, 2007. ©Springer-Verlag Berlin Heidelberg 2007. With kind permission of Springer Science+Business Media.*

```
function GnnRobust(gamma)
tspan=[0 10]; options=odeset(); n=size(MatrixA,1);
for i=1:8
    x0=4*(rand(n^2,1)-0.5*ones(n^2,1));
    [t,x]=ode45(@GnnRightHandSideImprecise,tspan,x0, ...
        options,gamma);
    for j=1:n^2
      k=mod(n*(j-1)+1,n^2)+floor((j-1)/n);
      subplot(n,n,k); plot(t,x(:,j)); hold on
    end
end
```

Similarly, we can show the computational error $\|X(t)-A^{-1}\|_F$ of GNN model (8.19) with large model-implementation errors. To do so, in the previously defined MATLAB function "NormError", we only need change "GnnRightHandSide" to "GnnRightHandSideImprecise". See Figure 8.13. Even with imprecise implementation, the perturbed neural network still works well, and its computational error $\|X(t)-A^{-1}\|_F$ is still bounded and very small. Moreover, as γ increases from 1 to 100, the convergence is expedited and the steady-state computational error is decreased. It is worth mentioning again that using power-sigmoid or sigmoid activation-functions has smaller steady-state residual error than using linear or power activation-functions. It is observed from other simulation data that when using power-sigmoid activation-functions, the maximum steady-state residual error is only 2×10^{-2} and 2×10^{-3} respectively for $\gamma = 100$ and $\gamma = 1000$. Clearly, compared to the case of using linear or pure power activation-functions, superior performance can be achieved by using power-sigmoid or sigmoid activation-functions under the same design specification. These simulation results have substantiated the theoretical results presented in previous sections and in [31].

8.4.4. The Singular Case

For the special case of $A \in \mathbb{R}^{n \times n}$ being singular, clearly it has no theoretical inverse, and any solutions generated by system (8.17) might be accepted provided that it is stable instead of divergence. By this singular-case study, we just wanted to know the system responses, and then see what hints could be given to the design of neural networks in applications, such

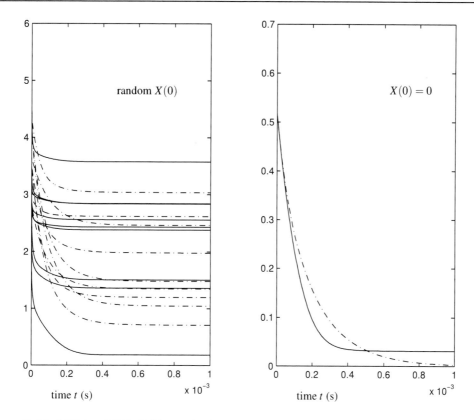

Figure 8.14. $\|X(t) - PINV(A)\|$ synthesized by (8.17) in the case of A being singular, where solid curves correspond to using the power-sigmoid function, while dashed-dotted curves correspond to the linear function. *Reproduced from Y. Zhang, Revisit the analog computer and gradient-based neural system for matrix inversion, Figure 9, Proceedings of 2005 IEEE International Symposium on Intelligent Control, pp. 1411-1416. ©[2005] IEEE. Reprinted, with permission.*

as, the inverse-kinematic robot control in the ensuing subsection. By following the analysis and results of [3,5,12,21,22,25,26,28–31,34–36], the marginal stability of (8.17) could be shown based on the singular value decomposition of A and $A^T A$ being positive semidefinite and having only zero and/or positive (real) eigenvalues.

A large number of numerical experiments also substantiate the typical situation as shown in Figure 8.14. The MATLAB "PINV" routine is used to provide a standard measure for comparing the convergence and distance of the solutions generated by neural system (8.17). As shown in the left plot of Figure 8.14, starting from any random initial states $X(0) \neq 0$, the neural system (8.17) is always convergent to an equilibrium, which is not necessarily the pseudoinverse solution. On the other hand, as shown in the right plot of Figure 8.14, starting from zero initial states $X(0) = 0$, the solutions of the neural system (8.17) are very similar or even equal to the PINV solution. Thus, as inspired from this study, for better consistency with prevailing numerical algorithms, we can always use the zero initial states $X(0) = 0$ to start the general matrix-inversion neural system (8.17); and, in the singular case, the linear activation-function is preferred.

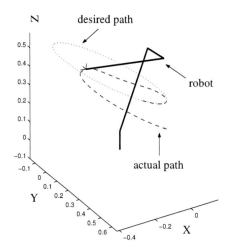

Figure 8.15. The end-effector moving along a circle of radius 30cm. *Reproduced from Y. Zhang, Revisit the analog computer and gradient-based neural system for matrix inversion, Figure 10, Proceedings of 2005 IEEE International Symposium on Intelligent Control, pp. 1411-1416. ©[2005] IEEE. Reprinted, with permission.*

8.4.5. An Application to Inverse Kinematics

The inverse-kinematic problem is one of the fundamental tasks in operating redundant manipulators [5–7,34,37–40]; i.e., to find the joint trajectories $\theta(t)$, given the trajectories of the end-effector, $r(t)$. The inverse-kinematic problem is usually handled at the velocity level via the following pseudoinverse/nullspace solution [37]:

$$\dot{\theta} = J^+ \dot{r} + (I - J^+ J) z \qquad (8.24)$$

where $J \in \mathbb{R}^{m \times n}$ is the Jacobian matrix with $m < n$, and $z \in \mathbb{R}^n$ is chosen as the negative gradient of any performance index to be minimized, like, for avoiding joint limits, configuration singularities, and/or environmental obstacles. By definition, the pseudoinverse $J^+ := J^T (JJ^T)^{-1}$ for J being full row rank. Thus, by defining $A = JJ^T$, we can re-exploit the general GNN system (8.17) to solve J^+ online. Specifically, $J^+ = J^T X$ where X is the state matrix generated by (8.17) in real-time.

The inverse kinematic control of robotic arms is like that of our human arm [41]: we just need know the position and orientation of our hands/palms/fingers, while at the same time our brain could command the corresponding joints to complete the task, without us knowing too much detail. In this sense, the brain seems to have a function of real-time matrix inversion, simulated by dynamic systems (8.17) and (8.24) [where the difference might be the accuracy, speed and flexibility].

The simulation based on Unimation PUMA560 robot arm [6,7] is given in Figure 8.15, where the desired motion of the end-effector (with only positioning considered) is a circle of radius 30cm. Note that this is an intentionally-designed "impossible" task that the desired circle actually exceeds the boundary of the end-effector workspace, and the matrix (JJ^T) might thus be singular. Even in this situation, as typically shown in Figures 8.16 and 8.17,

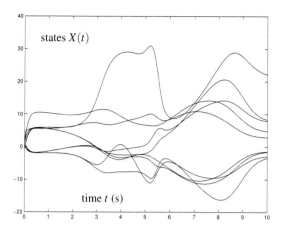

Figure 8.16. States $X(t)$ by (8.17), corresponding to $(JJ^T)^{-1}$, for robot. *Reproduced from Y. Zhang, Revisit the analog computer and gradient-based neural system for matrix inversion, Figure 11, Proceedings of 2005 IEEE International Symposium on Intelligent Control, pp. 1411-1416. ©[2005] IEEE. Reprinted, with permission.*

the performance is acceptably good, somewhat resembling human's decision-making. For a pure "possible" task such as a circle of radius 10cm, see Figure 7 of [30]. In that situation, simple numerical experiments show that increasing γ could improve the positioning accuracy, while using power-sigmoid function could also improve the accuracy as compared to the original linear model. Note that, due to space limitation, only the figures based on using the power-sigmoid function to synthesize (8.17) and (8.24) are presented.

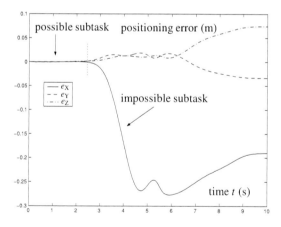

Figure 8.17. Positioning error on X, Y and Z axes at the end-effector. *Reproduced from Y. Zhang, Revisit the analog computer and gradient-based neural system for matrix inversion, Figure 12, Proceedings of 2005 IEEE International Symposium on Intelligent Control, pp. 1411-1416. ©[2005] IEEE. Reprinted, with permission.*

8.5. Conclusion

In this chapter, Zhang neural network is presented for linear equations solving of constant coefficients, which is depicted by an implicit dynamic-equation. It has been shown that the neural state $x(t)$ of such a ZNN model is globally convergent to the unique solution of (8.3). Moreover, the convergence rate could be expedited effectively by increasing the value of the design parameter γ. Compared to the previously-presented linear GNN model and its asymptotical convergence [11], a generalized GNN model and its global exponential convergence have been presented in this chapter for linear constant equations solving. We show that, by using linear activation-functions, exponential convergence can be achieved for the GNN model, and that, by using power-sigmoid activation-functions, the GNN model could perform much better than the former situation.

By considering different activation-functions and implementation errors, this chapter has also simulated the gradient neural network (8.16) for matrix inversion. A general model, (8.17), has thus been presented. Several important simulation techniques have also been introduced, i.e., Kronecker product of matrices and MATLAB routine "ode45". For superior convergence and robustness, the power-sigmoid activation-function is preferred. As for the singular case, zero initial states and linear activation-function are preferred. The application to PUMA560 robot control could further show a link among analog computer, neural networks, robot and human arm control. This has also provided immediate insights and answers to related problems of imprecise implementation [3, 5, 21, 22, 25, 26, 28, 29].

References

[1] Bazaraa, MS; Sherali, HD; Shetty, CM. *Nonlinear Programming - Theory and Algorithms*. New York: Wiley, 1993.

[2] Zhang, Y; Wang, J. A dual neural network for convex quadratic programming subject to linear equality and inequality constraints. *Physics Letters A,* 2002, vol. 298, no. 4, pp. 271-278.

[3] Steriti, RJ; Fiddy, MA. Regularized image reconstruction using SVD and a neural network method for matrix inversion. *IEEE Transactions Signal Processing,* 1993, vol. 41, no. 10, pp. 3074-3077.

[4] Sarkar, T; Siarkiewicz, K; Stratton, R. Survey of numerical methods for solution of large systems of linear equations for electromagnetic field problems. *IEEE Transactions on Antennas and Propagation,* 1981, vol. 29, no. 6, pp. 847-856.

[5] Sturges, Jr RH. Analog matrix inversion (robot kinematics). *IEEE Journal of Robotics and Automation,* 1988, vol. 4, no. 2, pp. 157-162.

[6] Zhang, Y; Wang, J. A dual neural network for constrained joint torque optimization of kinematically redundant manipulators. *IEEE Transactions on Systems, Man, and Cybernetics,* Part B, 2002, vol. 32, no. 5, pp. 654-662.

[7] Zhang, Y; Ge, SS; Lee, TH. A unified quadratic programming based dynamical system approach to joint torque optimization of physically constrained redundant manipulators. *IEEE Transactions on Systems, Man, and Cybernetics,* Part B, 2004, vol. 34, no. 5, pp. 2126-2132.

[8] Yu, X; Loh, NK; Miller, WC. New recursive algorithm for solving linear algebraic equations, *Electronics Letters,* 1992, vol. 28, no. 22, pp. 2069-2071.

[9] Wu, A; Duan, G. Explicit general solution to the matrix equation $AV + BW = EVF + R$. *IET Control Theory & Applications,* 2008, vol. 2, no. 1, pp. 56-60.

[10] Ding, F; Chen, T. Gradient based iterative algorithms for solving a class of matrix equations. *IEEE Transactions on Automatic Control,* 2005, vol. 50, no. 8, pp. 1216-1221.

[11] Wang, J. Electronic realisation of recurrent neural work for solving simultaneous linear equations. *Electronics Letters,* 1992, vol. 28, no. 5, pp. 493-495.

[12] Zhang, Y; Jiang, D; Wang, J. A recurrent neural network for solving Sylvester equation with time-varying coefficients. *IEEE Transactions on Neural Networks,* 2002, vol. 13, no. 5, pp. 1053-1063.

[13] Zhang, Y; Ge, SS. Design and analysis of a general recurrent neural network model for time-varying matrix inversion. *IEEE Transactions on Neural Networks,* 2005, vol. 16, no. 6, pp. 1477-1490.

[14] Pearlmutter, BA. Gradient calculations for dynamic recurrent neural networks: a survey. *IEEE Transactions on Neural Networks,* 1995, vol. 6, no. 5, pp. 1212-1228.

[15] Raida, Z. Improvement of convergence properties of Wang neural network. *Electronics Letters,* 1994, vol. 30, no. 22, pp. 1865-1866.

[16] Zhang, Y; Wang, J. Global exponential stability of recurrent neural network for synthesizing linear feedback control systems via pole assignment. *IEEE Transactions on Neural Networks,* 2002, vol. 13, no. 3, pp. 633-644.

[17] Zhang, Y. A set of nonlinear equations and inequalities arising in robotics and its online solution via a primal neural network. *Neurocomputing,* 2006, vol. 70, no. 1-3, pp. 513-524.

[18] Mead, C. *Analog VLSI and Neural Systems.* Reading, MA: Addison-Wesley, 1989.

[19] Ferreira, LV; Kaszkurewicz, E; Bhaya, A. Solving systems of linear equations via gradient systems with discontinuous righthand sides: application to LS-SVM. *IEEE Transactions on Neural Networks,* 2005, vol. 16, no. 2, pp. 501-505.

[20] Bhaya, A; Kaszkurewicz, E. A control-theoretic approach to the design of zero finding numerical methods. *IEEE Transactions on Automatic Control,* 2007, vol. 52, no. 6, pp. 1014-1026.

[21] Manherz, RK; Jordan, BW; Hakimi, SL. Analog methods for computation of the generalized inverse. *IEEE Transactions on Automatic Control,* 1968, vol. 13, no. 5, pp. 582-585.

[22] Carneiro, NCF; Caloba, LP. A new algorithm for analog matrix inversion. *Proceedings of the 38th Midwest Symposium on Circuits and Systems,* Brazil, 1995, vol. 1, pp. 401-404.

[23] Yeung, KS; Kumbi, F. Symbolic matrix inversion with application to electronic circuits. *IEEE Transactions on Circuits and Systems,* 1988, vol. 35, no. 2, pp. 235-238.

[24] El-Amawy, A. A systolic architecture for fast dense matrix inversion. *IEEE Transactions on Computers,* 1989, vol. 38, no. 3, pp. 449-455.

[25] Jang, J; Lee, S; Shin, S. *An Optimization Network for Matrix Inversion. In: Neural Information Processing Systems,* (Ed: Anderson, DZ), New York: American Institute of Physics, 1988, pp. 397-401.

References   227

[26] Luo, FL; Zheng, B. Neural network approach to computing matrix inversion. *Applied Mathematics and Computation,* 1992, vol. 47, no. 2-3, pp. 109-120.

[27] Cichocki, A; Unbehauen, R. Neural network for solving systems of linear equations and related problems. *IEEE Transactions on Circuits and Systems I: Fundamental Theory and Applications,* 1992, vol. 39, no. 2, pp. 124-138.

[28] Wang, J. A recurrent neural network for real-time matrix inversion. *Applied Mathematics and Computation,* 1993, vol. 55, no. 1, pp. 89-100.

[29] Song, J; Yam, Y. Complex recurrent neural network for computing the inverse and pseudo-inverse of the complex matrix. *Applied Mathematics and Computation,* 1998, vol. 93, no. 2-3, pp. 195-205.

[30] Zhang, Y; Ge, SS. A general recurrent neural network model for time-varying matrix inversion. *Proceedings of the 42nd IEEE Conference on Decision and Control,* USA, 2003, vol.6, pp. 6169-6174.

[31] Zhang, Y; Li, Z; Fan, Z; Wang, G. Matrix-inverse primal neural network with application to robotics. *Dynamics of Continuous, Discrete and Impulsive Systems,* Series A, 2007, vol. 14, pp. 400-407.

[32] Zhang, Y. Revisit the analog computer and gradient-based neural system for matrix inversion. *Proceedings of the IEEE International Symposium on Intelligent Control,* Cyprus, 2005, pp. 1411-1416.

[33] Anderson, JA; Rosenfeld, E. *Neurocomputing: Foundations of Research,* Cambridge, MA: The MIT Press, 1988.

[34] Wang, J; Zhang, Y. *Recurrent Neural Networks for Real-Time Computation of Inverse Kinematics of Redundant Manipulators. In: Machine Intelligence: Quo Vadis?* (Eds: Sincak, P; Vascak, J; Hirota K), Singapore: World Scientific, 2004, pp. 299-319.

[35] Zhang, Y; Heng, PA; Fu, AWC. Estimate of exponential convergence rate and exponential stability for neural networks. *IEEE Transactions on Neural Networks,* 1999, vol. 10, no. 6, pp. 1487-1493.

[36] Zhang, Y; Wang, J. Global exponential stability of recurrent neural networks for synthesizing linear feedback control systems via pole assignment. *IEEE Transactions on Neural Networks,* 2002, vol. 13, no. 3, pp. 633-644.

[37] Sciavicco, L; Siciliano, B. *Modelling and Control of Robot Manipulators.* Great Britain: Springer-Verlag London, 2000.

[38] Zhang, Y; Wang, J; Xu, Y. A dual neural network for bi-criteria kinematic control of redundant manipulators. *IEEE Transactions on Robotics and Automation,* 2002, vol. 18, no. 6, pp. 923-931.

[39] Zhang, Y; Wang, J; Xia, Y. A dual neural network for redundancy resolution of kinematically redundant manipulators subject to joint limits and joint velocity limits. *IEEE Transactions on Neural Networks,* 2003, vol. 14, no. 3, pp. 658-667.

[40] Zhang Y; Wang, J. Obstacle avoidance for kinematically redundant manipulators using a dual neural network. *IEEE Transactions on Systems, Man, and Cybernetics,* Part B, 2004, vol. 34, no. 1, pp. 752-759.

[41] Latash, ML. *Control of Human Movement.* Chicago: Human Kinetics Publishers, 1993.

Part IV

Final Comparisons and Discussions

Chapter 9

Unified Neural-Network Models

Abstract

ZNN and GNN models are unified and investigated in this chapter. Differing from the design of GNN model based on a nonnegative (or lower-bounded at least) scalar-valued norm-based energy function, the ZNN model is designed based on an indefinite matrix/vector-valued error function. For illustration and comparison of such two neural-network models, this chapter presents a cyclic-motion-generation (CMG) scheme (9.3)-(9.5) for redundant robot manipulators. Computer-simulation results based on three types of planar robot arms tracking a square path have substantiated again the efficacy of such a CMG scheme; moreover, theoretical analysis based on both gradient-descent and Zhang *et al.*'s neural-dynamic approaches have further shown the common effectiveness of the design methodologies.

9.1. Introduction

Due to the in-depth research in neural networks, numerous dynamic solvers based on recurrent neural networks have been developed and investigated [1–17]. Generally speaking, the methods reported in these references are related to the gradient descent algorithm in optimization, which can be summarized as follows: first, construct a cost function such that its minimal point is the solution of the given equation; then, a recurrent neural network is developed to evolve along a descent direction of this cost function until a minimum of the cost function is reached. A typical descent direction is defined by the negative gradient.

However, if the coefficients of the equation are time-varying, then gradient-based neural networks (GNN) may not work well. Because of the effects of the time-varying coefficients, the negative gradient direction can no longer guarantee the decrease of the cost function (or termed, energy function). Usually a neural network of much faster convergence in comparison to the time-varying coefficients is required for a real-time solution if the gradient-based method is adopted. The shortcomings of applying such a method to time-varying cases are two-fold: the much faster convergence is usually at the cost of the precision or with stringent restrictions on design parameters, and such method is not applicable to the case where the coefficients vary quickly or the case of large-scale complex control systems.

In this book, following the idea of using first-order time derivatives [6–8, 13], a special type of RNN model with implicit dynamics is developed, generalized and analyzed for solving online the time-varying problems [5–10, 13]. The resultant recurrent neural networks (or termed Zhang neural networks, ZNN) are elegantly introduced by defining the indefinite matrix-valued error functions (which means that the error function could be negative, positive, bounded or unbounded), instead of the usual positive (or lower bounded at least) scalar-valued norm-based cost functions such that computational errors can be made decreasing to zero exponentially (and globally). As noted, nonlinearity and errors always exist. Even if a linear activation function is used, the nonlinear phenomenon may appear in its hardware implementation. For superior convergence and better robustness, different kinds of activation functions (linear, sigmoid, power activation functions, and/or their variants, e.g., power-sigmoid functions) are investigated.

9.2. Unified Neural-Network Models

Differing from conventional gradient-based neural networks depicted in explicit dynamics, Zhang neural networks depicted in implicit dynamics are designed based on matrix/vector-valued error function, instead of the scalar-valued norm-based energy function. In this section, such two recurrent neural networks (i.e., ZNN and GNN) are developed and unified for final comparison and illustration.

9.2.1. Zhang Neural Networks

In this subsection, Zhang $et\ al.$'s neural-dynamic design method [5–10, 13] is presented, generalized and applied to the online solution of time-varying problems (and/or time-invariant ones, which could be viewed as a special case of time-varying ones). The design procedure could be formalized as follows.

Step 1. To monitor the process of problems solving, instead of a nonnegative scalar-valued norm-based energy-function usually associated with Hopfield-type and/or gradient-based networks [6–8], an indefinite matrix/vector-valued error function $E(t) = [e_{ij}(t)] \in \mathbb{R}^{m \times n}$ could be firstly defined and constructed (which means each element of the error function could be negative, positive, bounded or even unbounded), where the entry error $e_{ij}(t) \in \mathbb{R}$ is an element of the matrix/vector-valued error function $E(t)$, $\forall i = 1, 2, 3, \cdots, m$ and $j = 1, 2, 3, \cdots, n$.

Step 2. If such an error function $E(t)$ equals zero [i.e., each entry error $e_{ij}(t) \in \mathbb{R}$ of $E(t)$ is equal to zero], the neural state to be obtained [e.g., $X(t)$] could converge to the theoretical solution [e.g., $X^*(t)$] of the problem. Thus, the error-function time-derivative $\dot{E}(t)$ of $E(t)$ should be made such that every entry error $e_{ij}(t) \in \mathbb{R}$ of error-function $E(t) \in \mathbb{R}^{m \times n}$ converges to zero. In mathematics, we have to choose $\dot{e}_{ij}(t)$ such that $\lim_{t \to \infty} e_{ij}(t) = 0$.

Step 3. By following Zhang $et\ al.$'s design method [5–10, 13], a general form of the error-function time-derivative $\dot{E}(t)$ of $E(t)$ could be designed as follows (which is termed ZNN design formula for presentation convenience)

$$\dot{E}(t) := \frac{d(E(t))}{dt} = -\Gamma\Phi(E(t)), \tag{9.1}$$

where design parameter Γ and activation-function array $\Phi(\cdot)$ are described as follows.

- $\Gamma \in \mathbb{R}^{m \times m}$ is a positive-definite matrix, which is used to scale the convergence rate of the neural-network solution. For simplicity, we can use γI in place of Γ with $\gamma > 0 \in \mathbb{R}$. Γ (or γ), being a set of inductance parameters or reciprocals of capacitive parameters, should be set as large as the hardware permits (e.g., in analog circuits or VLSI [5,7,8,10,12]), or selected appropriately for simulative and experimental purposes.

- $\Phi(\cdot) : \mathbb{R}^{m \times n} \to \mathbb{R}^{m \times n}$ denotes a matrix activation-function array (or termed, a matrix activation-function mapping) of neural networks. A simple example of activation-function mapping $\Phi(\cdot)$ is the linear one, i.e., $\Phi(E) = E$. Different choices of $\Phi(\cdot)$ will lead to different performances. In general, any monotonically-increasing odd activation function $\phi(\cdot)$, being the ijth element of matrix array $\Phi(\cdot) \in \mathbb{R}^{m \times n}$, can be used for the construction of the neural network, where four basic types of activation functions $\phi(\cdot)$ are introduced in Section 1.5. and shown in Figure 1.3.

Step 4. Expanding the design formula (9.1) could lead to an implicit neural-dynamic equation, which is termed Zhang neural network (ZNN) for solving online time-varying problems. Then, starting from randomly-generated initial conditions [e.g., $X(0) := X_0 \in \mathbb{R}^{p \times q}$], the neural state matrix could (globally exponentially) converge to the theoretical solution of the problem. It is worth mentioning here that the time derivatives of the smoothly time-varying coefficients are assumed to be known or can be estimated, analytically or numerically.

9.2.2. Gradient Neural Networks

For comparison with the proposed ZNN method, it is also worth mentioning here that we can develop a gradient-based neural network to solve online the same time-varying problem as mentioned before. However, to the authors' best knowledge, similar to almost all numerical algorithms and neural-dynamic computational schemes, the gradient-based neural network is designed and developed intrinsically to perform exactly time-invariant (or say, static/stationary, constant) problems solving; i.e., of which the coefficients are constant (rather than time-varying) [6–8]. These algorithms and schemes are generally related to the gradient-descent method in optimization [2,4,6,17] described as the procedure below.

- Firstly, to solve for the neural-solution via gradient-based neural networks (GNN), a scalar-valued norm-based nonnegative (or lower-bounded at least) energy function, $\mathcal{E}(t) = \|E(t)\|^2 \in \mathbb{R}$ with some kind of matrix/vector norm $\|\cdot\|$, is defined. Note that, a minimum point of the energy function $\mathcal{E}(t)$ is achieved with $\mathcal{E}(t) = 0$, if and only if the neural-solution $X(t)$ is the exact/theoretical solution $X^*(t)$ of the constant problem solving.

- Secondly, a computational scheme could be designed to evolve along a descent direction of this energy function $\mathcal{E}(t)$, until the minimum point $X^*(t)$ is reached. It is worth mentioning here that a typical descent direction is the negative gradient of $\mathcal{E}(t)$, i.e., $-(\partial \mathcal{E}/\partial X) \in \mathbb{R}^{m \times n}$.

- Thirdly, the conventional gradient neural network design formula is defined and constructed as follows:

$$\dot{X}(t) = -\gamma \frac{\partial \mathcal{E}}{\partial X}. \tag{9.2}$$

Expanding the above GNN design formula (9.2) could lead to an explicit neural-dynamic equation, which is termed the conventional linear gradient-based neural network (GNN) for online solution of the constant problem (it may be a Hopfield-type neural network as well) [1, 3, 4, 11, 12], where design parameter γ is defined the same as that in the aforementioned ZNN models.

- Finally, as inspired by ZNN's design method, we could extend the above conventional linear gradient neural network to a general nonlinear form by employing an activation function array $\Phi(\cdot)$ [2, 4–6].

9.3. Comparisons and Differences

In this section, we would like to compare the two design methods and models of Zhang neural network and gradient neural network, which are exploited for the online solution of the same time-varying problem solving. The differences and novelties may lie in the following facts.

- The design of ZNN model is based on the elimination of every entry $e_{ij}(t) \in \mathbb{R}$ of the matrix/vector-valued indefinite error function $E(t) \in \mathbb{R}^{m \times n}$, with the value of $e_{ij}(t)$ being positive, negative, bounded or even unbounded. In contrast, the design of GNN model is based on the elimination of the scalar-valued norm-based energy function $\mathcal{E}(t) = \|E(t)\|^2$, which could only be positive or at least lower-bounded.

- ZNN model is depicted in an implicit dynamics, i.e., $M(t)\dot{X}(t) = \cdots$ (where $M(t)$ is the coefficient of the neural-dynamic system, and also termed the mass matrix in the MATLAB routine "ode45" [5, 6, 10, 13]), which might coincide well with systems in nature and in practice (e.g., in analogue electronic circuits and mechanical systems [7] owing to Kirchhoff's and Newton's laws, respectively). In contrast, GNN model is depicted in an explicit dynamics, i.e., $\dot{X}(t) = \cdots$, which is usually associated with classic Hopfield-type or gradient-based artificial neural networks [7–9, 13].

- ZNN model methodically and systematically exploits the time-derivative information of the time-varying coefficient(s) during its real-time solution process. In contrast, GNN model has not exploited such important information, and thus may not be effective on solving such time-varying problems.

- As discussed and analyzed in the preceding chapters, the neural state computed by ZNN model could (globally exponentially) converge to the theoretical time-varying solution $X^*(t)$. In contrast, GNN model could only generate an approximate result to the theoretical solution $X^*(t)$ with much larger steady-state residual errors.

- Belonging to a predictive approach, ZNN model and its design method make good use of the time-derivative information, and thus they could be more effective on the system convergence to a "moving" theoretical solution. In contrast, belonging to a conventional tracking approach, GNN model and its method act by adapting to the change of coefficient(s) in a posterior passive manner, and thus they theoretically can not catch the exact solution which is on the "move" (i.e., time-varying).

- As shown in [14] and [15], the connection from Newton iteration to ZNN models could be established. That is, Newton iteration for solving static problems appears to be a special case of the discrete-time ZNN models (by considering only the use of linear activation functions and fixing the step-size to be 1). Furthermore, this point may shows a new explanation to Newton iteration for static problems solving, which is evidently different from the traditional (or say, standard) explanations appearing in almost all literature and textbooks, e.g., via Taylor expansion [14, 15].

- The derivation of ZNN models might only need the less difficult knowledge of bachelors' mathematical course [e.g., see the scalar case of ZNN design formula (9.1)]. In contrast, the derivation of GNN models [such as (5.4) depicted in Subsubection 5.2.1.2.] requires more complicated mathematical knowledge of postgraduates' or even PhD's level.

9.4. An Illustrative Application

In this section, we focus on cyclic motion planning and online optimization techniques (specifically, quadratic programming, QP) for redundant manipulators. Computer simulations performed based on three types of multi-link planar robot arms, together with theoretical analysis based on gradient-descent and Zhang *et al*'s neural-dynamic methods, both substantiates the efficacy of the presented cyclic motion generation (CMG) scheme for redundant robots.

9.4.1. QP-Based Scheme-Formulation and Solver

As the cyclic motion (e.g., corresponding to a square path in [18]) is expected to generate, the minimization of the joint displacement $\|\theta(t) - \theta(0)\|_2^2$ between the current state $\theta(t) \in \mathbb{R}^n$ and the initial state $\theta(0) \in \mathbb{R}^n$ could be exploited, where $\|\theta\|_2^2 := \theta^T \theta$ denotes the two-norm of joint vector θ. This minimization can finally be derived as the minimization of quadratic performance index $\dot{\theta}^T \dot{\theta}/2 + c^T \dot{\theta}$ with $c = \beta(\theta(t) - \theta(0))$. Note that $\beta > 0 \in \mathbb{R}$ is used to adjust the magnitude of the manipulator's response to current joint displacement. In addition, such a minimization is subject to a periodical end-effector trajectory requirement, $\dot{r} = J(\theta)\dot{\theta}$, where $\dot{r} \in \mathbb{R}^m$ denotes the end-effector Cartesian-velocity vector, and $J(\theta)$ denotes the manipulator Jacobian matrix. It is worth mentioning that the degrees of redundancy is $n - m$.

Moreover, the avoidance of joint physical limits is important and necessary as well. Without considering these limits, physical damage may thus occur to robots. By considering joint limits $[\theta^-, \theta^+]$ and joint velocity limits $[\dot{\theta}^-, \dot{\theta}^+]$, the cyclic motion generation (CMG) scheme can be established as the following time-varying quadratic program in terms of joint velocity $\dot{\theta} \in \mathbb{R}^n$:

$$\text{minimize} \quad \frac{1}{2}\dot{\theta}^T\dot{\theta} + c^T\dot{\theta}, \tag{9.3}$$

$$\text{subject to} \quad c = \beta(\theta - \theta(0)),$$

$$J(\theta)\dot{\theta} = \dot{r}, \tag{9.4}$$

$$\xi^- \leq \dot{\theta} \leq \xi^+, \tag{9.5}$$

where the ith elements of unified low bound ξ^- and upper bound ξ^+ could be defined respectively as [19–22]:

$$\xi_i^- = \max\{\dot{\theta}_i^-, 2(\theta_i^- - \theta_i)\}, \ \xi_i^+ = \min\{\dot{\theta}_i^+, 2(\theta_i^+ - \theta_i)\}.$$

In CMG performance index (9.3), $c = \beta(\theta - \theta(0))$ might be called the CMG criterion, with β termed the CMG coefficient, used for eliminating the joint-angle-drift of redundant robot manipulators when moving repetitively (i.e., to generate cyclic motion).

Furthermore, we can solve the above quadratic program (9.3)-(9.5) by using MATLAB optimization routine "QUADPROG", but preferably by using recurrent neural nets due to their adaptive parallel-processing nature and convenience of circuits implementation [18–22]. It is worth mentioning here that, in the ensuing computer-simulations, QP (9.3)-(9.5) is solved actually by an LVI-based primal-dual neural network [19, 20].

9.4.2. Computer Simulation and Verification

In this subsection, the cyclic motion generation scheme (9.3)-(9.5) is simulated and applied based on 4-link, 5-link, and 6-link redundant planar robot arms. The end-effectors of such robot arms are expected to track squares, with task duration 80s.

9.4.2.1. 4-Link Planar Arm

The 4-link planar robot arm has four degrees of freedom (DOF) in the two-dimensional workspace, with two degrees of redundancy. Its joint physical limits are given as: $\theta^+ = -\theta^- = [\pi/3, \pi/3, \pi/3, \pi/3]^T$ in radians (in short, rad) and $\dot{\theta}^+ = -\dot{\theta}^- = [\pi, \pi, \pi, \pi]^T$ in radians per second (in short, rad/s). In addition, initial state $\theta(0) = [\pi/12, \pi/6, -\pi/4, \pi/8]^T$ rad [16]. Note that, in our simulations (including here and the ensuing subsections'), all joint angles and joint velocities are kept within their mechanical ranges, due to the inclusion of bound constraint $\xi^- \leq \dot{\theta} \leq \xi^+$ in our CMG scheme (9.3)-(9.5).

Firstly, the 4-link planar robot arm is expected to track a square path, which is synthesized without considering the CMG criterion (i.e., $\beta = 0$). Simulation results are shown in Figure 9.1 and Table 9.4.2.1., where J1T in the figure hereafter denotes the motion trajectory of joint 1, and so do J2T through J4T. In addition, as for the table, EF denotes the end-effector. Figure 9.1 illustrates the motion trajectories and joint-variable transients of

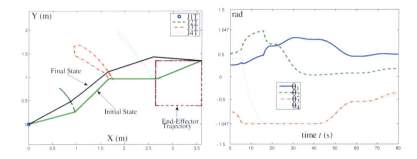

Figure 9.1. Noncyclic motion trajectories of 4-link planar robot arm performing a square path synthesized without considering CMG criterion (i.e., $\beta = 0$).

(Reproduced from K. Chen, L. Zhang et al., Cyclic motion generation of multi-link planar robot performing square end-effector trajectory analyzed via gradient-descent and Zhang et al's neural-dynamic methods, Figure 1, Proceedings of the 2nd International Symposium on Systems and Control in Aeronautics and Astronautics, pp. 1-6. ©[2008] IEEE. Reprinted, with permission).

Table 9.1. Joint-position change of 4-link robot arm when tracking a square path without considering the CMG criterion

#	Cartesian position $P(0)$ in meters	Cartesian position $P(80)$ in meters	$P(80) - P(0)$ in meters
Joint 1	(+0.0000000000, +0.0000000000)	(+0.0000000000, +0.0000000000)	(+0.0000000000, +0.0000000000)
Joint 2	(+0.9659258262, +0.2588190451)	(+0.8762474778, +0.4818613468)	(−0.0896783484, +0.2230423017)
Joint 3	(+1.6730326074, +0.9659258262)	(+1.6523355500, +1.1124858818)	(−0.0206970574, +0.1465600555)
Joint 4	(+2.6730326074, +0.9659258262)	(+2.6003140017, +1.4308203838)	(−0.0727186057, +0.4648945575)
EF	(+3.5969121399, +1.3486092586)	(+3.5969312498, +1.3486372915)	(+0.0000191099, +0.0000280328)

(Reproduced from K. Chen, L. Zhang et al., Cyclic motion generation of multi-link planar robot performing square end-effector trajectory analyzed via gradient-descent and Zhang et al's neural-dynamic methods, Table 1, Proceedings of the 2nd International Symposium on Systems and Control in Aeronautics and Astronautics, pp. 1-6. ©[2008] IEEE. Reprinted, with permission).

the robot arm when tracking the desired path. As shown in this figure, all joint motion trajectories are not closed, though the end-effector of the manipulator follows the desired square path successfully. In other words, the final state does not coincide with its initial state [i.e., $\theta_1(80) \neq \theta_1(0)$, $\theta_2(80) \neq \theta_2(0)$, $\theta_3(80) \neq \theta_3(0)$, $\theta_4(80) \neq \theta_4(0)$]. In addition, Table 9.4.2.1. shows the corresponding joint-position changes in terms of Cartesian coordinates. Evidently, the final joint-position change $P(80) - P(0)$ of θ_2, θ_3, and θ_4 are too large to be accepted in cyclic-motion tasks. Simply put, this simulation shows us a joint-angle-drift problem visibly.

For comparison, as the second computer-simulation in this 4-link example, the joint-angle-drift problem could be solved readily by considering the CMG criterion (i.e., with $\beta = 4$). Simulation results are shown in Figure 9.2 and Table 9.4.2.1.. As seen from Figure 9.2, all joints finally return to their initial values. Table 9.4.2.1. shows the corresponding joint-position changes in Cartesian coordinates, where the maximal position change is less

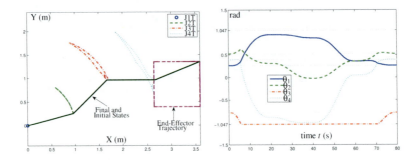

Figure 9.2. Cyclic motion trajectories of 4-link planar robot arm performing a square path synthesized by CMG scheme (9.3)-(9.5) with coefficient $\beta = 4$.

(Reproduced from K. Chen, L. Zhang et al., Cyclic motion generation of multi-link planar robot performing square end-effector trajectory analyzed via gradient-descent and Zhang et al's neural-dynamic methods, Figure 2, Proceedings of the 2nd International Symposium on Systems and Control in Aeronautics and Astronautics, pp. 1-6. ©[2008] IEEE. Reprinted, with permission).

Table 9.2. Joint-position change of 4-link robot arm when tracking a square path with the CMG criterion considered

#	Cartesian position $P(0)$ in meters	Cartesian position $P(80)$ in meters	$P(80) - P(0)$ in meters
Joint 1	$(+0.0000000000, +0.0000000000)$	$(+0.0000000000, +0.0000000000)$	$(+0.0000000000, +0.0000000000)$
Joint 2	$(+0.9659258262, +0.2588190451)$	$(+0.9659276581, +0.2588122084)$	$(+0.18318 \times 10^{-5}, -0.68366 \times 10^{-5})$
Joint 3	$(+1.6730326074, +0.9659258262)$	$(+1.6730527661, +0.9659006623)$	$(+2.01586 \times 10^{-5}, -2.51639 \times 10^{-5})$
Joint 4	$(+2.6730326074, +0.9659258262)$	$(+2.6730527660, +0.9659154406)$	$(+2.01585 \times 10^{-5}, -1.03856 \times 10^{-5})$
EF	$(+3.5969121399, +1.3486092586)$	$(+3.5969291931, +1.3486063700)$	$(+1.70531 \times 10^{-5}, -0.28885 \times 10^{-5})$

(Reproduced from K. Chen, L. Zhang et al., Cyclic motion generation of multi-link planar robot performing square end-effector trajectory analyzed via gradient-descent and Zhang et al's neural-dynamic methods, Table 2, Proceedings of the 2nd International Symposium on Systems and Control in Aeronautics and Astronautics, pp. 1-6. ©[2008] IEEE. Reprinted, with permission).

than 2.51640×10^{-5}m, very tiny and acceptable in practice. This acceptance is in view of the fact that the simulation was performed on a limited-accuracy digital computer with limited memory. As shown additionally in Figure 9.3, the end-effector positioning error is very tiny as well. This substantiates the efficacy of our CMG scheme (9.3)-(9.5) on cyclic motion generation of robot manipulators [16].

9.4.2.2. 5-Link Planar Arm

In this subsubsection, the presented CMG scheme is applied to a five-link planar robot arm, which has five DOF and works in two-dimensional workspace with three-DOF redundancy. The physical limits of joint angles and joint velocities are given as $\theta^+ = -\theta^- = [\pi/3, \pi/3, \pi/3, \pi/3, \pi/3]^T$rad and $\dot{\theta}^+ = -\dot{\theta}^- = [\pi, \pi, \pi, \pi, \pi]^T$rad/s, respectively. In addition, initial state $\theta(0) = [-\pi/12, \pi/12, \pi/6, -\pi/4, \pi/3]^T$rad. As seen from Figure 9.4 and

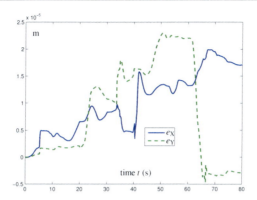

Figure 9.3. End-effector positioning error of 4-link planar robot arm when tracking a square path synthesized with the CMG criterion considered.

(Reproduced from K. Chen, L. Zhang et al., Cyclic motion generation of multi-link planar robot performing square end-effector trajectory analyzed via gradient-descent and Zhang et al's neural-dynamic methods, Figure 3, Proceedings of the 2nd International Symposium on Systems and Control in Aeronautics and Astronautics, pp. 1-6. ©[2008] IEEE. Reprinted, with permission).

Table 9.3, closed trajectories may not be obtained in the joint space when we do not consider the CMG criterion (i.e., with $\beta = 0$), though the end-effector performs the desired square path in the workspace. In contrast, Figure 9.5 and Table 9.4.2.2. show that the final state could fit well with the initial state by using our CMG criterion (i.e., with $\beta = 4$), where the position change is very tiny, less than 10^{-5} in meters (i.e., 10^{-2} in millimeters) [16]. Simulation results based on this 5-link planar robot arm also substantiate the efficacy of CMG scheme (9.3)-(9.5), especially the CMG measure index (9.3).

9.4.2.3. 6-Link Planar Arm

The six-link planar robot arm has six DOF, with four degrees of redundancy. In this subsubsection, it is employed as well to verify the effectiveness and efficiency of our presented CMG scheme (9.3)-(9.5). Similar to previous examples, the limits of all joint-variables and velocities are given as $\pm\pi/3$rad and $\pm\pi$rad/s, respectively. The initial state $\theta(0) = [-\pi/12, \pi/6, \pi/12, -\pi/6, \pi/4, -\pi/4]^T$rad. Figures 9.6 and 9.7 illustrate the non-cyclic and cyclic motion trajectories of such a six-link planar robot arm synthesized via our CMG scheme (9.3)-(9.5). On one hand, Figure 9.6 shows that, without considering the CMG criterion, some joint-trajectories are not closed in the sense that $\theta_2(80) \neq \theta_2(0)$, $\theta_3(80) \neq \theta_3(0)$, $\theta_4(80) \neq \theta_4(0)$, $\theta_5(80) \neq \theta_5(0)$. On the other hand, as seen from Figure 9.7, all final joint states could coincide well with their initial ones by using CMG scheme (9.3)-(9.5) (specifically, using the CMG criterion with $\beta = 4$) [16].

In summary, computer simulation and comparison based on three types of planar robot arms all have demonstrated the efficacy of the presented cyclic-motion-generation scheme-formulation (9.3)-(9.5) on drift-free redundancy resolution of robot manipulators [16]. Finally, before ending this subsection, it is worth pointing out that the presented QP-based CMG scheme is not only effective on planar robot arms, but also effective on three-

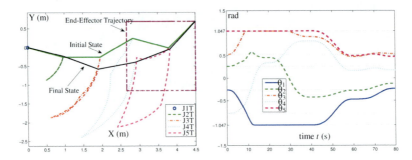

Figure 9.4. Noncyclic motion trajectories of 5-link planar robot arm performing a square path synthesized without considering CMG criterion (i.e., $\beta = 0$).

(Reproduced from K. Chen, L. Zhang et al., Cyclic motion generation of multi-link planar robot performing square end-effector trajectory analyzed via gradient-descent and Zhang et al's neural-dynamic methods, Figure 4, Proceedings of the 2nd International Symposium on Systems and Control in Aeronautics and Astronautics, pp. 1-6. ©[2008] IEEE. Reprinted, with permission).

dimensional redundant manipulators such as PA10 and PUMA560 [19, 21, 22].

9.4.3. CMG Performance-Index Analysis

As the preceding subsection illustrates successful simulation results of CMG scheme applied to different types of planar robot manipulators, in this subsection we investigate theoretically the reason why such a CMG scheme [specifically, CMG performance index (9.3)] succeeds. The analysis could be established in the following two ways.

9.4.3.1. Via Gradient-Descent Method

To show the correctness and effectiveness of the CMG performance index (9.3), we could take the following steps via the so-called gradient-descent method.

Firstly, to achieve cyclic motion generation of both end-effector and joint angles [i.e., $\theta(T) = \theta(0)$, where T denotes the final time of the cyclic motion], it is natural to define the following position-change function of scalar-valued two-norm based form: $\mathcal{E}(t) = \|\theta(t) - \theta(0)\|_2^2/2$, which is to be minimized over the cyclic task duration $[0,T]$. Evidently, the cyclic purpose is achieved, if the minimum zero value of this position-change function $\mathcal{E}(t)$ can be obtained at final time instant $t = T$ [i.e., if $\theta(T) - \theta(0) = 0$].

Secondly, using the gradient-descent method [17, 18], we could set $\dot{\theta} = -\beta(\partial \mathcal{E}(t)/\partial \theta)$; i.e., $\dot{\theta} = -\beta(\theta - \theta(0))$. Evidently, the gradient-descent method originates from the elimination of the scalar-valued norm-based position-change function $\mathcal{E}(t)$ with an exponential convergence rate β. In addition, the above can be rewritten as $\dot{\theta} + \beta(\theta - \theta(0)) = 0$.

Finally, as other factors (e.g., end-effector motion requirement and avoidance of joint physical limits) have to be considered in the cyclic-motion-generation scheme, the dynamic equation $\dot{\theta}(t) + \beta(\theta(t) - \theta(0)) = 0$ could be achieved only theoretically. Thus, minimizing $\|\dot{\theta}(t) + \beta(\theta(t) - \theta(0))\|_2^2/2$ appears to be more feasible in practice for the cyclic motion generation of redundant manipulators. It follows that expanding $\|\dot{\theta}(t) + \beta(\theta(t) - \theta(0))\|_2^2/2$

Table 9.3. Joint-position change of 5-link robot arm when tracking a square path without considering the CMG criterion

#	Cartesian position $P(0)$ in meters	Cartesian position $P(80)$ in meters	$P(80) - P(0)$ in meters
Joint 1	$(+0.0000000000, +0.0000000000)$	$(+0.0000000000, +0.0000000000)$	$(+0.0000000000, +0.0000000000)$
Joint 2	$(+0.9659258262, -0.2588190451)$	$(+0.9728285382, -0.2315267481)$	$(+0.0069027120, +0.0272922969)$
Joint 3	$(+1.9659258262, -0.2588190451)$	$(+1.9103777676, -0.5793793641)$	$(-0.0555480586, -0.3205603190)$
Joint 4	$(+2.8319512300, +0.2411809548)$	$(+2.8938802152, -0.3984845451)$	$(+0.0619289851, -0.6396655000)$
Joint 5	$(+3.7978770563, -0.0176380902)$	$(+3.8308623386, -0.0491072411)$	$(+0.0329852822, -0.0314691509)$
EF	$(+4.5049838375, +0.6894686909)$	$(+4.5050345285, +0.6894669652)$	$(+0.0000506909, -0.0000017257)$

(Reproduced from K. Chen, L. Zhang et al., Cyclic motion generation of multi-link planar robot performing square end-effector trajectory analyzed via gradient-descent and Zhang et al's neural-dynamic methods, Table 3, Proceedings of the 2nd International Symposium on Systems and Control in Aeronautics and Astronautics, pp. 1-6. ©[2008] IEEE. Reprinted, with permission).

Table 9.4. Joint-position change of 5-link robot arm when tracking a square path with the CMG criterion considered

#	Cartesian position $P(0)$ in meters	Cartesian position $P(80)$ in meters	$P(80) - P(0)$ in meters
Joint 1	$(+0.0000000000, +0.0000000000)$	$(+0.0000000000, +0.0000000000)$	$(+0.0000000000, +0.0000000000)$
Joint 2	$(+0.9659258262, -0.2588190451)$	$(+0.9659242978, -0.2588247494)$	$(-0.15284 \times 10^{-5}, -0.57042 \times 10^{-5})$
Joint 3	$(+1.9659258262, -0.2588190451)$	$(+1.9659242978, -0.2588276884)$	$(-0.15284 \times 10^{-5}, -0.86433 \times 10^{-5})$
Joint 4	$(+2.8319512300, +0.2411809548)$	$(+2.8319475941, +0.2411759616)$	$(-0.36359 \times 10^{-5}, -0.49932 \times 10^{-5})$
Joint 5	$(+3.7978770563, -0.0176380902)$	$(+3.7978748796, -0.0176376373)$	$(-0.21766 \times 10^{-5}, +0.04528 \times 10^{-5})$
EF	$(+4.5049838375, +0.6894686909)$	$(+4.5049830557, +0.6894677489)$	$(-0.07817 \times 10^{-5}, -0.09420 \times 10^{-5})$

(Reproduced from K. Chen, L. Zhang et al., Cyclic motion generation of multi-link planar robot performing square end-effector trajectory analyzed via gradient-descent and Zhang et al's neural-dynamic methods, Table 4, Proceedings of the 2nd International Symposium on Systems and Control in Aeronautics and Astronautics, pp. 1-6. ©[2008] IEEE. Reprinted, with permission).

yields

$$(\dot{\theta} + \beta(\theta(t) - \theta(0)))^T (\dot{\theta} + \beta(\theta(t) - \theta(0)))/2,$$

with $c := \beta(\theta(t) - \theta(0))$. Furthermore, it could be proved readily that minimizing the above expression is equivalent to minimizing the performance index $\dot{\theta}^T \dot{\theta}/2 + c^T \dot{\theta}$, which is exactly the same as CMG performance index (9.3).

9.4.3.2. Via Zhang et al's Neural-Dynamic Method

For comparative and more sufficient purposes, we could introduce an alternative approach for showing the correctness and effectiveness of CMG performance index (9.3). The so-called Zhang *et al.*'s neural-dynamic approach originates from the online neural solution of time-varying matrix and/or vector algebra problems [6–8,16]. We could derive the quadratic CMG performance index (9.3) via this method as follows.

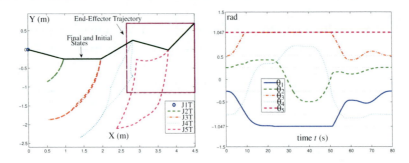

Figure 9.5. Cyclic motion trajectories of 5-link planar robot arm performing a square path synthesized by CMG scheme (9.3)-(9.5) with coefficient $\beta = 4$.

(Reproduced from K. Chen, L. Zhang et al., Cyclic motion generation of multi-link planar robot performing square end-effector trajectory analyzed via gradient-descent and Zhang et al's neural-dynamic methods, Figure 5, Proceedings of the 2nd International Symposium on Systems and Control in Aeronautics and Astronautics, pp. 1-6. ©[2008] IEEE. Reprinted, with permission).

Firstly, to generate the cyclic motion for redundant manipulators, instead of using the scalar-valued norm-based position-change function $E(t)$ of gradient-descent method in the preceding subsubsection, we could define alternatively a vector-valued position-change function $E(t) := \theta(t) - \theta(0) \in \mathbb{R}^n$ in this subsubsection.

Secondly, to eliminate every entry $e_j(t)$ of the vector-valued position-change function $E(t)$ over the task duration $[0,T]$, we can follow Zhang's method and simply set [6–8, 16]:

$$\dot{E}(t) = -\beta E(t), \qquad (9.6)$$

where $\beta > 0 \in \mathbb{R}$ is used to adjust the exponential-convergence rate of E to zero. In addition, we know that the solution to (9.6) is evidently $E(t) = \exp(-\beta t)E(0)$, and that, within the time-period of $4/\beta$ seconds, $|e_j(t)|$ would be less than 1.85% of $|e_j(0)|$, $\forall j \in \{1, 2, \cdots, n\}$ [8, 21].

Thirdly, expanding (9.6) with definition $E(t) = \theta(t) - \theta(0)$ yields the same result as that in the second step of Subsubsection 9.4.3.1.: $\dot{\theta}(t) = -\beta(\theta(t) - \theta(0))$. That is, through Zhang et al.'s neural-dynamic method, we could achieve the same results as those by the gradient-descent method. Similar to the final step of Subsubsection 9.4.3.1., it follows that quadratic performance index (9.3), $\dot{\theta}^T\dot{\theta}/2 + c^T\dot{\theta}$, is the very one that could lead to the cyclic motion of redundant arms, theoretically at least.

In summary, via two different approaches, we have both shown the theoretical correctness and effectiveness of performance index $\dot{\theta}^T\dot{\theta}/2 + c^T\dot{\theta}$ employed in cyclic motion generation of redundant robot manipulators.

9.5. Conclusion

In this chapter, ZNN and GNN models have been unified and investigated for the online solution of time-varying problems. From the formulation and design method of these two

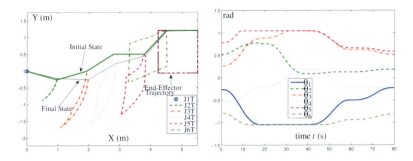

Figure 9.6. Noncyclic motion trajectories of 6-link planar robot arm performing a square path synthesized without considering CMG criterion (i.e., $\beta = 0$). *Reproduced from K. Chen, L. Zhang et al., Cyclic motion generation of multi-link planar robot performing square end-effector trajectory analyzed via gradient-descent and Zhang et al's neural-dynamic methods, Figure 6, Proceedings of the 2nd International Symposium on Systems and Control in Aeronautics and Astronautics, pp. 1-6. ©[2008] IEEE. Reprinted, with permission.*

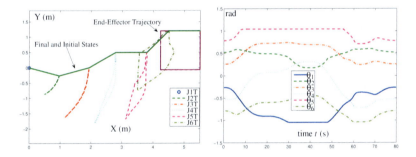

Figure 9.7. Cyclic motion trajectories of 6-link planar robot arm performing a square path synthesized by CMG scheme (9.3)-(9.5) with coefficient $\beta = 4$. *Reproduced from K. Chen, L. Zhang et al., Cyclic motion generation of multi-link planar robot performing square end-effector trajectory analyzed via gradient-descent and Zhang et al's neural-dynamic methods, Figure 7, Proceedings of the 2nd International Symposium on Systems and Control in Aeronautics and Astronautics, pp. 1-6. ©[2008] IEEE. Reprinted, with permission.*

neural networks, it is obviously shown that ZNN makes full use of the time-derivative information, while GNN does not utilize such important information. This point may show the reason why ZNN could solve time-varying problems effectively and exactly, and GNN could only approximately approach the exact solution with large lagging-behind error. This chapter has also presented a cyclic-motion-generation (CMG) scheme (9.3)-(9.5) for redundant robot manipulators, which could incorporate the avoidance of joint physical limits as well. Computer-simulation results based on three types of planar robot arms tracking a square path have substantiated again the efficacy of such a CMG scheme. Moreover, theoretical analysis based on both gradient-descent and Zhang *et al.*'s neural-dynamic approaches have shown the common effectiveness of the two design methodologies.

References

[1] Jang, J; Lee, S; Shin, S. An optimization network for matrix inversion. *Neural Information Processing and Computing,* 1988, pp. 397-401, New York: American Institute of Physics.

[2] Wang, J. A recurrent neural network for real-timematrix inversion. *Applied Mathematics and Computation,* 1993, vol. 55, no. 1, pp. 89-100.

[3] Manherz, RK; Jordan, BW; Hakimi, SL. Analog methods for computation of the generalized inverse. *IEEE Transactions on Automatic Control,* 1968, vol. 13, no. 5, pp. 582-585.

[4] Zhang, Y. Revisit the analog computer and gradient-based neural system for matrix inversion. *Proceedings of IEEE International Symposium on Intelligent Control,* 2005, pp. 1411-1416.

[5] Zhang, Y; Chen, K; Ma, W. MATLAB simulation and comparison of Zhang neural network and gradient neural network for online solution of linear time-varying equations. *Proceedings of International Conference on Life System Modeling and Simulation,* 2007, pp. 450-454.

[6] Zhang, Y; Chen, K. Comparison on Zhang neural network and gradient neural network for time-varying linear matrix equation $AXB = C$ solving. *Proceedings of IEEE International Conference on Industrial Technology,* 2008, pp. 1-6.

[7] Zhang, Y; Jiang, D; Wang, J. A recurrent neural network for solving Sylvester equation with time-varying coefficients. *IEEE Transactions on Neural Networks,* 2002, vol. 13, no. 5, pp. 1053-1063.

[8] Zhang, Y; Ge, SS. Design and analysis of a general recurrent neural network model for time-varying matrix inversion. *IEEE Transactions on Neural Networks,* 2005, vol. 16, no. 6, pp. 1477-1490.

[9] Zhang, Y; Peng, H. Zhang neural network for linear time-varying equation solving and its robotic application. *Proceedings of the Sixth International Conference on Machine Learning and Cybernetics,* 2007, pp. 3543-3548.

[10] Zhang, Y; Guo, X; Ma, W. Modeling and simulation of Zhang neural network for online linear time-varying equations solving based on MATLAB Simulink. *Proceedings*

of the 7th International Conference on Machine Learning and Cybernetics, 2008, pp. 805-810.

[11] Steriti, RJ; Fiddy, MA. Regularized image reconstruction using SVD and a neural network method for matrix inversion. *IEEE Transactions on Signal Processing,* 1993, vol. 41, no. 10, pp. 3074-3077.

[12] Wang, J; Electronic realisation of recurrent neural network for solving simultaneous linear equations. *Electronics Letters,* 1992, vol. 28, no. 5, pp. 493-495.

[13] Zhang, Y; Yang, Y. Simulation and comparison of Zhang neural network and gradient neural network solving for time-varying matrix square roots. *Proceedings of International Symposium on Intelligent Information Technology Application,* 2008, vol. 2, pp. 966-970.

[14] Zhang, Y; Ma, W; Yi, C. The link between Newton iteration for matrix inversion and Zhang neural network (ZNN). *Proceedings of IEEE International Conference on Industrial Technology,* 2008, pp. 1-6.

[15] Zhang, Y; Cai, B; Liang, M; Ma, W. On the variable step-size of discrete-time Zhang neural network and Newton iteration for constant matrix inversion. *Proceedings of International Symposium on Intelligent Information Technology Application,* 2008, vol. 1, pp. 34-38.

[16] Chen, K; Zhang, L; Zhang, Y. Cyclic motion generation of multi-link planar robot performing square end-effector trajectory analyzed via gradient-descent and Zhang et al's neural-dynamic methods. *Proceedings of the 2nd International Symposium on Systems and Control in Aeronautics and Astronautics,* 2008, pp. 1-6.

[17] Zhang, Y; Chen, K. Global exponential convergence and stability of Wang neural network for solving online linear equations. *Electronics Letters,* 2008, vol. 44, no. 2, pp. 145-146.

[18] Zhang, Y; Lv, X; Li, Z; Yang, Z; Zhu, H. Effective neural remedy for drift phenomenon of planar three-link robot arm using quadratic performance index. *Electronics Letters,* 2008, vol. 44, no. 6, pp. 435-437.

[19] Zhang, Y; Lv, X; Yang, Z; Li, Z. Repetitive motion planning of kinematically redundant manipulators using LVI-based primal-dual neural network. *Proceedings of IEEE International Conference on Mechatronics and Automation,* 2007, pp. 3138-3143.

[20] Zhang, Y; Ge, SS; Lee, TH. A unified quadratic-programming based dynamical-system approach to joint torque optimization of physically constrained redundant manipulators. *IEEE Transactions on System, Man, and Cybernetics,* Part B, 2004, vol. 34, no. 5, pp. 2126-2132.

[21] Zhang, Y. *Analysis and Design of Recurrent Neural Networks and their Applications to Control and Robotic Systems,* Ph.D. Dissertation, Chinese University of Hong Kong, 2002.

[22] Zhang, Y; Wang, J; Xia, Y. A dual neural network for redundancy resolution of kinematically redundant manipulators subject to joint limits and joint velocity limits. *IEEE Transactions on Neural Networks,* 2003, vol. 14, no. 3, pp. 658-667.

Appendix A

In order to lay a basis for further discussion, investigation and analysis on Zhang neural dynamics, some preliminaries related to ZNN such as matrix fundamental operations (e.g., matrix multiplication, matrix inversion and Kronecker product), concepts and definitions, are introduced in this appendix.

A.1. Computation of Matrix and Vector

The problem of linear/nonlinear matrix equations solving (e.g., matrix inversion) is considered to be a very fundamental issue widely encountered in science and engineering fields such as automatical control [1–3], optimization [4–6], robotic inverse kinematics [7–9], and signal processing [10–12]. In this section, we would introduce the concepts and properties of matrix and vector, as well as the basic matrix and/or vector operation.

A.1.1. Notation of Matrix and Vector

A matrix is made of a set of elements. The elements of the matrix could be real numbers, as well as functions. Generally speaking, the matrix is rectangular.

Definition A.1.1. [13] Let $\mathbb{R}$ denote the set of real numbers. All m-by-n real matrices in the vector space are defined as follow

$$A = [a_{ij}] \in \mathbb{R}^{m \times n} \Longleftrightarrow A = \begin{bmatrix} a_{11} & a_{12} & \cdots & a_{1n} \\ a_{21} & a_{22} & \cdots & a_{2n} \\ \vdots & \vdots & \ddots & \vdots \\ a_{m1} & a_{m2} & \cdots & a_{mn} \end{bmatrix},$$

where $a_{ij} \in \mathbb{R}$ denotes the ith-row and jth-column element of the matrix A with $i = 1, 2, \cdots, m$ and $j = 1, 2, \cdots, n$.

In general, if a capital letter is used to denote a matrix (e.g., A, B, and C), then the corresponding lower-case letter with subscript ij refers to the ith-row and jth-column element (e.g., a_{ij}, b_{ij}, and c_{ij}). In addition, we could use the notation $[A]_{ij}$ and $A(i, j)$ to designate the matrix element [13].

Definition A.1.2. [13] All real n-vectors in the vector space are defined as follow

$$x \in \mathbb{R}^n \Longleftrightarrow x = \begin{bmatrix} x_1 \\ x_2 \\ \vdots \\ x_n \end{bmatrix},$$

where $x_i \in \mathbb{R}$ denotes the ith element of the vector x with $i = 1, 2, \cdots, n$.

Similarly, we could also use the notation $[x]_i$ and $x(i)$ to designate the ith vector element. It is worth pointing out that, in this book, $\mathbb{R}^n$ and $\mathbb{R}^{n \times 1}$ are mathematically equivalent, and so $x \in \mathbb{R}^n$ is termed as a *column* vector. On the other hand, the *row* vector could be defined as follows

$$x \in \mathbb{R}^{1 \times n} \Longleftrightarrow x = \begin{bmatrix} x_1 & x_2 & \cdots & x_n \end{bmatrix},$$

where $x_i \in \mathbb{R}$ denotes the ith *row* element of the vector x with $i = 1, 2, \cdots, n$. Note that, if x is a *column* vector, then $y = x^T$ is a *row* vector (where subscript ^{T} denotes the transpose of a matrix or vector which will be introduced in the ensuing part). According to the above definitions, we could observe that *column* vector is one of a special case of matrix, so is a *row* vector. In other words, a *column* vector could be obtained when $n = 1$ in $A \in \mathbb{R}^{m \times n}$ and a *row* vector could be obtained when $m = 1$ in $A \in \mathbb{R}^{m \times n}$.

A.1.2. Matrix/Vector Basic Operation

Basic matrix operations include matrix addition/subtraction, scalar-matrix multiplication and matrix-matrix multiplication, matrix transposition [13]:

- Assume that matrices $A \in \mathbb{R}^{m \times n}$, $B \in \mathbb{R}^{m \times n}$ and $C \in \mathbb{R}^{m \times n}$, then the addition/subtraction of A and B could be formulated as

$$C = A \pm B \Longleftrightarrow c_{ij} = a_{ij} \pm b_{ij}.$$

- Assume that matrices $A \in \mathbb{R}^{m \times n}$, $B \in \mathbb{R}^{m \times n}$, and the scalar $\alpha \in \mathbb{R}$, then the scalar-matrix multiplication of α and A could be formulated as

$$B = \alpha A \Longleftrightarrow b_{ij} = \alpha a_{ij}.$$

- Assume that matrices $A \in \mathbb{R}^{m \times p}$, $B \in \mathbb{R}^{p \times n}$ and $C \in \mathbb{R}^{m \times n}$, then the matrix-matrix multiplication of A and B could be formulated as

$$C = AB = \left[\sum_{k=1}^{p} a_{ik} b_{kj} \right]_{m \times n} \Longleftrightarrow c_{ij} = \sum_{k=1}^{p} a_{ik} b_{kj}.$$

On the other hand, basic *column* vector and *row* vector operations are quite similar to the presented operation except the dot product (or termed, inner product) between two vectors.

Assume that matrices $A \in \mathbb{R}^{m \times n}$ and $B \in \mathbb{R}^{n \times m}$, then the transposition of A could be defined as [13]

$$B = A^T \Longleftrightarrow b_{ij} = a_{ji},$$

where subscript ^{T} denotes the transpose of a matrix and/or vector. In addition, the basic properties of the matrix transposition include

1) $(A^T)^T = A, \ \forall A \in \mathbb{R}^{m \times n}$

2) $(A + B)^T = A^T + B^T, \ \forall A, B \in \mathbb{R}^{m \times n}$

3) $(AB)^T = B^T A^T, \ \forall A \in \mathbb{R}^{m \times n}, B \in \mathbb{R}^{n \times p}$.

Assume $x \in \mathbb{R}^n$ and $y \in \mathbb{R}^n$, then the dot product (inner product) could be defined below

$$< x, y >= x^T y = y^T x = < y, x >= \sum_{i=1}^{n} y_i x_i \in \mathbb{R},$$

and the outer product could be defined as follows:

$$xy^T = \begin{bmatrix} x_1 \\ x_2 \\ \vdots \\ x_n \end{bmatrix} \begin{bmatrix} y_1 & y_2 & \cdots & y_n \end{bmatrix} = \begin{bmatrix} x_1 y_1 & x_1 y_2 & \cdots & x_1 y_n \\ x_2 y_1 & x_2 y_2 & \cdots & x_2 y_n \\ \vdots & \vdots & \ddots & \vdots \\ x_n y_1 & x_n y_2 & \cdots & x_n y_n \end{bmatrix} \in \mathbb{R}^{n \times n}.$$

In general, the order of matrices' positions is not communicative in these combined operations. For example, if $A \in \mathbb{R}^{m \times p}$ and $B \in \mathbb{R}^{p \times n}$, then the multiplication $C = AB \in \mathbb{R}^{m \times n}$; but the multiplication $D = BA$ does not exist. Moreover, the multiplication AB and the multiplication BA are not equivalent in most case. In mathematics, $AB \neq BA$ (which implies that matrix-multiplication is quite different from scalar-multiplication). In other words, matrix-matrix multiplication does not always satisfy the commutative law of multiplication.

A.2. Matrix Inversion, Trace and Kronecker Product

Acting as a closely-related topic of linear equation solving, the problem of matrix inversion is usually viewed as an essential part of many solutions [11, 14–16].

Definition A.2.1. [17] For a given square matrix $A \in \mathbb{R}^{n \times n}$, the matrix $B \in \mathbb{R}^{n \times n}$ that satisfies the conditions $AB = I$ and $BA = I$ is called the inverse of matrix A, and is denoted by $B = A^{-1}$, where $I \in \mathbb{R}^{n \times n}$ denotes the identity matrix.

It is worth pointing out here that not all of the square matrices are invertible (e.g., the zero matrix and some nonzero matrices whose determinants equal zero). A given square matrix $A \in \mathbb{R}^{n \times n}$ is said to be *invertible* if and only if its determinant is nonzero (or say, its rank is equal to the matrix-dimension n). On the other hand, an invertible matrix is said to be *nonsingular*, and a square matrix with no inverse is called a *singular matrix* [17]. Take the above-presented linear matrix equation $Ax = b$ as an example, there exists a unique feasible solution if and only if the coefficient matrix A is *nonsingular*. In addition, the following properties on matrix inversion could be obtained.

Proposition A.2.1. *[13, 17] For the given invertible matrices $A, B \in \mathbb{R}^{n \times n}$, we could have*

- $(A^{-1})^{-1} = A$

- $(AB)^{-1} = B^{-1} A^{-1}$

- $(A^T)^{-1} = (A^{-1})^T$. □

Definition A.2.2. [17] The trace of the square matrix $A = [a_{ij}] \in \mathbb{R}^{n \times n}$ is defined to be the sum of the entries lying on the main diagonal of A, $\forall i, j = 1, 2, 3, \cdots, n$. That is,

$$\text{trace}(A) = \text{tr}(A) = a_{11} + a_{22} + \cdots + a_{nn} = \sum_{i=1}^{n} a_{ii}.$$

Moreover, the basic properties of the matrix trace include

1. $\text{tr}(A) = \text{tr}(A^T)$, $\forall A \in \mathbb{R}^{n \times n}$

2. $\text{tr}(AA^T) = \text{tr}(A^T A)$, $\forall A \in \mathbb{R}^{m \times n}$

3. $\text{tr}(AB) = \text{tr}(BA) = \text{tr}(A^T B^T) = \text{tr}(B^T A^T)$ $\forall A \in \mathbb{R}^{n \times m}, B \in \mathbb{R}^{m \times n}$

4. $\text{tr}(\alpha A + \beta B) = \alpha \text{tr}(A) + \beta \text{tr}(B)$ $\forall A, B \in \mathbb{R}^{n \times n}, \alpha, \beta \in \mathbb{R}$

5. $\text{tr}(A) = \sum_{i=1}^{n} a_{ii} = \sum_{i=1}^{n} \lambda_i$, where λ_i is the ith eigenvalue of square matrix $A \in \mathbb{R}^{n \times n}$ with $i = 1, 2, \cdots, n$.

Generally speaking, many linear matrix equations could be transformed to linear system of equations via Kronecker product and/or vectorization technique. For example, differential matrix equation could be transformed to differential equation in the form of vectors by using Kronecker product and/or vectorization techniques.

Definition A.2.3. [16,17] The Kronecker product (also known as the *tensor product* or the *direct product*) of two matrices $A = [a_{ij}] \in \mathbb{R}^{m \times n}$ ($\forall i = 1, 2, 3, \cdots, m$ and $j = 1, 2, 3, \cdots, n$) and $B \in \mathbb{R}^{p \times q}$ is defined as follows:

$$A \otimes B := \begin{bmatrix} a_{11}B & a_{12}B & \cdots & a_{1n}B \\ a_{21}B & a_{22}B & \cdots & a_{2n}B \\ \vdots & \vdots & \ddots & \vdots \\ a_{m1}B & a_{m2}B & \cdots & a_{mn}B \end{bmatrix} \in \mathbb{R}^{mp \times nq}.$$

In general, $A \otimes B \neq B \otimes A$. The basic properties of the Kronecker product include

1) $(A \otimes B)^T = A^T \otimes B^T$, $\forall A \in \mathbb{R}^{m \times n}, B \in \mathbb{R}^{p \times q}$

2) $A \otimes (B + C) = A \otimes B + A \otimes C$, $\forall A \in \mathbb{R}^{m \times n}, B, C \in \mathbb{R}^{p \times q}$

3) $(A + B) \otimes C = A \otimes C + B \otimes C$, $\forall A, B \in \mathbb{R}^{m \times n}, C \in \mathbb{R}^{p \times q}$

4) $A \otimes (B \otimes C) = (A \otimes B) \otimes C$, $\forall A \in \mathbb{R}^{m \times n}, B \in \mathbb{R}^{p \times q}, C \in \mathbb{R}^{r \times s}$

5) $(A \otimes B)(C \otimes D) = (AC) \otimes (BD)$, $\forall A \in \mathbb{R}^{m \times n}, B \in \mathbb{R}^{p \times q}, C \in \mathbb{R}^{n \times r}, D \in \mathbb{R}^{q \times s}$.

Besides, with matrix $A = [a_{ij}] \in \mathbb{R}^{m \times n}$, we could associate the vector $\text{vec}(A) \in \mathbb{R}^{mn \times 1}$ defined by

$$\text{vec}(A) = [a_{11}, \cdots, a_{m1}, a_{12}, \cdots, a_{m2}, \cdots, a_{1n}, \cdots, a_{mn}]^T.$$

It is worth mentioning here that the matrix equation could be transformed to the vector form by using the Kronecker product and the above $\text{vec}(\cdot)$ operations. For instance, let X be unknown, given $A \in \mathbb{R}^{m \times n}, B \in \mathbb{R}^{p \times q}, C \in \mathbb{R}^{m \times q}$, the matrix equation $AXB = C$ could be vectorized as the system of $(B^T \otimes A)\text{vec}(X) = \text{vec}(C)$.

A.3. Vector/Matrix Norms

Norms serve the same purpose on vector spaces that absolute value does on the real line: they furnish a measure of distance. More precisely, $\mathbb{R}^n$ together with a norm on $\mathbb{R}^n$ defines a metric space [13].

Definition A.3.1. [13] A vector norm on $\mathbb{R}^n$ is a function $f : \mathbb{R}^n \to \mathbb{R}$ that satisfies the following properties:

1) $f(x) \geq 0, \ x \in \mathbb{R}^n; \ f(x) = 0$ iff $x = 0$;

2) $f(x+y) \leq f(x) + f(y), \ x, y \in \mathbb{R}^n$;

3) $f(\alpha x) = |\alpha| f(x), \ \alpha \in \mathbb{R}, x \in \mathbb{R}^n$.

Then we could denote such a function with a double bar notation: $f(x) = ||x||$. It is worth mentioning that subscripts on the double bar are used to distinguish between various norms. A useful class of vector norms are the $p-$norms, which could be defined as follows:

$$\|x\|_p = (|x_1|^p + \cdots + |x_n|^p)^{\frac{1}{p}} = \left(\sum_{k=1}^{n} |x_k|^p \right)^{\frac{1}{p}}, \quad 1 \leq p \leq +\infty.$$

Of these the 1, 2 and ∞ norms are the most important:

- $\|x\|_1 = |x_1| + \cdots + |x_n|$,

- $\|x\|_2 = (|x_1|^2 + \cdots + |x_n|^2)^{1/2}$,

- $\|x\|_\infty = \max_{1 \leq i \leq n} |x_i|$.

Specially, a *unit* vector with respect to the norm $\| \cdot \|$ is a vector x that satisfies $\|x\| = 1$.

The following important lemma concerning p-norms is the *Hölder inequality* [13].

Lemma 1.3.1. *Consider the two vectors $x, y \in \mathbb{R}^n$, there exists*

$$|\langle x, y \rangle| = |x^T y| \leq \|x\|_p \|y\|_q,$$

where $p > 1$, $q > 1$, and $\frac{1}{p} + \frac{1}{q} = 1$.

A very important special case of the above inequality is the *Cauchy-Schwartz* inequality:

$$|x^T y| \leq \|x\|_2 \|y\|_2.$$

In the same way, the following definition on matrix norms could be obtained.

Definition A.3.2. [13] A matrix norm on $\mathbb{R}^{m \times n}$ is a function $f : \mathbb{R}^{m \times n} \to \mathbb{R}$ that satisfies the following four properties:

1) $f(A) \geq 0, \ A \in \mathbb{R}^{m \times n}; \ f(A) = 0$ iff $A = 0$;

2) $f(A + B) \leq f(A) + f(B), \ A, B \in \mathbb{R}^{m \times n}$;

3) $f(\alpha A) = |\alpha| f(A), \; \alpha \in \mathbb{R}, A \in \mathbb{R}^{m \times n}$;

4) $f(AB) \leq f(A)f(B), \; A \in \mathbb{R}^{m \times n}, B \in \mathbb{R}^{n \times m}$.

As with the presented vector norms, we could also use a double bar notation with subscripts to designate matrix norm, i.e., $\|A\| = f(A)$.

The most frequently used matrix norms in numerical linear algebra are the Frobenius norm (or simply termed F-norm)

$$\|A\|_F = \sqrt{\sum_{i=1}^{m} \sum_{j=1}^{n} |a_{ij}|^2} = \sqrt{\text{trace}(A^T A)},$$

and the matrix 2-norm which induced by the Euclidean vector norm

$$\|A\|_2 = \max_{\|x\|_2=1} \|Ax\|_2 = \sqrt{\lambda_{\max}},$$

where $A = [a_{ij}] \in \mathbb{R}^{m \times n}$, $x \in \mathbb{R}^{n \times 1}$ and $\lambda_{\max}$ is the largest eigenvalue of the matrix $A^T A$.

In addition, the 1 and ∞ matrix norms induced by the vector 1-norm and ∞-norm could be defined as follows:

$$\|A\|_1 = \max_{\|x\|_1=1} \|Ax\|_1 = \max_j \sum_i |a_{ij}|$$

$$\|A\|_\infty = \max_{\|x\|_\infty=1} \|Ax\|_\infty = \max_i \sum_j |a_{ij}|.$$

In other words, the matrix 1-norm $\|A\|_1$ is the largest absolute *column* sum; while the matrix ∞-norm $\|A\|_\infty$ is the largest absolute *row* sum.

Moreover, the Frobenius and the 1-, 2-, and ∞-norms satisfy certain inequalities that are frequently used in the analysis of matrix computation [13]. For example, given a matrix $A = [a_{ij}] \in \mathbb{R}^{m \times n}$, we have

- $\|A\|_2 \leq \|A\|_F \leq \sqrt{n}\|A\|_2$,

- $\max_{i,j} |a_{ij}| \leq \|A\|_2 \leq \sqrt{mn} \max_{i,j} |a_{ij}|$,

- $\frac{1}{\sqrt{n}}\|A\|_\infty \leq \|A\|_2 \leq \sqrt{m}\|A\|_\infty$,

- $\frac{1}{\sqrt{m}}\|A\|_1 \leq \|A\|_2 \leq \sqrt{n}\|A\|_1$.

A.4. Matrix Differentiation

When a dynamic system is discussed and analyzed, the matrix differentiation may be used to simplify the presentation and solution processing. Thus, the matrix differentiation is introduced in this section.

Definition A.4.1. [18] If $A(t) = [a_{ij}(t)] \in \mathbb{R}^{m \times n}$, for $i = 1, 2, 3, \cdots, m$ and $j = 1, 2, 3, \cdots, n$, then $\partial A / \partial t$ is defined to be $[\partial a_{ij}(t)/\partial t]$; that is, the derivative of $A(t)$ is obtained by differentiating each element of A.

Unless specified otherwise (e.g., A is symmetric), we assume that the elements of all the matrices differentiated are functionally independent (i.e., unconstrained). From definition A.4.1, given appropriately-dimensioned matrices, we have

i) $\frac{\partial\{AX(t)B\}}{\partial t} = A\frac{\partial X(t)}{\partial t}B,$

ii) $\frac{\partial\text{vec}(X(t))}{\partial t} = \text{vec}\left(\frac{\partial X(t)}{\partial t}\right),$

iii) $\frac{\partial\{A(t)B(t)C(t)\}}{\partial t} = \frac{\partial A(t)}{\partial t}B(t)C(t) + A(t)\frac{\partial B(t)}{\partial t}C(t) + A(t)B(t)\frac{\partial C(t)}{\partial t},$

iv) $\frac{\partial\{A(t)\otimes B(t)\}}{\partial t} = \frac{\partial A(t)}{\partial t}\otimes B(t) + A(t)\otimes\frac{\partial B(t)}{\partial t}.$

Definition A.4.2. [18] Let $f(X)$ be a scalar function of the elements x_{ij} of the matrix $X \in \mathbb{R}^{m\times n}$. Then the derivative of f with respect to X, written as df/dX, is the matrix with the ijth element $\partial f/\partial x_{ij}$, that is,

$$\frac{df}{dX} = \left[\frac{\partial f}{\partial x_{ij}}\right]_{m\times n} = \begin{bmatrix} \frac{\partial f}{\partial x_{11}} & \frac{\partial f}{\partial x_{12}} & \cdots & \frac{\partial f}{\partial x_{1n}} \\ \frac{\partial f}{\partial x_{21}} & \frac{\partial f}{\partial x_{22}} & \cdots & \frac{\partial f}{\partial x_{2n}} \\ \vdots & \vdots & \ddots & \vdots \\ \frac{\partial f}{\partial x_{m1}} & \frac{\partial f}{\partial x_{m2}} & \cdots & \frac{\partial f}{\partial x_{mn}} \end{bmatrix}.$$

Especially, we have the following derivative of f with respect to the vector $x \in \mathbb{R}^n$

$$\frac{df}{dx} = \left[\frac{\partial f}{\partial x_1}, \frac{\partial f}{\partial x_2}, \cdots, \frac{\partial f}{\partial x_n}\right]^T.$$

For example, let matrix $A = [a_{ij}] \in \mathbb{R}^{n\times n}$ and vector $x \in \mathbb{R}^n$. Assume that $f(x) = x^T A x$, then we have

$$\frac{df}{dx} = \begin{bmatrix} \frac{\partial f}{\partial x_1} \\ \vdots \\ \frac{\partial f}{\partial x_n} \end{bmatrix} = \begin{bmatrix} \sum_{k=1}^n a_{k1}x_k + \sum_{p=1}^n a_{1p}x_p \\ \vdots \\ \sum_{k=1}^n a_{kn}x_k + \sum_{p=1}^n a_{np}x_p \end{bmatrix} = (A^T + A)x.$$

In addition, if matrix A is symmetric, then $df/dx = 2Ax$.

For readers' convenience, given appropriately-dimensioned matrices A, B and C, the following three main properties of matrix differentiation are listed:

- $\frac{\partial}{\partial A}\text{trace}(BAC) = B^T C^T,$

- $\frac{\partial}{\partial A}\text{trace}(BA^T C) = CB,$

- $\frac{\partial}{\partial A}\text{trace}(ABA^T) = AB^T + AB.$

References

[1] Zhang, Y. *Analysis and Design of Recurrent Neural Networks and their Applications to Control and Robotic Systems*, Ph.D. Dissertation, Chinese University of Hong Kong; 2002.

[2] Zhang, Y; Wang, J. Recurrent neural networks for nonlinear output regulation. *Automatica*, 2001, vol. 37, no. 8, pp. 1161-1173.

[3] Zhang, Y; Wang, J. Global exponential stability of recurrent neural networks for synthesizing linear feedback control systems via pole assignment. *IEEE Transactions on Neural Networks*, 2002, vol. 13, no. 3, pp. 633-644.

[4] Xia, Y; Wang, J. A general projection neural network for solving monotone variational inequalities and related optimization problems. *IEEE Transactions on Neural Networks*, 2004, vol. 15, no. 2, pp. 318-328.

[5] Zhang, Y; Ge, SS; Lee, TH. A unified quadratic-programming based dynamical system approach to joint torque optimization of physically constrained redundant manipulators. *IEEE Transactions on Systems, Man, and Cybernetics*, Part B, 2004, vol. 34, no. 5, pp. 2126-2132.

[6] Guo, W; Qiao, Y; Hou, H. BP neural network optimized with PSO algorithm and its application in forecasting. *Proceedings of IEEE International Conference on Information Acquisition*, 2006, pp. 617-621.

[7] Zhang, Y; Wang, J; Xu, Y. A dual neural network for bi-criteria kinematic control of redundant manipulators. *IEEE Transactions on Robotics and Automation*, 2002, vol. 18, no. 6, pp. 923-931.

[8] Zhang, Y; Wang, J; Xia, Y. A dual neural network for redundancy resolution of kinematically redundant manipulators subject to joint limits and joint velocity limits. *IEEE Transactions on Neural Networks*, 2003, vol. 14, no. 3, pp. 658-667.

[9] Zhang, Y; Wang, J. A dual neural network for constrained joint torque optimization of kinematically redundant manipulators. *IEEE Transactions on Systems, Man, and Cybernetics*, part B, 2002, vol. 32, no. 5, pp. 654-662.

[10] Tank, DW; Hopfield, JJ. Simple neural optimization networks: an A/D converter, signal decision circuit, and a linear programming circuit. *IEEE Transactions on Circuits and Systems*, 1986, vol. 33, no. 5, pp. 533-541.

[11] Steriti, RJ; Fiddy, MA. Regularized image reconstruction using SVD and a neural network method for matrix inversion. *IEEE Transactions on Signal Processing,* 1993, vol. 41, no. 10, pp. 3074-3077.

[12] Kung, SY. *Digital Neural Network.* New Jersey: Prentice Hall, Englewood Cliffs, 1993.

[13] Golub, GH; Van LCF. *Matrix Computations, 3rd ed.* Baltimore and London: The Johns Hopkins University Press, 1996.

[14] Zhang, Y; Ge, SS. Design and analysis of a general recurrent neural network model for time-varying matrix inversion. *IEEE Transactions on Neural Networks,* 2005, vol. 16, no. 6, pp. 1477-1490.

[15] Zhang, Y; Ma, W; Cai, B. From Zhang neural network to Newton iteration for matrix inversion. *IEEE Transactions on Circuits and Systems I: Regular Papers,* 2009, vol. 56, no. 7, pp. 1405-1415.

[16] Zhang, Y; Jiang, D; Wang, J. A recurrent neural network for solving sylvester equation with time-varying coefficients. *IEEE Transactions on Neural Networks,* 2002, vol. 13, no. 5, pp. 1053-1063.

[17] Meyer, CD. *Matrix Analysis and Applied Linear Algebra Book and Solutions Manual (Hardcover).* Philadelphia: Society for Industrial and Applied Mathematics, 2001.

[18] Seber, GAF. *A Matrix Handbook for Statisticians.* Hoboken, New Jersey: Wiley-Interscience, 2008.

Index

activation function, 9
approximate result, 159
artificial neural networks, 1

bipolar-sigmoid activation function, 10
block diagram, 21
BP neural network, 3

Cartesian, 37
circuit schematics, 93
coefficients, 28
column, 250
column element, 249
computational error, 29
constant coefficients, 51
constant/static, 6
convergence time, 96

descent direction, 4
design parameter, 21
differential equation, 177
discrete-time, 159
dot product, 250
dynamic-equation, 92

electronic circuits, 203
electronic implementation, 204
end-effector, 37
energy function, 4
entry, 93
entry error, 75
entry-to-entry, 157
equilibrium point, 7
error back-propagation, 3
error function, 5
error-free, 127
error-function derivative, 21

exact solution, 159

feedback, 2
feedforward, 2
finite difference, 83
Frobenius norm , 254
function blocks, 32

generalized inverse, 103
globally asymptotically stable, 7
globally convergent, 7
globally exponentially convergent, 7
globally exponentially stable, 8
gradient algorithm, 4
gradient neural network, 22
gradient-based dynamic, 175

Hessian matrix, 42
Hopfield-neural-network, 4

implicit dynamic equation, 21
inner product, 250
inverse, 69
inverse-kinematic, 36
invertibility condition, 70
invertible, 251

Jacobian matrix, 37
joint-space, 102
joints, 104

Kronecker product, 77

left pseudoinverses, 98
linear activation function, 10
linear equation, 204
linear matrix equation, 138
linear time-invariant, 202

local minimum point, 193
locally asymptotically stable, 7
locally exponentially stable, 7
locally stable, 7
lower-bounded, 158
Lyapunov equation, 126
Lyapunov function candidate, 75, 77
Lyapunov stability theory, 75

mass matrix, 26
MATLAB routine, 119
matrix, 249
matrix derivatives, 25
matrix differentiation, 254
matrix inversion, 69
matrix norm, 253
matrix pseudoinverse, 97
matrix square root, 156
matrix-form, 81
matrix-valued, 73
minimum point, 116
model-implementation error, 23
monotonically-increasing odd functions, 75
Moore-Penrose, 103
moving, 159
multiple root, 189
multiplication, 250
multiplication term, 192

negative gradient, 51
negative-definite, 157
network architecture, 204
neural dynamics, 181
neuron, 92
Newton iteration, 159
nonlinear equation, 176
nonlinear-matrix-equation, 156
nonsingular, 251
norm-based, 44

ODE routine, 159
ode45, 25
ordinary differential equation, 4

perturbed model, 23

PINV, 221
positive-definite, 156
power activation function, 10
power-sigmoid activation function, 10
predictive approach, 159
PUMA560, 37

realization errors, 24
recurrent neural networks, 3
redundant manipulator, 102
reshape, 141
right pseudoinverse, 98
robot arm, 37
Robustness, 29
row, 250

scalar-valued, 44
simulation techniques, 25
Simulink modeling techniques, 32
singular, 220
sinusoidal-matrix, 84
square-based, 176
state feedback, 4
state matrix, 93
static problems, 51
steady-state, 80
steady-state errors, 52
superior convergence, 76
Sylvester equation, 114
symbolic object, 27

Taylor series, 75
theoretical solution, 115
time derivative, 27
time-varying, 6
time-varying convex quadratic programming problem, 47
time-varying linear-equality constraints, 47
time-varying quadratic minimization, 42
Toeplitz matrix, 90
trace, 252
tracking approach, 159
trajectories, 90
transpose, 250
two-norm, 206

unitary matrix, 94

vector element, 250
vector norm, 253
vector space, 249
vector-form, 82
vector-valued, 21, 43
vectorization techniques, 81
velocity-level, 37

work space, 105
wrong solution, 193

Zhang dynamics, 176
Zhang neural network, 21
ZNN-design formula, 44